SURFACE SCIENCE
VOL. II

INTERNATIONAL CENTRE FOR THEORETICAL PHYSICS, TRIESTE

SURFACE SCIENCE

LECTURES PRESENTED AT
AN INTERNATIONAL COURSE
AT TRIESTE FROM 16 JANUARY TO 10 APRIL 1974
ORGANIZED BY THE
INTERNATIONAL CENTRE FOR THEORETICAL PHYSICS, TRIESTE

In two volumes

VOL. II

INTERNATIONAL ATOMIC ENERGY AGENCY
VIENNA, 1975

THE INTERNATIONAL CENTRE FOR THEORETICAL PHYSICS (ICTP) in Trieste was established by the International Atomic Energy Agency (IAEA) in 1964 under an agreement with the Italian Government, and with the assistance of the City and University of Trieste.

The IAEA and the United Nations Educational, Scientific and Cultural Organization (UNESCO) subsequently agreed to operate the Centre jointly from 1 January 1970.

Member States of both organizations participate in the work of the Centre, the main purpose of which is to foster, through training and research, the advancement of theoretical physics, with special regard to the needs of developing countries.

SURFACE SCIENCE, IAEA, VIENNA, 1975
STI/PUB/396
ISBN 92−0−130375−0

Printed by the IAEA in Austria
November 1975

FOREWORD

The International Centre for Theoretical Physics, pursuing its objective of research and training with a comprehensive and synoptic coverage in various disciplines, has already presented three courses on solid-state physics: Theory of Condensed Matter (1967); Theory of Imperfect Crystalline Solids (1970); and Electrons in Crystalline Solids (1972). These Proceedings have been published by the International Atomic Energy Agency.

The present Proceedings constitute a fourth international course, held from 16 January to 10 April 1974, devoted to surface science, a new topic, holding the centre of current interest in solid-state physics. The directors of the course were V. Celli and G. Chiarotti (Italy), F. García-Moliner (Spain), S. Lundqvist (Sweden), N.H. March and J.M. Ziman (United Kingdom).

Generous grants from the Swedish International Development Authority and the United Nations Development Programme are gratefully acknowledged.

Abdus Salam

EDITORIAL NOTE

CONTENTS OF VOLUME II

Part III

ATOMS AND MOLECULES OF SURFACES

THEORY OF HYDROGEN
CHEMISORPTION ON METALS

W. BRENIG
Physics Department,
Technical University Munich,
Garching/Munich,
Federal Republic of Germany

Abstract

THEORY OF HYDROGEN CHEMISORPTION ON METALS.
1. Introduction. 2. Local density of states. 3. Discussion.

1. INTRODUCTION

Although an isolated hydrogen atom has only one electron, chemisorbed hydrogen can exhibit strong many-body effects. Owing to the coupling between the chemisorbed atom and the metal, charge can be transferred between atom and metal. Even though the correlation effects may be small in the metal because of screening, the intra-atomic Coulomb potential at the adatom remains.

Intra-atomic correlation effects can be studied in a simple model [1] containing essentially three parameters: the band width B of the metal, the coupling strength V between metal and atom and the strength U of the intra-atomic Coulomb repulsion. A typical situation is shown in Fig.1. In this figure the Fermi level E_f and the work function ϕ of the metal, as well as the ionization energy $\epsilon_a = -I$ and the electron affinity $\epsilon_a + U = -A$ of the isolated H atom, are shown.

2. LOCAL DENSITY OF STATES

The local single-particle density of states $\rho(\epsilon)$ at the adatom can be read off from the single-particle Green's function g as $\rho(\epsilon) = \text{Im } g(\epsilon - i0)/\pi$. $\rho(\epsilon)$ is closely related to the change of the photo emission [2] and field emission [3] spectrum by chemisorption. It can be used [4-6] to calculate the electronic part of the binding energy ΔE and the adsorbate charge q which is related to the work function change $\Delta\phi$ during adsorption.

From Fig.1 the Green's function for the isolated atom can be read off:

$$g_\sigma(z) = \frac{1-n}{z - \epsilon_a} + \frac{n}{z - \epsilon_a - U} \tag{1}$$

Here we have introduced $z = \epsilon - i0$ and $n = \langle n_{-\sigma} \rangle$ the number of electrons of spin $-\sigma$ in the ground state. The quantity $\rho(\epsilon)$ introduced above is then the density of states of spin σ electrons at the adatom.

FIG.1.   Energy levels for a hydrogen atom far from the surface.

If the atom is brought close to the metal surface the two poles of g(z) are shifted and broadened due to the coupling V. This coupling can be taken into account easily in the limit of vanishing U. Then a simple golden rule calculation yields

$$g_o(z) = \frac{1}{z - \epsilon_a - \Gamma(z)} \tag{2}$$

where H_s = substrate Hamiltonian,

$$\Gamma(z) = V^2 \langle b | \frac{1}{z - H_s} | b \rangle \tag{3}$$

and $|b\rangle$ is essentially that single-particle state of the metal which has maximum overlap with the screened Coulomb potential of the adatom. The imaginary part of $\Gamma(z)$ is thus the local density of states in the metal 'near the adatom' and, roughly speaking, has a width B.

The properties of Eq.(2) depend crucially on the ratio of the two parameters B and V. For $V \ll B$ (weak coupling) one just finds a small shift and width of order V^2 of the original level ϵ_a. For $V \gg B$(strong coupling) the state $|a\rangle$ of the adatom is coupled strongly to the state $|b\rangle$ of the metal forming the bonding and antibonding state of a 'surface molecule'. The two resulting states then couple weakly to the rest of the metal. In the strong coupling case one thus finds essentially a splitting of the original peak at ϵ_a into two peaks even for U = 0.

If both U and V are taken into account one expects in general four peaks as the combined effect of Coulomb splitting and bonding-antibonding splitting.

The treatment of these effects is complicated by the fact that V does not only shift, broaden and split the two levels of Eq.(1) separately but also leads to a coupling of the two levels.   The result of a generalized golden rule calculation [7] can be written as follows. Equation (1) is first rewritten as

$$g_o(z) = \frac{1}{z - \epsilon_a - Un - \dfrac{U^2 n(1-n)}{z - \epsilon_a - U(1-n)}} \tag{1'}$$

The coupling V then leads to an additional width function in each of the two denominators:

$$g_o(z) = \frac{1}{z - \epsilon_a - Un - \Gamma - \dfrac{U^2 n(1-n)}{z - \epsilon_a - \Gamma - U(1-n) - m(z)/n(1-n)}} \tag{4}$$

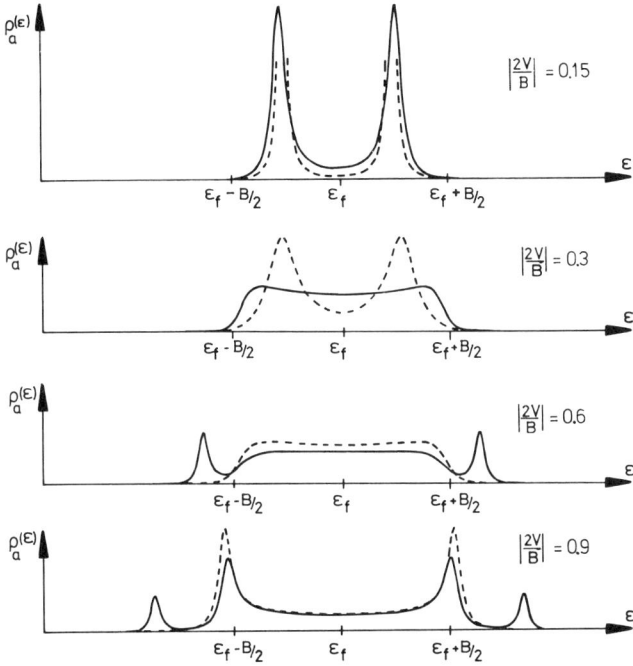

FIG. 2.   Spectral density $\rho(\epsilon)$ in the symmetric case $2\epsilon_a + U = 2E_f$ for different V values in the weak coupling theory (full line) and the Hartree-Fock approximation (broken line).

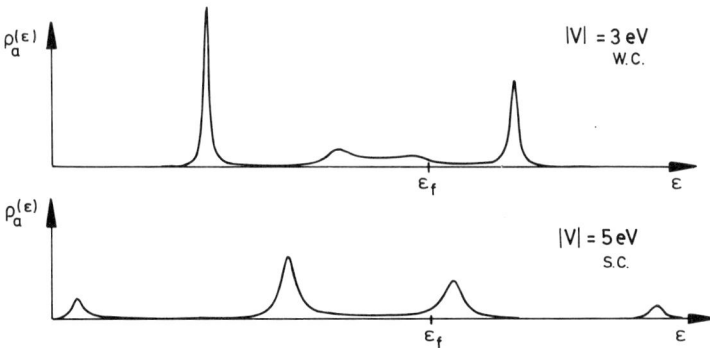

FIG. 3.   Spectral density $\rho(\epsilon)$ for hydrogen on nickel. V = 3 eV (weak coupling theory) and V = 5 eV (strong coupling theory).

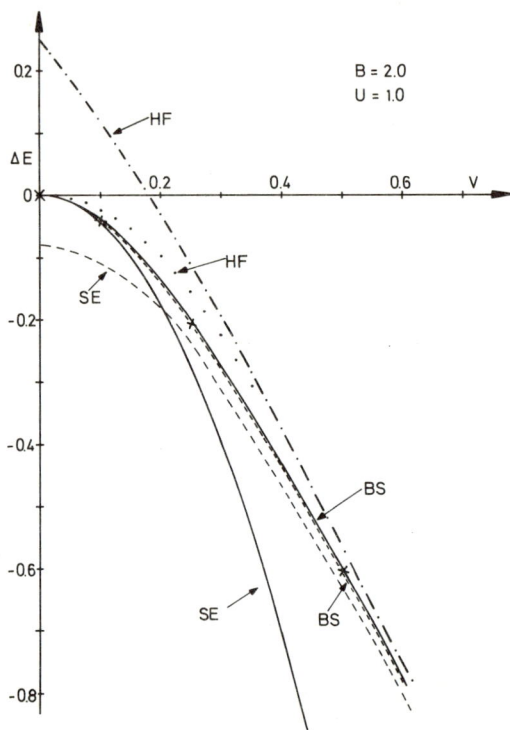

FIG.4.   Binding energy for a short metal 'chain'.
———BS: our weak coupling theory;   ——— SE: Schrieffer's weak coupling theory;
- - - - BS: our strong coupling theory;   — — -SE: Schrieffer's strong coupling theory;
- . — . — : restricted HF theory;   · · · · · : unrestricted HF theory;
  x : exact computer results

The quantity m(z) has a simple form only in the two limiting cases of strong and weak coupling, which can be combined into

$$m(z) = \left(\frac{1}{2} - 2 \langle \vec{S}_a \cdot \vec{S}_b \rangle\right)\Gamma(z) + \Delta \tag{5}$$

where $\langle \vec{S}_a \cdot \vec{S}_b \rangle$ is the ground-state spin correlation at $|a\rangle$ and $|b\rangle$ and is zero for weak coupling and $-3/4$ for strong coupling. Δ is a principal value integral which is a small real quantity and can be neglected in most cases.

If m(z) were to be neglected each of the atomic levels would be shifted and broadened by $\Gamma(z)$ separately. m(z), however, leads to an additional coupling and can produce large changes in the widths of the final levels [7].

Figures 2 and 3 show various results for the local density of states calculated from Eq.(4). Figure 4 shows results for the binding energy of a short metal 'chain' compared with various other calculations. Figure 5 shows the charge q at the adatom compared with the Hartree-Fock results.

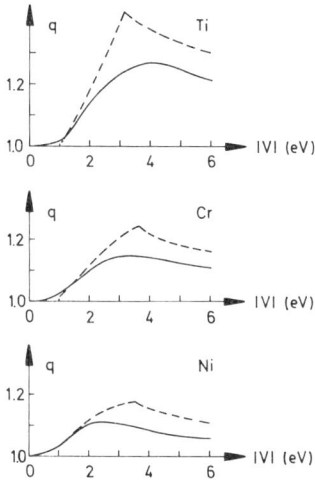

FIG.5.   Charge q of the adsorbate for H on Ti, Cr and Ni.   Full line: weak coupling theory;
broken line: Newns' results.

3. DISCUSSION

The expression (4) was obtained from a generalized second-order
calculation in V. It is thus correct in the large U, small V, limit. On the
other hand, if one lets $U \to 0$, V arbitrary, then, neglecting the U^2 term in (4),
one obtains the familiar Hartree-Fock solution. Thus (4) is also correct in
the small U, large V, limit. Furthermore, one can consider the limiting case
$B \to 0$, and U,V arbitrary. In this case the exact solution can easily be found
and compared with (4). It turns out that (4) is correct in the so-called
symmetry case when ϵ_a and $\epsilon_a + U$ have equal distance from the Fermi level E_f.
These results, together with the good agreement of the binding energies in
Fig.4 with exact results, suggest that (4) is an approximation which inter-
polates rather well between various limiting situations. Defects of (4) therefore will be due largely to the defects of the original model.

There are important electronic interaction effects not contained in the
Anderson model: in particular, the Coulomb interaction between electrons
at the adatom and in the metal. The direct part of this interaction leads to
the familiar image charge effects. These effects have a tendency to lower
the affinity level and to raise the ionization level by $\approx e^2/4d$ where d is the
distance from the metal surface. This corresponds to a screening of U. On
the other hand, the exchange terms have the opposite effect and can be of
comparable magnitude at the equilibrium distance. Both effects can be taken
into account by an appropriate renormalization of the parameters of the
Anderson model.

There are, however, important dynamical effects associated with the
extra-atomic Coulomb interaction. They lead to the occurrence of plasmon
satellites [8] and paramagnon satellites [9] of the single-particle spectrum.
This corresponds to an additional broadening of the original level which is
missing in the Anderson model.

ACKNOWLEDGEMENTS

The details of the calculations reported here can be found in Ref.[7].
The author is indebted to L. Cederbaum, W. Domcke, W. Götze, D. Menzel,
H. Schmidt and R. Schrieffer for interesting discussions.

REFERENCES

[1] ANDERSON, P.W., Phys.Rev. 124 (1961) 41.

[2] PENN, D.R., Phys.Rev.Lett. 28 (1972) 1041.

[3] PENN, D., GOMER, R., COHEN, M.H., Phys.Rev.Lett. 27 (1971) 26; Phys.Rev. B5 (1972) 768.

[4] GRIMLEY, T.B., Proc.Phys.Soc. 90 (1967) 751; 92 (1967) 776; J.Phys.C 3 (1970) 1934.

[5] NEWNS, D.M., Phys.Rev. 178 (1969) 1123.

[6] KJÖLLERSTRÖM, B., SCALAPINO, D.J., SCHRIEFFER, J.R., Phys.Rev. 148 (1966) 665.

[7] BRENIG, W., SCHÖNHAMMER, K., Z.Phys. 267 (1974) 201.

[8] HEWSON, A.C., NEWNS, D.M., "Effect of the image force in chemisorption", Imperial College,
London, preprint.

[9] BRENIG, W., Z.Phys. B20 (1975) 55.

ATOMIC SCATTERING

H. NAHR
Physics Institute IV,
University of Erlangen,
Erlangen,
Federal Republic of Germany

Abstract

ATOMIC SCATTERING.
    The scattering of atoms with thermal energy is a method for investigating the topmost monolayer
of a solid surface. The scattering process is first treated classically and the influence of the properties
of both the gas atom and the solid is discussed by use of simple models. Whereas the classical treatment
is suitable for inelastic scattering processes with many phonons involved, elastic and few-phonon inelastic
scattering processes are covered by the quantum-mechanical scattering theory, which is briefly summarized.
Recent experiments are discussed, where alkali halide and metal single crystals are investigated by atomic
beam scattering. By using light gas atoms (H, H_2, He) diffraction occurs, and the structure of adsorbed
layers, the surface Debye temperature, and the gas-solid interaction potential can be studied.

1. INTRODUCTION

Surface properties can be investigated by analysing particles emitted
in some way from the surface, therefore carrying information about the
surface itself. Probing particles are photons, electrons, atoms, molecules
and ions. Emission processes are thermal emission (desorption), stimulated
emission and scattering.

Many combinations of particle and emission process have proved to
be useful methods for studying surface properties as well as particle-
surface interaction. Some of them are listed below together with the kind
of information they can give. But it should be noted that at present each
method is restricted to the study of some few properties, and it is necessary to combine several methods to get a sufficiently complete picture of
the surface.

(a) Chemical composition

Infra-red spectroscopy (IRS), X-ray appearance spectroscopy (XPS,
APS), photon and electron stimulated emission of electrons (ESCA
and AES), thermal and electron stimulated desorption of neutrals
(flash desorption and EID), field desorption of ions, and secondary
ion mass spectroscopy (SIMS).

(b) Structure

Field electron microscopy (FEM) and field ion microscopy (FIM),
low-energy electron diffraction (LEED), and atomic scattering.

(c) Debye temperature, phonon spectrum

LEED and atomic scattering.

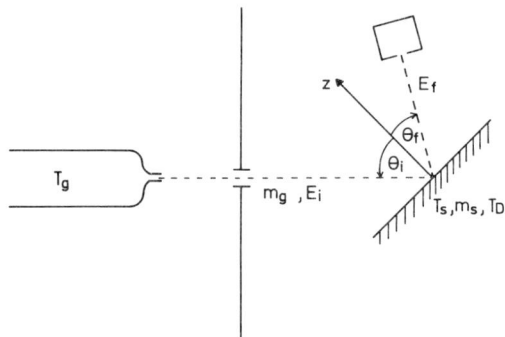

FIG. 1.   Atomic beam scattering from solid surfaces (schematic).

(d) Adsorption

   LEED, ASCA, AES, work-function measurement, flash desorption,
   ellipsometry, and atomic scattering.

(e) Gas-surface interaction

   Atomic scattering.

   Each surface-probing technique gives rise to a different point of view
about the surface, a trivial reason being the information depth. If the
information depth exceeds the topmost monolayer (ESCA, AES) the emitted
particle is simultaneously affected by the surface and the bulk. This is
also a difficulty in LEED. The interference of electrons scattered in deeper
regions (2-5 monolayers) with those scattered from the first monolayer
may be confusing. So the use of non-penetrating particles in scattering,
which thermal energy atoms are, gives us a basic advantage. On the other
hand, neutral atoms are not as easy to handle as electrons. Beam production, as well as energy analysis and detection, are more complicated.
   Even atomic scattering can give different 'information depth' in the
following meaning. An atom with high kinetic energy — let us say helium
at 10 eV — will come closer to a surface atom than will a thermal energy
atom of 50 meV. Therefore the fast atom will probe a greater range of
the interaction potential and will mostly be influenced by the high-energy
repulsive part, whereas the slow atom is sensitive to the attractive part
of the potential. Furthermore, a large atom such as xenon has a greater
probability of coming into contact with more than one surface atom, so
that (i) the potential is smeared out over the surface, and (ii) multiple
scattering becomes likely.
   Another difference, which is not so obvious, arises from the fact that
an atom may be scattered classically, i.e. that a kind of 'billiard ball'
mechanics is applicable to describe the scattering data; or it may be
diffracted and its behaviour can only be understood by applying quantum-
mechanical models. With a plane-wave incident on a surface we must
drop the picture of probing particle and closest approach. Of course, this
different behaviour exists only in our conception of the scattering process.

It is satisfying to see that the results from the different experimental methods together with proper theoretical models begin to combine into a consistent physical picture of the surface. The contribution of atomic scattering to this picture consists in successful investigation of structure and thermodynamic properties of some surfaces.

However, the domain of atomic scattering is gas-surface interaction and the questions related to it.

The motivation for the first scattering experiments from well defined surfaces was to prove the wave character of slowly moving light particles which was postulated by the early quantum theory. In these experiments, Estermann, Frisch and Stern (1931) scattered He and H_2 from LiF and found diffraction maxima according to diffraction from a two-dimensional grid of like ions. Furthermore, they found peculiar minima in the angular distribution. They could be explained by Lennard-Jones and Devonshire (1936, 1937a) as selective adsorption (see Section 4).

A second period began about 1955 with the scattering of atoms and molecules from a variety of polycrystalline and amorphous surfaces. They were all unintentionally contaminated by residual gases and oil vapour. The intention was to study energy and momentum transfer at the gas-solid interface in order to find the boundary conditions for the Boltzmann equation and to calculate drag and lift of bodies in the high atmosphere.

Some modern questions of general interest are the understanding of the trapping process and the chemical interaction between gas and surface atoms and between gas atoms on the surface[1], the study of the scattering process itself, and the interaction potential. Some aims for the future are the evaluation of the surface phonon spectrum from scattering data, electron distribution on a metal surface, and spin-dependent processes in the gas-solid interaction.

The basic arrangement consists of a source producing a well-collimated beam, the surface and a detector (see Fig.1). The surface, usually a single crystal cut in a plane with low Miller indices, can be rotated to adjust the angle of incidence (denoted by θ_i or θ_0) and the azimuthal angle γ between the projection of the incoming beam and an arbitrary crystal direction. The surface temperature is an important parameter in atomic scattering. Values between 10 K and 2200 K have been used. The angular distribution of the scattered atoms is usually measured by rotating the detector within the plane of incidence (the plane containing the beam axis and the surface normal z). A position of the detector outside this plane is denoted by the angle ϕ.

All gaseous species can, in principle, be used as beam particles. Common projectiles are the noble gases, hydrogen and deuterium, both atomic and molecular, and sometimes atmospheric gases and vaporized alkali metals.

In the following, the energy of incidence will be restricted to below 20 eV. Beyond this energy, sputtering and penetration into the surface become possible, requiring a completely different theoretical treatment.

---

[1] "Atoms" may stand for "molecules" so long as confusion is not possible.

Emphasis will be given to the thermal energy range 10 - 200 meV. This
is also the range of phonon energies; thus phonon events are easily
identified. The wavelength of light thermal atoms is in the range of the
lattice spacing:

$$T_g = 300 \text{ K} \rightarrow 2 \text{ k } T_g = 51.7 \text{ meV}$$
$$m_g = 1 \text{ (hydrogen)} \rightarrow \lambda = 1.15 \text{ Å}$$
Lattice spacing in LiF: d = 2.84 Å

Thus diffraction is possible (but not necessary, as we shall see later).

Recent developments in thermal beam scattering, both theoretical
and experimental, will be discussed in greater detail. Earlier work has
been reviewed by Stickney (1967) and Beder (1967). Recent reviews have
been given by Goodman (1971b), Smith (1973) and Toennies (1974).

2.   GENERAL DESCRIPTION OF THE SCATTERING PROCESS

Classical and quantum-theoretical models are used in surface
scattering. The decision of which model should be able to describe a
given gas-solid interaction is usually made in a rather qualitative manner.
High energy and momentum of the incident gas atom usually give rise to
a scattering process where many phonons are involved. When many phonons
are involved, the quantization becomes unimportant and a classical description should be valid. On the other hand, if we pass to a scattering process
with a light atom of small kinetic energy (on some arbitrary scale), the
number of phonons involved should be small and classical models must fail.

We like to distinguish between the classical and the quantum range on
a more quantitative basis. We use as characteristic the Debye-Waller
factor (DWF) of the system under study.

2.1. The Debye-Waller factor

The DWF is well known in neutron scattering and Mössbauer effect.
For a certain scattering process the DWF gives the probability that the
scattering is elastic. The DWF is written as

$$\exp(-2W) \quad \text{where} \quad 2W = \langle (\Delta \vec{k} \cdot \vec{u})^2 \rangle \tag{2.1}$$

For the surface $\Delta \vec{k} = \vec{k}_f - \vec{k}_i$ is the total momentum transfer to the surface,
and $\vec{u}$ is the displacement vector. Instead of taking the thermal average
in (2.1), $\vec{u}$ is usually averaged independently and only the z components
are kept:

$$2W = [\Delta k_z(\theta_i, \theta_f)]^2 \langle u_z^2(T_S) \rangle \tag{2.2}$$

Following a suggestion of Beeby (1971), the momentum transfer Δk_z is
taken after the atom has passed the attractive well. For specular reflection
this is

$$(\Delta k_z)^2 = \frac{8m_g E_i}{\hbar^2} \left(\cos^2 \theta_i + \frac{D}{E_i} \right) \tag{2.3}$$

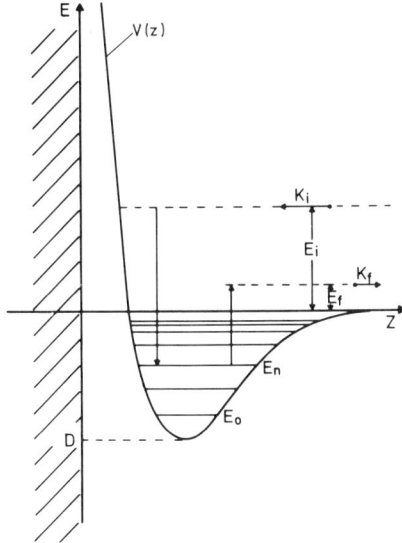

FIG. 2.   Energy diagram of the gas-solid interaction.

In the high-temperature limit the mean square displacement of the surface atoms is given by

$$\langle u_z^2 \rangle = \frac{3\hbar^2 \, T_S}{M_S \, k_B \, T_D^2} \tag{2.4}$$

with M_S = mass number of the solid atoms
      T_D = surface Debye temperature.

Equation (2.4) is valid for $T_{SF} \gtrsim \frac{1}{4} \, T_D$. For low temperatures the zero-point motion of the surface atoms must be taken into account.
    With (2.4) and (2.3) we get from (2.2)

$$2W = \frac{24 \, \mu \, E_i \, T_S}{k_B \, T_D^2} \left(\cos^2 \theta_i + \frac{D}{E_i} \right) \tag{2.5}$$

where $\mu = m_g/M_S$ mass ratio between gas and solid atom. The total specular intensity is given by

$$I_{sp} = K \, I_i \, e^{-2W} \tag{2.6}$$

The factor K gives the ratio of the specular scattered intensity to the total elastic scattered intensity. K = 1 in the absence of diffraction peaks. Equation (2.6) can be straightforwardly used for scattering angles other than specular.
    By calculating the DWF we get the total elastic-scattered flux and can decide whether to regard it as large or as negligible. As present

quantum theories are restricted in general to zero-phonon processes or one-phonon processes, whereas classical models cannot explain elastic scattering with reasonable assumptions on the solid, this is at the same time a decision on whether we can apply quantum models or classical models.

2.2. One-dimensional description of the interaction process

Figure 2 shows a schematic diagram of the interaction potential in the direction perpendicular to the surface. In the classical language, a gas atom approaches the surface with kinetic energy E_i and velocity v_i. It is first accelerated by the attractive part of the potential, then slowed down by the repulsive part until it reaches the turning point where $V(z) = E_i$. $V(z)$ is time dependent because of (i) the thermal motion of the lattice atoms and (ii) the response of the lattice to the impact of the gas atom.

Computer simulation shows that oscillations of the lattice atom which acts as collision partner results almost only in oscillations of the repulsive wall.

In our simple picture, the gas atom gains or loses energy if the wall moves during the collision in positive or negative z direction, respectively. If there is an energy loss $> E_i$ the gas atom is trapped in the attractive well.

In the quantum-mechanical description an incident wave with wave vector $\vec{k_i}$ is reflected from the potential. The derivative of the potential allows for interaction with phonons in the lattice. This results in a transition between continuum states (C ← C) where the wave vectors are related to the phonon energies in the usual way:

$$\frac{\hbar^2}{2\,m_g}\,(k_i^2 - k_f^2) = \sum_{m=1}^{M} (\pm\hbar\omega_m) \tag{2.7}$$

The positive sign stands for phonon creation.

Transitions into bound states (B ← C) occur when

$$E_n = \frac{\hbar^2}{2\,m_g}\,k_i^2 - \sum_{m=1}^{M} (\pm\hbar\omega_m) < 0 \tag{2.8}$$

Transitions between bound states (B ← B) or into continuum states (C ← B) are similarly connected with phonon processes.

It should be mentioned here that all these transitions can also occur elastically by reciprocal lattice vectors but, as atomic scattering occurs on the first monolayer, only reciprocal lattice vectors parallel to the surface (denoted by $\vec{G}$) are important and they are not included in this simple one-dimensional picture.

2.3. Interaction potential

The interaction potential is the essential quantity governing the gas-solid collision. It may be separated into an attractive and a repulsive part. The repulsive part is due to overlapping non-bonding electron wave functions.

The attractive part may be due to 'chemical' or 'physical' bonding between the two atoms. Chemical potentials are usually rather strong, with well-depth energy $D > 1$ eV, and exclusive in character. They give rise to chemical reactions on the surface or with the surface rather than to direct scattering (see e.g. Smith and Palmer (1972) and references given there).

In common gas-solid scattering systems the potential is of the Van der Waals (VdW) type. The well depth is usually much smaller than 1 eV (He-LiF: $D = 7.6$ meV; H-LiF: $D = 17.9$ meV; Ar-C (graphite): $D \approx 100$ meV). The VdW potential between the gas atom and the solid is written as a sum over the pairwise potentials:

$$V(\vec{r}) = \sum_{j=1}^{N} v_{ij} \left(|\vec{r} - r_j| \right) \tag{2.9}$$

Equation (2.9) implies that the VdW forces are additive and by use of (2.9) we neglect three-body forces. (For calculation of VdW pairwise potentials including three-body terms, see Dalgarno and Davison (1966).)

The pairwise potential is written in one of the following well-known analytical forms:

Lennard-Jones (12,6):

$$v_{ij} = D \left[\left(\frac{r_m}{r_{ij}} \right)^{12} - 2 \left(\frac{r_m}{r_{ij}} \right)^{6} \right] \tag{2.10}$$

Morse potential:

$$v_{ij} = D \left[\exp \{ -2\alpha (r_{ij} - r_m) \} - 2 \exp \{ -\alpha (r_{ij} - r_m) \} \right] \tag{2.11}$$

where r_m = internuclear distance at the minimum of the potential.

Both Morse potential and Lennard-Jones (12,6) potential have been modified in many ways (Lennard-Jones (m,n), Kihara, Buckingham). Especially a combination of the exponential repulsive part of the Morse potential and the attractive part of L-J (12,6) seems reasonable. For a discussion of pairwise potentials, see common textbooks.

We note that in (2.10) and (2.11) only two parameters are relevant: D and r_m in (2.10) and D and α in (2.11), because r_m in (2.11) only shifts the potential without changing its form.

With the assumptions made so far (which are not severe) we can calculate $V(\vec{r})$ in terms of the pairwise potential parameters. However, these parameters are only known in a few cases. A priori calculations (Hartree-Fock, etc.) are restricted to few-electron systems and must otherwise be regarded as approximate. The extent of the electron wave functions (the 'radius' of the electron shell) influences very heavily the well-depth energy D. Even if the wave function is precise to within 5%, the calculated D may be affected with an error of $> 50\%$. Therefore experimental parameters must be used for the heavier systems. Among

them, the best-known potentials are those of the noble gases. In all the
cases where pairwise potentials are unknown, approximate methods must
be applied, as follows.

The repulsive and the attractive terms of the potential are considered
separately:

$$v_{ij} = R_{ij} + A_{ij} \qquad\qquad (2.12)$$

The attractive part is calculated from the well-known London (or
Kirkwood) formula by using the electronic polarizabilities p of atom i
and atom (or ion) j, respectively:

London:

$$A_{ij} = - \frac{2}{3} \frac{p_i p_j}{r_{ij}^6} \frac{\Delta_i \Delta_j}{\Delta_i + \Delta_j} \qquad\qquad (2.13)$$

where Δ is the energy term of the first excited electronic state;

Kirkwood:

$$A_{ij} = - \frac{2}{3} \frac{e\hbar}{r_{ij}^6 \sqrt{m_e}} \frac{p_i p_j}{\sqrt{p_i/N_i} + \sqrt{p_j/N_j}} \qquad\qquad (2.14)$$

where e, m_e = electronic charge and mass
$\qquad\;\; N_i$ = number of electrons in the outer shell.

As both formulas give an attractive potential too low, in (2.13) Δ is
replaced by the ionization potential U, and N in (2.14) is replaced by the
total number of electrons in the atom (Mavroyannis and Stephen (1962)).

The repulsive part is calculated by using empirical combination
rules:

$$R_{ij} = (R_{ij} \cdot R_{ij})^{\frac{1}{2}} \qquad\qquad (2.15)$$

If the exponential shape is used for R, we get

$$R = B \exp(-\alpha r) \longrightarrow \quad \begin{array}{l} B_{ij} = (B_{ii} \cdot B_{jj})^{\frac{1}{2}} \\[2mm] \alpha_{ij} = \frac{1}{2}(\alpha_{ii} + \alpha_{jj}) \end{array} \qquad\qquad (2.16)$$

The combination rules allow us to extrapolate the potential of the atom
under study from a suitable combination with other atoms, and one pairwise potential between like atoms.

With the limitations in mind we can now calculate v_{ij} from Eqs (2.12)
and (2.13) or (2.14) and (2.15). However, the repulsive potential determined
for the atom or ion under study cannot be exact, as the combination rules
are only approximate (the 'radii' of the alkali and halogen ions in the
alkalihalides, for example, differ appreciably from species to species,
due to the different configurations). On the other hand, the attractive
part is approximately determined too, as (2.13) and (2.14) are not exact
and — more important — the electronic polarizabilities found in the

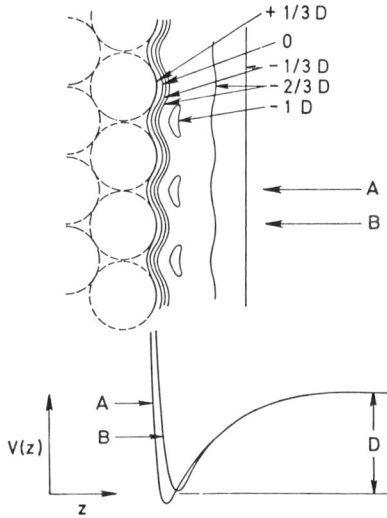

FIG. 3. Schematic diagram of the interaction potential on a strongly periodic surface (Toennies, 1974).
The equipotential lines are given in units of the well-depth energy D. The potential V(z) in the lower
part is shown for two different positions, A and B, with respect to the surface.

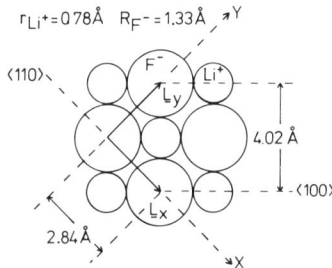

FIG. 4. Schematic cross-section through the (001) plane of LiF. The radii of the ions are taken from
Addison (1961).

literature show remarkable scatter. This scatter is partly due to experimental uncertainty and partly to different chemical environment and to
different frequency ranges used in the experiments (usually the extrapolation $p_0 = p(\nu : \nu \rightarrow 0)$ is made). The enumeration of sources for
systematic errors could be continued and makes the whole system
questionable.

The method becomes reliable, however, if we can measure the interaction potential on one surface precisely. Then we can match the pairwise
potential parameter in order that $V(\vec{r})$ fits the experimental result. With
the parameters now fixed we can extrapolate to the unknown potential of
another gas atom on the same surface (or of the same atom on a modified
surface). This is another reason to measure the interaction potential in
some cases, and probably the most precise method is selective adsorption
(Section 4).

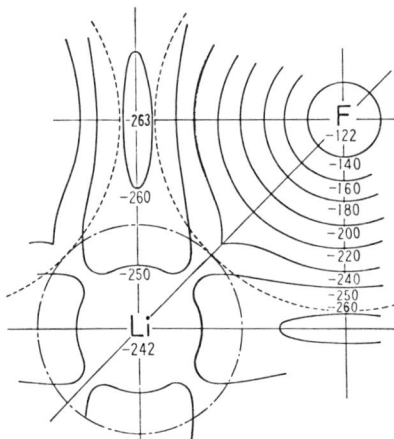

FIG. 5.   Equipotential curves calculated for He on LiF (Tsuchida, 1969) by summation over pairwise
potentials. The distance from the surface plane is z = 2. 5 Å.

2.4. Schematic diagram of the interaction potential

A schematic diagram of the interaction potential is shown in Fig.3,
where a strongly periodic potential has been assumed. In the lower part,
two cross-sections are shown, indicating that the periodicity mainly
influences the repulsive part of the potential.

The interaction potential for He on LiF has been calculated by
Tsuchida (1969) by summing over pairwise potentials. A schematic cross-
section through the (001) plane of LiF is shown in Fig.4. The equipotential curves calculated for a distance z = 2.5 Å above this plane are shown
in Fig.5. Tsuchida used the Lennard-Jones (12,6) pairwise potential and
replaced the (unknown)[2] He-Li$^+$ and He-F$^-$ parameters by those for the
pairs He-He and He-Ne, respectively.

In the quantum-mechanical scattering theories (see below), the
interaction potential $V(\vec{r})$ is expanded in a Fourier series:

$$V(\vec{r}) = \sum_{\vec{G}} V_{\vec{G}}(z) \exp(i\vec{G}\vec{R}) \tag{2.17}$$

The Fourier components $V_{\vec{G}}$ are represented by a Morse potential for
$\vec{G} = \vec{0}$:

$$V_{\vec{0}}(z) = D \{\exp[-2\alpha(z-z_m)] - 2 \exp[-\alpha(z-z_m)]\} \tag{2.18}$$

---

[2] Experimental potential parameters He-Li$^+$, measured at higher kinetic energies, cannot simply
be extrapolated to the thermal energy range.

and as exponential repulsive for $\vec{G} \neq 0$:

$$V_{\vec{G}}(z) = \kappa_{\vec{G}} D \exp [- 2\alpha (z - z_m)] \quad \text{for} \quad \vec{G} \neq \vec{0} \tag{2.19}$$

Here D, α are well-depth energy and range parameter of the Morse potential, respectively, and $\vec{G}$ is a reciprocal lattice vector. We note that for $\kappa_{\vec{G}} = 0$ we have a flat surface, whereas for $\kappa_{\vec{G}} \gg 0$ we have a surface with strong periodicity (structured surface). Examples of a flat surface are the close-packed (111) metal surfaces, whereas alkali halides are typically structured.

2.5. Dimensionless parameters

Before discussing scattering experiments, some further parameters are defined:

$$\epsilon = \frac{E_i}{D} \tag{2.20}$$

This parameter indicates whether or not the attractive part of the interaction potential is important. For $\epsilon \gg 1$ the potential can be approximated as purely repulsive, resulting in considerable simplification of the calculation. If $\epsilon \lesssim 1$ the well is important for the scattering process, and trapping into the well becomes likely.

$$\delta = \frac{E_i}{\hbar \omega_m} \tag{2.21}$$

where ω_m is the maximum modal frequency. δ is also significant for trapping. It indicates whether transition into a bound state is possible by a one-phonon process, $\delta \leq 1$, or whether a multiphonon process is necessary: $\delta > 1$. δ also indicates how drastically the kinetic energy of the incident particle can be changed in a single-phonon inelastic collision.

$$\xi = \frac{v_i \alpha}{\omega_m} \tag{2.22}$$

where α is the range parameter of a Morse potential. ξ relates the period of a lattice vibration to the time that the atom travels within the range of the potential. (We keep in mind, however, that neither the extent of the potential nor the velocity during the collision is a fixed quantity.) If $\xi \gg 1$, the solid atoms are effectively stationary during the collision process, so far as they do not participate in the encounter. The interaction between the surface atoms (the 'springs') may be ignored. For $\xi \ll 1$ the solid atoms make many oscillations during the collision time, and energy can be transported into the lattice. The elastic properties of the solid becomes important.

$$\mu = \frac{m_g}{m_s} \tag{2.23}$$

is the mass ratio already defined. For $\mu \ll 1$ the net energy transfer to the solid is small. This may be seen from a binary head-on collision,

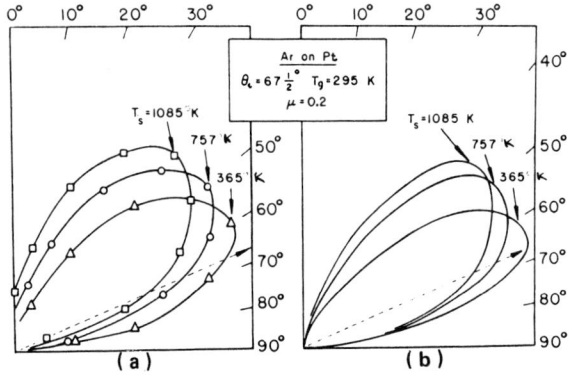

FIG. 6.   Comparison of the dependence of experimental and theoretical scattering patterns on the surface temperature T_S (Stickney, 1967). (a) Experimental results for Ar on Pt; (b) Corresponding theoretical results based on the hard-cube model.

where the energy transfer ΔE from an atom at energy E_i to a stationary free atom m_s is given by

$$\frac{\Delta E}{E_i} = 4\mu\,(1+\mu)^2 \approx 4\mu \quad \text{for} \quad \mu \ll 1$$

μ is closely related to ξ as $\mu\xi^2 \propto E_i$. It also gives an indication of how many surface atoms are involved in the collision. For $\mu \ll 1$ a single collision is sufficient to reverse the momentum of the gas atom. For $\mu \gtrsim 1$ a single collision may not be sufficient and multiple collision becomes likely.

3.   CLASSICAL SCATTERING

In this section gas-surface scattering under experimental conditions is discussed, where the DWF is small and elastic (coherent) scattering may therefore be ignored. Inelastic one- and multi-phonon processes will predominate, and classical models are appropriate. This region is relatively widespread with respect to the experimental parameters m_g, E_i, T_S and T_D. This may be illustrated by an example: Ar-W. We calculate the DWF by using

$$\mu \approx 0.2; \quad D \approx 80 \text{ meV}; \quad T_D \approx \frac{T_{D\,\text{bulk}}}{\sqrt{2}} \approx 270 \text{ K} \,\hat{=}\, 23.2 \text{ meV}/k_B$$

We choose

$$E_i = 10 \text{ meV}; \quad T_S = 90 \text{ K} \approx T_D/3; \quad \theta_i = 0°$$

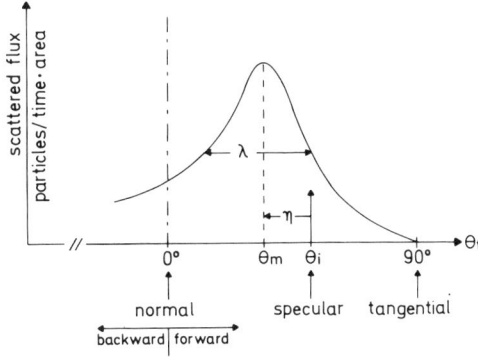

FIG. 7.   Lobular angular distribution (schematic).

Then we get

$$2W = 24\mu \, \frac{E_i T_S}{k_B T_D^2} \left(\cos^2 \theta_i + \frac{D}{E_i} \right) \approx 6$$

$$\frac{I_{sp}}{I_0} = e^{-2W} \approx e^{-6} \approx 2 \times 10^{-3}$$

Although the experimental conditions are chosen to favour elastic scattering, this amounts to less than 1% (at $\theta_i = \theta_r \approx 90°:0.5\%$). Indeed, elastic scattering has never been observed in this system. The typical angular distribution for rare gases scattered from metal surfaces shows a lobular structure, indicative for inelastic scattering.

Lobes of essentially similar shape were often reported in the period before 1970, and a typical example is shown in Fig.6(a). Here the angular distribution of the scattered flux is plotted in polar co-ordinates. The Ar beam had a Maxwellian velocity distribution, which is given by

$$\frac{dI}{dv} = \frac{2I_0}{v_h} \left(\frac{v}{v_h} \right)^3 \exp \left[- \left(\frac{v}{v_h} \right)^2 \right] \tag{3.1}$$

where

$$v_h = \left(\frac{2k_B T}{m_g} \right)^{\frac{1}{2}}$$

is the most probable velocity in the oven, which is at temperature T. The data are taken for three surface temperatures of the Pt target, whose surface condition is uncertain.

Typical features of these lobes are the following (see also Fig.7):

(a) The maximum is shifted more or less against the specular direction.

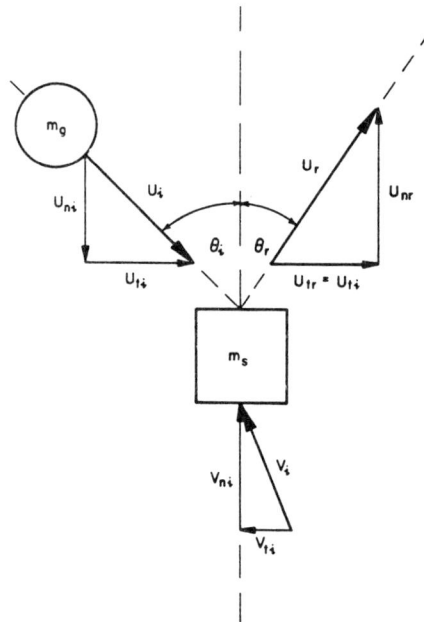

FIG. 8.   Schematic representation of the hard-cube model (Stickney, 1967).

(b) The lobes are broad. The width λ at half maximum height is
     larger than the incoming beam, even if this is nearly monoenergetic.
(c) The shift of the maximum, $\eta = \theta_i - \theta_m$, depends in a characteristic
     manner on θ_i, E_i, T_s and μ.

A very simple classical model can give a qualitative explanation for these
findings.

3.1. Hard-cube model

This model was formulated by Logan and Stickney (1966) (see also
Stickney (1967)). It is mentioned here only because of its simplicity. The
following assumptions have been made:

(a) The surface is effectively flat. The tangential component of the
     gas atom momentum $\vec{P}$ is unchanged. (All V_g of Eq.(2.19) are
     equal to zero.)
(b) The interaction potential is that acting between hard bodies:

$$V(z) = \begin{cases} 0 & \text{for } z > z_0 \\ \infty & \text{for } z < z_0 \end{cases}$$

(c) The solid is represented by a collection (i.e. gas) of hard cubes
     confined within the volume of the solid by a square-well potential.
     Each gas atom experiences a single collision with one of the hard
     cubes.

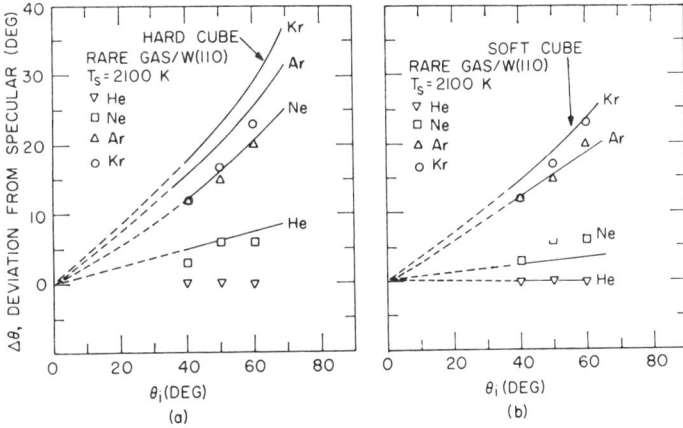

FIG. 9.   Scattering of rare gases from W(110).   Dependence of the angular position of the lobe maximum on angle of incidence.   The experimental data are compared with corresponding calculations based on the hard-cube model (a) and on the soft-cube model (b), respectively (Yamamoto and Stickney, 1970).

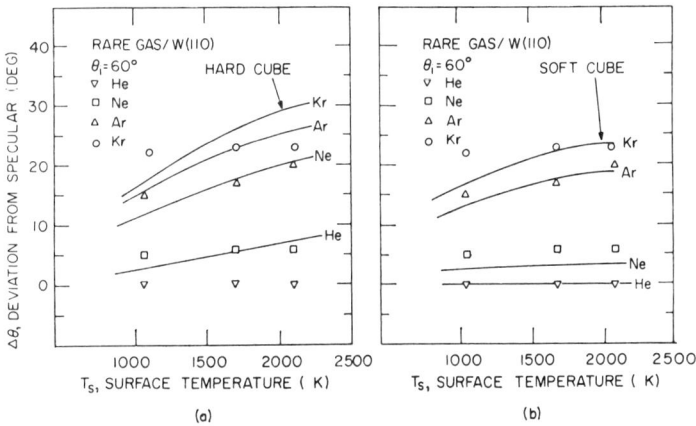

FIG. 10.   As Fig. 9, but for several surface temperatures and angle of incidence fixed at $\theta_i = 60°$.

Because of assumption (b) the interaction time vanishes and the atoms of the solid may be regarded as free particles, i.e. uncoupled from each other, during the gas-solid collision.   The situation is shown in Fig. 8. With assumptions (a), (b) and (c) the calculation becomes very simple. The conservation laws for energy and momentum are used for the normal components of velocity of gas and surface atom, respectively.   The tangential motion of the gas atom is unchanged; that of the solid atom has no influence.   By integrating over the velocity distributions for gas and solid atoms, the differential cross-section is calculated:

$$\frac{1}{\overline{v_{zi}}} \frac{dR}{d\theta_R} = \frac{3}{4} B_2 \left(1 + \frac{B_1}{\cos\theta_i}\right) \left(\frac{m_s T_g}{m_g T_s}\right)^{\frac{1}{2}} \left(1 + B_1^2 \frac{m_s T_g}{m_g T_s}\right)^{-5/2} \qquad (3.2)$$

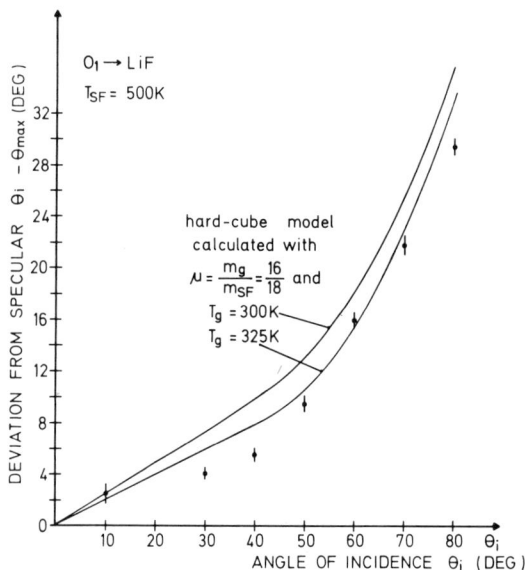

FIG. 11.   Atomic oxygen scattered from LiF(001).   The deviation of the lobe maximum from the specular direction is plotted against angle of incidence.   Experimental data (Hoinkes et al., 1973) are compared with calculations based on the hard-cube model. $T_B \gtrsim 300°K$ (estimated).

with

$$B_1 = \frac{1+\mu}{2} \sin \theta_i \, \cot \theta_r - \frac{1+\mu}{2} \cos \theta_i$$

$$B_2 = - \frac{1+\mu}{2} \frac{\sin \theta_i}{\sin^2 \theta_r}$$

The result for Ar on Pt is shown in Fig.6(b). The lobular structure is represented quite well, and so is the shift of the maximum with T_S.

Detailed comparisons with experimental data have shown, however, that the model has many shortcomings. Backward scattering is not included; the angular distribution does not fall off to zero at $\theta_f \to 90°$, especially if μ is large, and the lobe width λ comes out wrong.

The position of the lobe maximum, however, is reproduced qualitatively well. This can be seen from Figs 9(a) and 10(a) (Yamamoto and Stickney (1970)). Here the cube-model results are compared with experimental η values for several rare gases on W(110). η is plotted against θ_i and T_S, respectively. These results (and many others not given here) show, however, that the interaction is exaggerated in most cases. A conclusion would be that the assumptions of the models should be, at least qualitatively, correct − among them the surface flatness. This conclusion can be misleading.

Alkali halides, known to be strongly periodic from diffraction experiments, can show lobular scattering, too. This is the case for metal vapours,

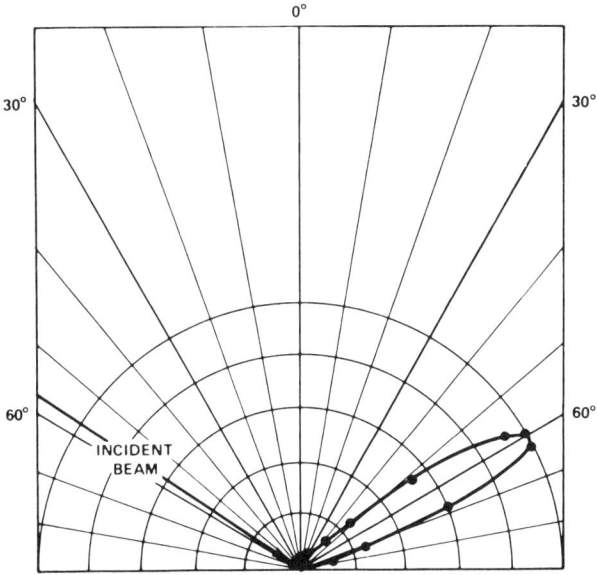

FIG. 12. Polar plot of the scattering distribution of a nearly monoenergetic Ar beam of E_i = 2.7 eV from epitaxially grown Ag(111). Surface temperature was approximately 570 K; θ_i = 58°. FWHM of the scattered beam is 17° (Hays et al., 1972).

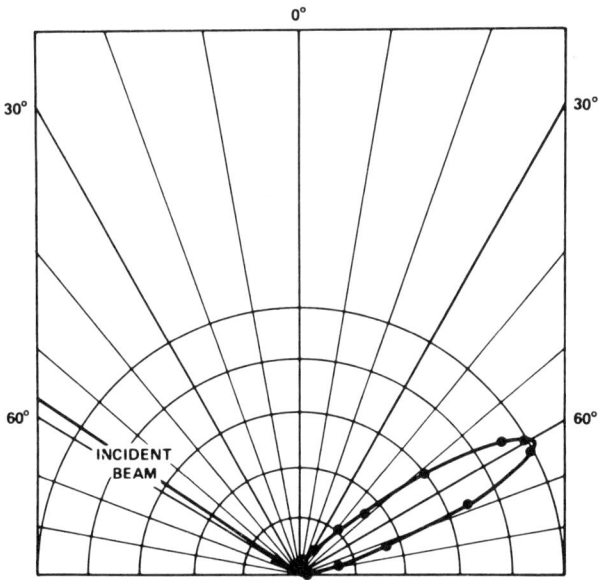

FIG. 13. As Fig. 12, but for a 18.1 eV argon beam. FWHM is 20°, showing little dependence on E_i.

e.g. Hg on NaCl (Zahl and Ellet (1931)). This is also the case for Xe on
LiF(001) (O'Keefe et al. (1971)), and for atomic oxygen on LiF(001)
(Hoinkes et al. (1973)).

The lobular structure in the scattering distribution of atomic oxygen
is superimposed on a large cosine distribution, which is due to initial
trapping and subsequent desorption. After subtraction of the cosine distribution, rather broad lobes remain; these result from direct scattering.
They are shifted from the specular direction towards the surface normal
in a similar manner as with the rare gases on metal surfaces. The deviation η is plotted against θ_i in Fig.11 together with the prediction of the
hard-cube model. Two slightly different beam temperatures are assumed,
showing the dependence on beam temperature.

Even in this case there is a qualitative agreement between the hard-
cube model and experimental data. It becomes clear that the appearance
of lobes in a scattering experiment is a rather common feature. Lobes
have also been observed at higher beam energies. In Fig.12 a polar plot
is shown for a 2.7 eV argon beam scattered from Ag(111) (Hays et al.
(1972)). The angular distribution becomes narrow again, as we have large
values of ϵ, δ and ξ (see Eqs (2.20) - (2.22)). In particular, $\xi \gg 1$ indicates
that the solid atoms may be regarded as stationary.

The angular distribution does not change very much if E_i becomes
still larger (Fig.13). Inelastic processes associated with higher energy
losses (plasmons, electronic excitation of the gas atom) are possible, but
not indicated in the angular distribution.

We saw in previous figures that the hard-cube model can give a
qualitative description of both the appearance of lobes and their angular
position. But an improvement of the model is desirable for two reasons:
(a) the agreement with experiment has to be quantitative; and (b) as the
assumptions of the model are so simple, we can get almost no information
about the surface or the interaction potential. As we have seen, not even
the flatness of the surface can be concluded from the comparison of this
model with experiment.

So it was quite obvious that the model would be improved by introducing a more realistic interaction potential, consisting of two terms: an
attractive term and a 'soft' repulsive term. Consequently, the interaction
time becomes finite and the surface properties become more important.
During the interaction time, energy can be exchanged between the surface
atom, upon which the gas atom collides, and the surrounding lattice. Thus,
the elastic properties of the surface are introduced.

3.2. Soft-cube model

The soft-cube model was developed by Logan and Keck (1968) as a
successor to the hard-cube model. The basic assumptions of the hard-cube
model are kept: flat surface and single collision with one surface atom.
The interaction potential is assumed to consist of a stationary attractive
part and an exponential repulsive part moving with the surface atom involved
in the collision. This is indicated in the schematic diagram, Fig.14. As
the attractive potential is stationary, its shape is arbitrary in this classical
treatment and a step is assumed. The surface atoms are linear harmonic
oscillators (the 'spring' in Fig.14) and have equilibrium distribution of
energy at the temperature of the solid.

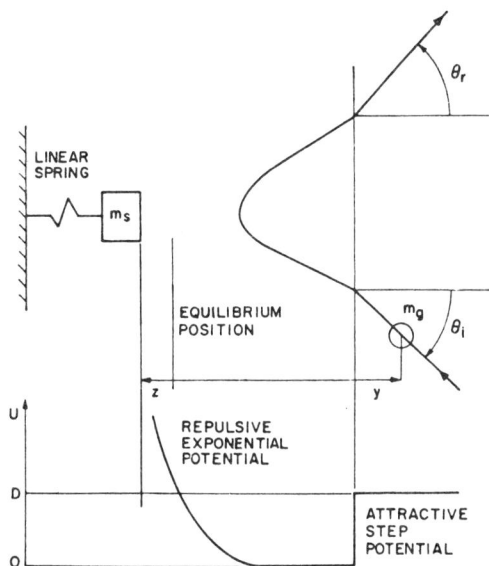

FIG. 14. Schematic representation of the soft-cube model (Logan and Keck, 1968).

The equations of motion become more complicated than with the hard-cube model and a computer has to be used for calculating the differential cross-section. However, the cross-section can still be expressed in analytical form. It is given by Eq.(48) in the original paper of Logan and Keck (please note that there is a misprint in Eq.(B2) of Appx.B). The model has three parameters: D, α and ω. D and α are Morse parameters. ω is related to a characteristic temperature of the solid by $\hbar\omega = k_B T_c$, and T_c is assumed to be the surface Debye temperature T_D.

In Fig.9(b) the result of a fit procedure is shown, together with the same experimental data as in Fig.9(a). The agreement is much better than with the hard-cube model (which is to be expected, as additional free parameters have been introduced). In principle, all three parameters could be evaluated by fitting the experimental data; but this would be an overinterpretation of the experiment. Therefore, two parameters, D and T_D, are taken from other experiments or are estimated. D was taken from estimates of the heat of adsorption, and T_D was replaced by the bulk Debye temperature. So only α was subject to the fit procedure. The result for the four noble gases He, Ne, Ar, Kr ranges from $\alpha = 6.7$ Å^{-1} for Kr to $\alpha = 8.33$ Å^{-1} for Ne and Ar (He lying in between). These values are remarkably large, about a factor of two over the noble-gas pairwise interaction parameters, and they are not significantly dependent on the gas atom. This gives us little trust in the validity of the soft-cube model. The simple classical models are rather discouraging as soon as quantitative information about the interaction is expected. This is also the case with other simple classical theories, which will not be discussed here. They are, however, able to illustrate the qualitative feature of the scattering process and can explain in general the lobular structure.

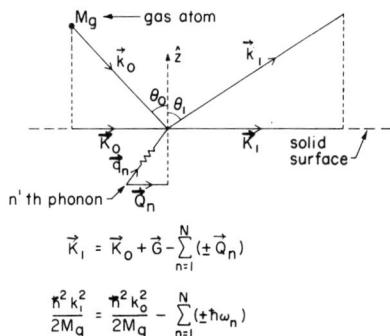

$$\vec{K}_1 = \vec{K}_0 + \vec{G} - \sum_{n=1}^{N}(\pm \vec{Q}_n)$$

$$\frac{\hbar^2 k_1^2}{2M_g} = \frac{\hbar^2 k_0^2}{2M_g} - \sum_{n=1}^{N}(\pm \hbar\omega_n)$$

FIG. 15.   Quantum-mechanical analogy to the cube models (Goodman, 1970).

Goodman (1970) pointed out that there is a quantum-mechanical basis for the cube models. In the vector diagram shown in Fig.15 a flat surface is equivalent to vanishing $\vec{G}$ and $\vec{Q}_n$, so that only phonons with wave vector perpendicular to the surface are kept. Then phonon annihilation increases the normal velocity of the gas atom. The outgoing angle deviates from the specular direction towards the surface normal; the atom gains energy from the solid. Phonon creation has the opposite effect: the atom loses energy to the solid. Both effects together result in a broadening of the angular distribution (which would be purely specular in the absence of phonon processes). If phonon annihilation predominates over phonon creation, the lobe will deviate to the surface normal, and vice versa.

In the next section, we drop the oversimplifying assumptions that had to be made in order to get closed-form expressions for the scattering distributions.

3.3.   Computer simulation models

In a classical theory of greatest generality, calculation of the differential cross-section must take into account surface properties and the interaction potential. As soon as the surface cannot be treated as a continuum, but the individual motion of the solid atoms is important, this is a many-body problem. For the solution of this problem, a Monte-Carlo technique is applied. The classical trajectory of a gas atom is calculated during the collision process by summing up the pairwise potentials of the gas atom with a certain number of solid atoms. By averaging over many trajectories, the (classical) collision process is simulated.

Calculations of this type have been performed by Oman (1968), Lorenzen and Raff (1971), and McClure (1970, 1972a, b). Several models have been developed; they differ from one another by the number of solid atoms taken into account, the type of interaction potential, the representation of the solid, and so on. Pairwise interaction potentials of the Lennard-Jones (12,6) type (Eq.(2.10)) are used by Oman and McClure, whereas Lorenzen and Raff use a Morse potential (Eq.(2.11)). The interaction between the solid atoms is either represented by springs or ignored.

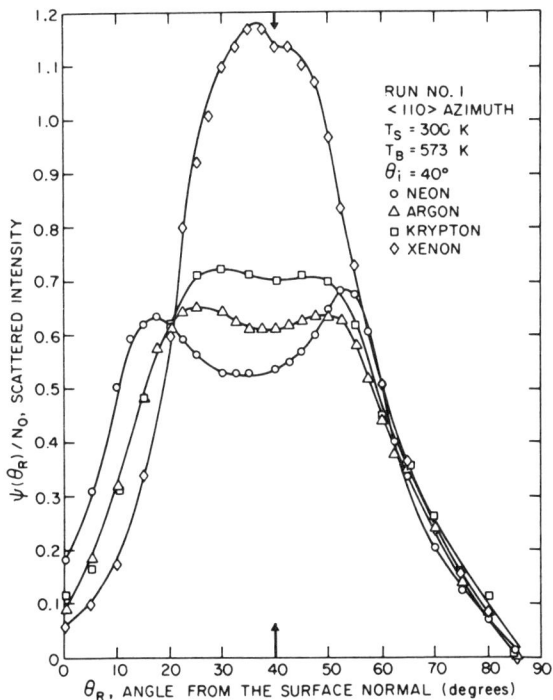

FIG. 16.   Scattering distribution of rare gases from LiF(001). The scattered intensity is given in units of (atoms/second SR)/(incident atoms/second). The gases Ne, Ar and Kr show rainbows (from O'Keefe et al., 1971).

Both 'cold' and 'warm' lattices are used. The thermal motion of the solid is simulated either by motion of each surface atom at random phase and amplitude or by coupling to phonon modes.

The trajectories are calculated with the 'aiming points' on the surface either uniformly distributed or chosen statistically. A large number of trajectories must be calculated to obtain a small statistical error in the distribution function $F(E_f, \theta_f, \phi_f)$, even at angles or energies where the differential cross-section is small. In practice 10^4 to 10^5 trajectories must be calculated and this is rather time-consuming. By keeping the collision mechanics relatively simple, angular distributions with a resolution of a few degrees can be calculated in a reasonable time.

Calculations of this type are, of course, strongly dependent on the pairwise interaction parameters. It was mentioned in Section 2 that it would be advantageous to have experimental data of some kind to which the pairwise potential parameters can be fitted. For this purpose a classical structure effect called 'surface rainbow' has proved useful.

3.4. Rainbow scattering

Rainbow scattering is a special case of structure scattering. It is characterized by a double lobular structure. The lobes occur at positions

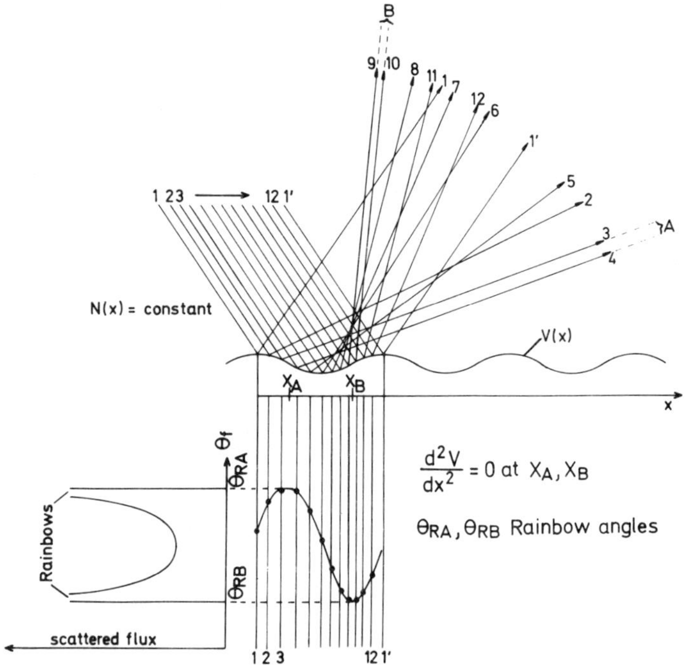

FIG. 17.   Formation of surface rainbows. Schematic diagram for a surface with one-dimensional structure.

where no diffraction maxima are expected and they shift with angle of incidence and surface temperature other than predicted by the cube models.

They were first observed experimentally by O'Keefe et al. (1971) for Ne, Ar and Kr scattered from LiF(001) (see Fig.16). They are also found in calculations of McClure et al. (1969) and in earlier calculations of Oman.

The formation of surface rainbows can be seen from a simplified model, where the interaction potential is assumed to be periodic, time dependent, and stepwise repulsive: the surface is a 'corrugated mirror' (see Fig.17).

Atoms incident on this surface under angle of incidence θ_i will be scattered into outgoing angles smaller than specular ($\theta_f < \theta_i$) if they hit the surface within the region, where $dV/dx > 0$ (atoms 6, 7, 8, 9, 10, 11 and 12 in Fig.17). Trajectories with aiming points within the region $dV/dx < 0$ are associated with scattering angles $\theta_f > \theta_i$ (atoms 2, 3, 4 and 5). The maximum scattering angle, denoted by θ_{RA} belongs to atoms 3 and 4, having aiming points around x_A. Atoms 9 and 10, being scattered around x_B, have minimum scattering angle θ_{RB} .

A plot of the resulting scattering angle as function of the position of the aiming point x shows an accumulation around θ_{RA} and θ_{RB} . This is because the potential $V(x)$ has turning points at x_A and x_B. θ_{RA} and θ_{RB}

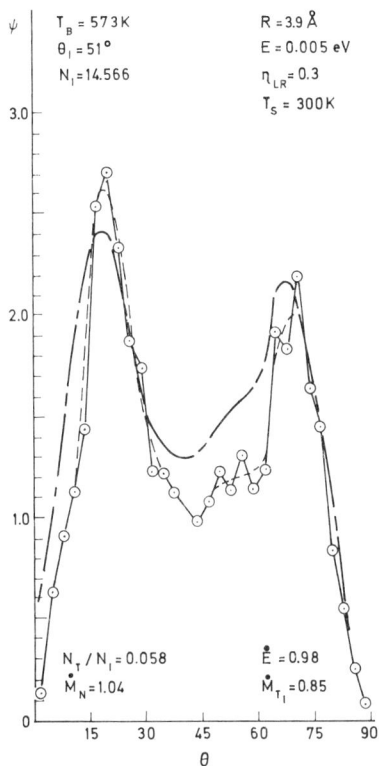

FIG. 18.   Rainbow scattering of Ne from LiF(001).   Comparison of numerical calculations of McClure (1972b) with an experimental flux distribution (broken line) (Smith et al., 1970).

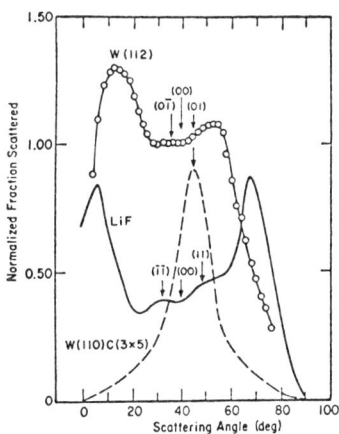

FIG. 19.   Rainbow scattering of Ne from W(112). T_S = 1170 K; T_g = 300 K (Maxwellian beam); θ_i = 40°. Arrows indicate the calculated positions of diffraction peaks (- o - o - o -).   For comparison, the angular distribution of Ne from LiF(001) in the $\langle 100 \rangle$ azimuth (———), and Ne from W(110) C (3 × 5) at θ_i = 45° (------) is shown (Stoll and Merrill, 1973).

are called the rainbow angles. A generalization to the case of two-dimensional periodicity of the surface is straightforward. The flux distribution on a real surface is smeared out due to the thermal motion of the solid atoms.

McClure (1972a, b) made a comparison with the experimental results of Smith et al. (1970) for Ne on LiF(001) with the $\langle 100 \rangle$ azimuth in the plane of incidence. McClure fitted the pairwise potential parameters so that the rainbow angles are reproduced. The comparison between calculated and experimental angular distribution is shown in Fig.18. The agreement is similarly good at other angles of incidence and beam temperatures. The interaction potential, which leads to this agreement, has a well-depth energy of D = 16 meV. This is very close to the value estimated by Hoinkes et al. (1972c), who found D = 15 meV.

Recently, surface rainbows have also been reported for Ne on the strongly periodic surface of W(112) by Stoll and Merrill (1973). This is the first classical structure effect found on a metal surface. In Fig.19 this result is shown together with the distribution for Ne on LiF of Smith, O'Keefe and Palmer. The authors report that the surface rapidly contaminates below T_S = 1170 K, even at a background pressure of 5×10^{-10} Torr.

4.   QUANTUM SCATTERING

In this section processes are discussed where either the scattered atoms interfere with each other, i.e. the wave functions are coherent and superpose each other, or single phonons are resolved. Coherent superposition of the atomic wave functions is not restricted to elastic processes. It can also occur in inelastic scattering, when phonons with fixed $\vec{q}$ are created or annihilated.

In this section the term diffraction will often be used. Exactly speaking, this term may only be used for the non-specular beams, which are — in analogy to optics — diffracted away from the specular direction. But in general the specular beam is included in the diffracted beams, if not otherwise stated.

In Section 4.1 quantum-mechanical elastic scattering theories are briefly discussed, and in Section 4.2 a geometrical method is used to explain the kinematics of the diffraction process.

4.1. Quantum theory of scattering

The first quantum theory of the gas-solid interaction was formulated by Lennard-Jones, Devonshire and Strachan (LJDS) in a series of papers (1935 - 1937). (For a review, see Goodman (1971a).) LJDS used the Morse-type potential (Eqs (2.17) - (2.19)). For elastic scattering (LJD (1936, 1937a, b)), a distorted-wave Born approximation was used (by DWBA). This theory was generalized when the essential restriction DWBA was eliminated by Cabrera, Celli, Goodman and Manson (1970), referred to as CCGM. DWBA is valid only if the total scattered flux is fundamentally specular. This condition is met only in a few cases (like H-LiF). In many other cases, among them the most intensively studied system, He-LiF, the total diffracted flux greatly exceeds the specular flux. So the elimination of DWBA now allows us to calculate diffracted intensities of any kind.

The following notation has become common:

$$\vec{r} = (x, y, z) = (\vec{R}, z) \tag{4.1}$$

$$\vec{k} = (k_x, k_y, k_z) = (\vec{K}, k_z) \tag{4.2}$$

and

$$\vec{G} = n_1 \vec{G}_1 + n_2 \vec{G}_2 \tag{4.3}$$

where

$$\vec{G}_1 = 2\pi \frac{\vec{L}_2 \times \vec{z}}{(\vec{L}_1 \times \vec{L}_2) \vec{z}} \quad \text{and} \quad \vec{G}_2 = 2\pi \frac{\vec{L}_1 \times \vec{z}}{(\vec{L}_1 \times \vec{L}_2) \vec{z}}$$

with $\vec{L}_1$, $\vec{L}_2$ two fundamental translational vectors of the lattice.

In the elastic scattering theory we have to use the thermally averaged potential $\langle V(\vec{r}, \vec{u}) \rangle = v(\vec{r})$. We assume that $V(\vec{r}, \vec{u})$ is obtained by summation of pairwise potentials over all lattice atoms. Then it can be shown (CCGM) that we can get $v(\vec{r})$ from $V(\vec{r})$ (Eqs (2.17) - (2.19)) by the substitution:

$$\kappa'_{\vec{G}} = \kappa_{\vec{G}} \exp\left(-\frac{1}{2} \vec{G}^2 \langle u_x^2 \rangle\right) \tag{4.4}$$

$$D' = D \exp\left(-\alpha^2 \langle u_z^2 \rangle\right) \tag{4.5}$$

$$z'_m = z_m + \frac{3}{2} \alpha \langle u_z^2 \rangle \tag{4.6}$$

We get

$$v(\vec{r}) = \sum_{\vec{G}} v_{\vec{G}}(z) \exp(i\vec{G}\vec{R}) \tag{4.7}$$

with

$$\frac{v_{\vec{0}}(z)}{D'} = \exp\left[-2\alpha (z - z'_m)\right] - 2 \exp\left[-\alpha (z - z'_m)\right] \tag{4.8}$$

$$\frac{v_{\vec{G}}(z)}{D'} = \kappa'_{\vec{G}} \exp\left[-2\alpha (z - z'_m)\right] \text{ for } \vec{G} \neq \vec{0} \tag{4.9}$$

We write down the Schrödinger equation for the gas-atom wave function $\psi(\vec{r})$ in the potential $v(\vec{r})$:

$$\left[-\frac{\hbar^2}{2m} \nabla + v(\vec{r})\right] \psi(\vec{r}) = \frac{\hbar^2 k^2}{2m} \psi(\vec{r}) \tag{4.10}$$

Because of the periodicity of $v(\vec{r})$ the wave function can be expanded into partial waves using the Bloch theorem:

$$\psi(\vec{r}) = \sum_{\vec{G}} \psi_{\vec{G}}(z) \exp[i(\vec{G}+\vec{K})\vec{R}] \tag{4.11}$$

The summation goes over all $\vec{G}$ including $\vec{G} = \vec{0}$. We insert (4.11) and (4.7) into (4.10) and get for the $\psi_{\vec{G}}(z)$

$$\left[\frac{d^2}{dz^2} + k_{\vec{G}z}^2 - \frac{2m}{\hbar^2} v_0(z)\right]\psi_{\vec{G}}(z) = \frac{2m}{\hbar^2} \sum_{\vec{G}'\neq\vec{G}} v_{\vec{G}-\vec{G}'}(z)\,\psi_{\vec{G}'}(z) \tag{4.12}$$

$k_{\vec{G}z}$ is defined in analogy with (4.2):

$$k_{\vec{G}z}^2 = k^2 - (\vec{K}+\vec{G})^2 \tag{4.13}$$

We solve the homogeneous equation from (4.12) for $v_{\vec{G}-\vec{G}'} = 0$ and get the complete set of eigenstates $\phi_\alpha(z)$ to the potential $v_0(z)$. The Schrödinger equation defining $\phi_\alpha(z)$ is

$$\left[\frac{d^2}{dz^2} + \alpha^2 - \frac{2m}{\hbar^2} v_0(z)\right]\phi_\alpha(z) = 0 \tag{4.14}$$

with the eigenvalues $E_\alpha = (\hbar^2\alpha^2)/2m$.

Now $\psi_{\vec{G}}$ is written as linear combination in ϕ_α:

$$\psi_{\vec{G}}(z) = \sum_\alpha C_{\vec{G}\alpha}\,\phi_\alpha(z) \tag{4.15}$$

We obtain an equation defining the $C_{\vec{G}\alpha}$ by inserting (4.15) into (4.12):

$$C_{\vec{G}\alpha}(k_{\vec{G}z}^2 - \alpha^2) = \frac{2m}{\hbar^2} \sum_{\vec{G}'\neq\vec{G}} \sum_\beta C_{\vec{G}'\beta} \langle\beta|v_{\vec{G}-\vec{G}'}|\alpha\rangle \text{ for } k_{\vec{G}z} \neq \alpha \tag{4.16}$$

The ambiguity for $k_{\vec{G}z}^2 = \alpha^2$ is resolved by demanding that $\psi(\vec{r})$ describe an incoming plane wave $\vec{k}$ and outgoing scattered waves. We obtain

$$C_{\vec{G}\alpha} \exp(-i\xi_0) = \delta_{\alpha,\vec{0}}\,\delta_{\vec{G},\vec{0}} + \frac{2m}{\hbar^2} \frac{\langle\vec{G}\alpha|t|\vec{0}k_z\rangle}{k_{\vec{G}z}^2 - \alpha^2 + i\epsilon} \tag{4.17}$$

where the matrix elements are given by (4.16):

$$\langle\vec{G}\alpha|t|\vec{0}k_z\rangle = \exp(-i\xi_0) \sum_{\vec{G}'\neq\vec{G}} \sum_\beta C_{\vec{G}'\beta} \langle\alpha|v_{\vec{G}-\vec{G}'}|\beta\rangle \tag{4.18}$$

We insert (4.16), (4.17) and (4.18) into (4.15) and get for $\psi_{\vec{G}}(z)$

$$\psi_{\vec{G}}(z) \, \exp\left(-i\xi_0\right) = \phi_0 \, \delta_{\vec{G},\,\vec{0}} + \frac{2m}{\hbar^2} \sum_{\substack{\alpha \\ Gz}} \frac{\langle \vec{G}\alpha \, | \, t \, | \, \vec{0} \, k_z \rangle}{k_z^2 - \alpha^2 + i\epsilon} \, \phi_\alpha(z) \tag{4.19}$$

From (4.19) and (4.11) we can calculate $\psi(\vec{r})$.
   $\psi(\vec{r})$ is essentially determined by the $\psi_{\vec{G}}$ as $\exp[i(\vec{G}+\vec{K})\,R]$ in Eq.(4.11) is only a phase factor. So we discuss some special cases of $\psi_{\vec{G}}$ for $z \to \infty$:

(a) $\psi_{\vec{0}}(z) \propto \exp\left(-ik_z z\right) + A[1 - 2i\,D_{\vec{0}}^{\vec{0}}]\,\exp(ik_z z)$ \hfill (4.20)

   describes the specular beam.

(b) $\psi_{\vec{G}}(z) \propto B\,\exp[ik_{\vec{G}z}\,z]\,D_{\vec{G}}^{\vec{G}}$ \hfill (4.21)

   is the diffracted beam accompanied by $\vec{G}$.

(c) $\psi_{\vec{G}}^{m}(z) \propto D_{\vec{G}}^{m}\phi_m(z)$ \hfill (4.22)

   are the wave functions describing a bound state in the interaction
   potential $v_0(z)$. $\phi_m(z)$ are the eigenfunctions of (4.14) belonging to
   negative α^2 with the eigenvalues E_n.

   The energy of an atom going into a bound state E_n is larger than E_n
by an amount equal to the kinetic energy associated with its motion
parallel to the surface. This total energy is given by

$$E_{n\vec{G}} = E_n + \frac{\hbar^2}{2m}\,(\vec{K}+\vec{G})^2 \tag{4.23}$$

This energy is not necessarily equal to the energy of incidence,
$E = \hbar^2 k^2/2m$, because the lifetime in the bound state is finite. If, however, within the Heisenberg uncertainty relation, both energies are equal,
we have a resonant transition from the continuum state into a bound state:

$$E = \frac{\hbar^2 k^2}{2m} = E_n + \frac{\hbar^2}{2m}\,(\vec{K}+\vec{G})^2 \tag{4.24}$$

(selective adsorption).
   The intensities in the outgoing beams are given by

$$\frac{I_{\vec{F}}}{I_{\vec{0}}} = R_{\vec{F}} = |\delta_{\vec{F},\,\vec{0}} - 2i\,D_{\vec{F}}^{F}|^2 \tag{4.25}$$

where $\vec{G}$ is replaced by $\vec{F}$ in order to indicate that $k_{\vec{G}z}^2 > 0$ and that $\vec{G} = \vec{F}$
is associated with a final diffracted state. The matrix elements are
given by

$$D_{\vec{F}}^{\vec{F}} \propto \langle F\,k_{\vec{F}z} \, | \, t \, | \, \vec{0}\,k_z \rangle \tag{4.26}$$

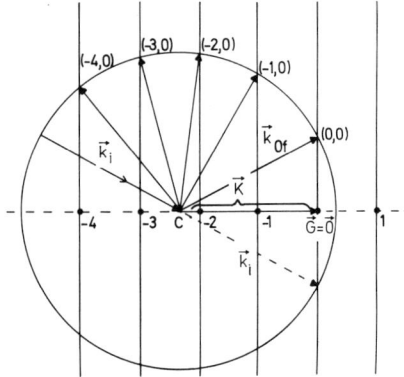

FIG. 20.   Ewald construction in the plane of incidence.   The angular position of the in-plane diffracted beams is given by the intersection of the circle round C with the rods.

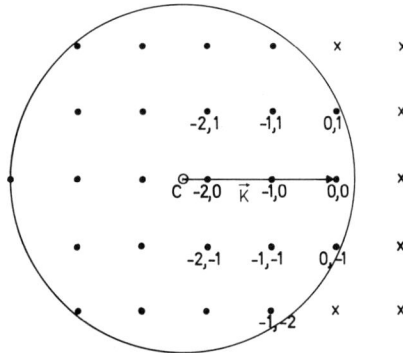

FIG. 21.   Ewald construction in the plane of the surface.   No diffracted beams outside the circle are possible.   The reciprocal lattice vectors are denoted by the order of the diffracted beam with which they are associated.

It can be shown that, as a consequence of energy resonance (4.24), $D^0_{\vec{0}}$ vanishes (CCGM) and therefore

energy resonance: $R_{\vec{0}} = 1$

Resonant transition into bound states E_n should therefore show in a maximum in the elastic-scattered beam and a sharp decrease in the diffracted beams.   In the experiments, however, a sharp decrease is found both in specular and diffracted beams, although the reason for this is not clear.

4.2.   Ewald construction

The Ewald construction is a geometrical method to find the angular positions of diffracted beams.   It is also generally applicable in X-ray

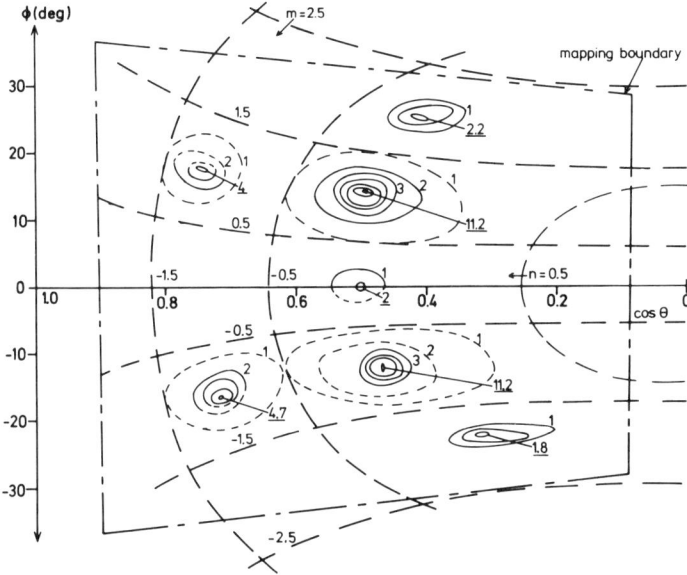

FIG. 22.   Diffraction of He from LiF(001).   The intensity contours are plotted in the cos θ_f, ϕ_f plane. The solid and the short dashed lines are lines of equal flux.   The boundary between each diffracted beam of integer order is indicated by a longer dashed line.   Experimental conditions are: E_i = 58 meV ± 10%; T_S = 300 K; θ_i = 60°; γ = 45° = $\langle 110 \rangle$ azimuth (Gillerlain and Fisher, 1971).

and neutron diffraction in solids.   For surface scattering it is used in a modification for two-dimensional problems.

The reciprocal lattice vectors $\vec{G}$, as defined by Eq.(4.3), span out a plane parallel to the x-y plane.   Each $\vec{G}$ is associated with a point in this plane.   The array of these points is the reciprocal lattice.   If the real lattice is quadratic with spacing d, it is transformed into the reciprocal lattice, also quadratic, with spacing g = $2\pi/d$, and x and y axes are exchanged.   The advantage of this transformation is that we can add wave vectors and $\vec{G}$ vectors because they both have the same dimension: length^{-1}.

A cross-section through a reciprocal lattice is shown in Fig.20 together with the vector diagram of the incoming and outgoing wave vectors.   The diffracted beams are found as intersection of the circle round C with the rods going through the reciprocal lattice points.   It should be noted here that the attractive part of the potential does not influence this construction, as the coherent superposition is made at z → ∞, where the interaction potential vanishes.   A top view of the reciprocal lattice is shown in Fig.21.   Lattice vectors, which are accompanied by final diffracted beams, lie within the circle around C.   No diffracted beams outside the circle are possible so long as the scattering is elastic.

4.3.   Diffraction from alkali halides

Diffractive scattering has been studied in detail on the (001) surfaces of LiF and NaF.   Since the experiments of Estermann, Frisch and Stern

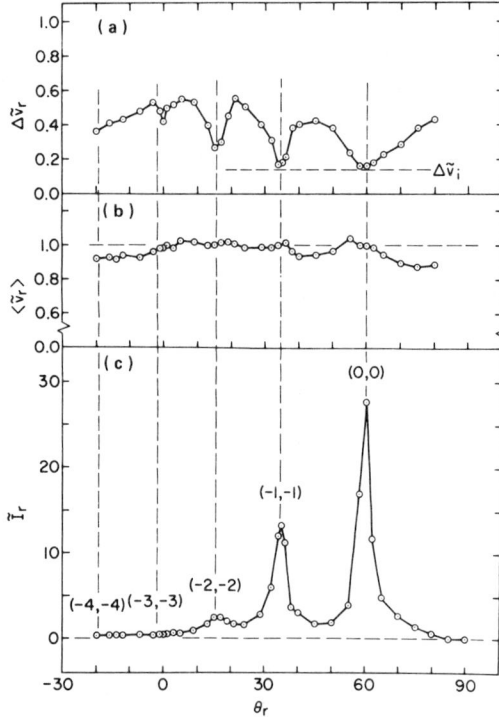

FIG. 23.   Scattering of He on LiF(001) in the $\langle 100 \rangle$ azimuth. Intensity, mean speed and speed distribution half-width versus scattering angle θ_r. Vertical dashed lines are drawn at theoretically predicted nominal diffraction angles (Fisher and Bledsoe, 1972).

(1931), LiF especially has received much attention. Refined experimental techniques have improved the significance of the experimental data within the last years. A scattering experiment with He on LiF(001) performed by Gillerlain and Fisher (1971) is shown in Fig.22. The flux distribution of the diffracted helium atoms is plotted in the θ, ϕ plane. The plot is similar to Fig.21. The position of the diffracted beams, as indicated by the contours of equal intensity, is in good agreement with the expected values. The specular beam is rather small, whereas the (1,0) and (-1,0) beams are the most intense. The experimental parameters were:

E_i = 58 meV, ΔE_i = ± 10%   (He nozzle beam)

θ_i = 60°, T_S = 300 K, γ = 0°

An angular distribution in the plane of incidence is shown in Fig.23. This experiment was done by Fisher and Bledsoe (1972). The experimental conditions are as described above, except γ = 45° and ϕ = 0°, so that the detector scans along the (0,0), (-1,-1), (-2,-2) $\cdots$ peaks. The surface cleanliness was not, either in this or in any other scattering experiment

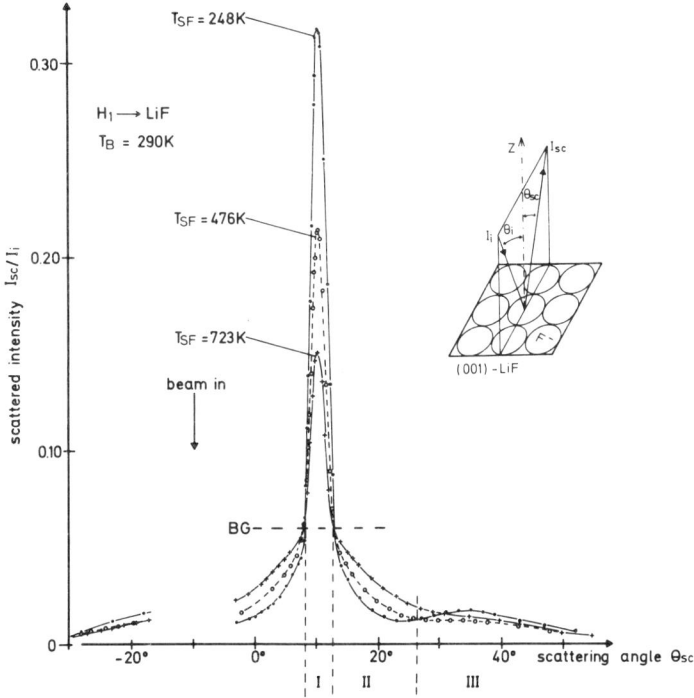

FIG. 24.   Angular distribution of atomic hydrogen scattered from LiF(001) in the $\langle 110 \rangle$ azimuth for three surface temperatures. Region I: specular beam I_{00}; Region II: inelastic events; Region III: first-order diffraction peak I_{10}. Inelastic background subtraction (BG) shown for T_S = 732 K (Hoinkes et al., 1972b).

on alkali halides, controlled by independent methods and this is a short-coming. There has been great experience, however, with these materials, especially with LiF. Surfaces both cleaved in air and cleaved in ultrahigh vacuum show the same properties. Ellipsometric studies show that LiF(001) is stable, even if it is exposed to water vapour up to 10^{-2} Torr (Bayh (1973)). Evidence for surface cleanliness is drawn in this experiment from the fact that, by increasing the surface temperature, the diffracted intensity decreases. This is in agreement with the trend expected on the basis of the DWF. Surface contamination would result in the opposite effect: the diffracted intensities would increase as the contaminants desorb.

   The temperature dependence of the diffracted intensity has been studied by Hoinkes et al. (1972a,b) with H on LiF. The angular distribution of the scattered flux density at various surface temperatures is shown in Fig. 24.

   A beam with Maxwellian velocity distribution (T_g = 290 K) is incident on LiF(001) at θ_i = 10°. The $\langle 110 \rangle$ direction was in the plane of incidence. At the lowest surface temperature, T_S = 248 K, the specular peak is very pronounced and amounts to up to 33% of the incoming beam. The first-order diffraction peak also occurs at this temperature (region III). It is

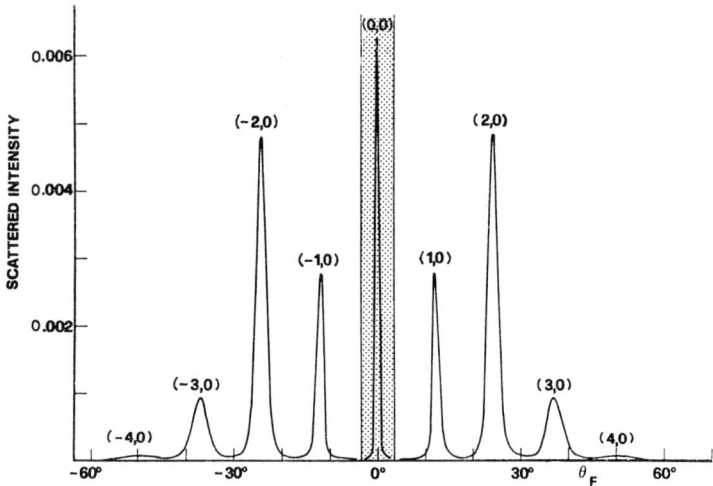

FIG. 25.   Diffraction of He from LiF(001) at T_S = 10 K. θ_i = 0°.   The detector scans along the $\langle 110 \rangle$
azimuth.   The specular intensity is extrapolated from measurements at small θ_i.   A nozzle beam with
narrow velocity distribution was used. $E_i \approx 58$ meV.   (From unpublished results of Boato et al., 1974.)

smeared out due to the spread in velocity (wavelength). At higher surface
temperature the diffracted intensity is reduced and an inelastic shoulder
grows round the specular peak (region II).

By going to lower surface temperatures, the DWF increases until the
mean square displacement of the solid atoms is determined essentially
by the zero-point motion. This results in considerable increase in
diffractive scattering over inelastic scattering. This can be seen from
a recent experiment by Boato et al. (1974), shown in Fig.25. The diffraction pattern is shown at normal incidence (the specular beam is extrapolated from an experiment at small θ_i). The in-plane distribution shows
peaks up to the order (4,0). Almost no inelastic background is seen between
the peaks. This is for two reasons:

(a)   The phonon density is very low at T_S = 10 K, so that phonon
       annihilation is negligible and phonon creation greatly reduced.
(b)   A bolometer detector was used, which is sensitive to the energy-
       flux density multiplied by the energy-accommodation coefficient,
       rather than to the particle-flux density. So if atoms had lost
       energy to the surface they would give a reduced detector signal.

The intensities in the diffraction peaks are quite different. There
is no monotonic decay from the specular peak to the higher order. The
intensity falls off to the (1,0) peak, rises again at the (2,0) peak and then
finally falls along (3,0) and (4,0) to zero.

Such behaviour is still more pronounced in the diffraction of Ne from
LiF, as shown in Fig.26. T_S was kept at 80 K to prevent contamination
from the beam. The $\langle 110 \rangle$ direction was oriented in the plane of incidence.

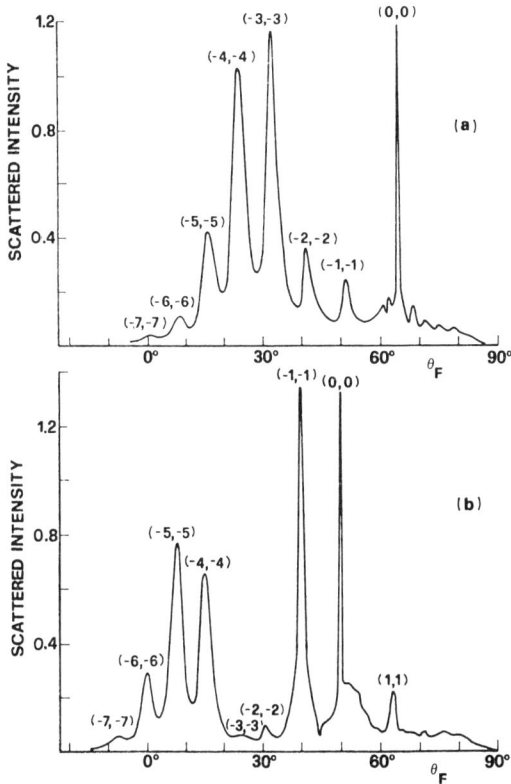

FIG. 26.   Diffraction of Ne from LiF(001) at T_S = 80 K in the $\langle 100 \rangle$ azimuth. (a) θ_i = 65°; (b) θ_i = 50°.
Ne nozzle beam, $E_i \approx 58$ meV. (From unpublished results of Boato et al., 1974.)

In Fig.26(a) θ_i was 65°. Maxima until the order $(-7,-7)$ are resolved. The
specular peak is of lower intensity than the higher-order peaks, as these
are broadened due to the velocity distribution in the beam. An envelope
over the peak heights has a maximum at the $(-3,-3)$ peak and a second
one at the specular peak.

In Fig.26(b) θ_i is smaller (50°) and the $(-1,-1)$ peak is now the highest
one. An envelope over the peaks again has two maxima, one around θ_r = 10°
and a second around θ_r = 45°. The results have not yet been analysed by
using the CCGM theory because they are rather new. So the authors pro-
pose a simplified model developed by Garibaldi et al. (1974), which is
called the 'quantum theory of surface rainbow'. We remember that the
occurrence of two intensity maxima is a typical feature of surface rainbows, and the gross intensity distribution may be regarded as due to
rainbow scattering, whereas the subsequent superposition of the coherent
wave functions may lead to a separation into diffraction peaks.

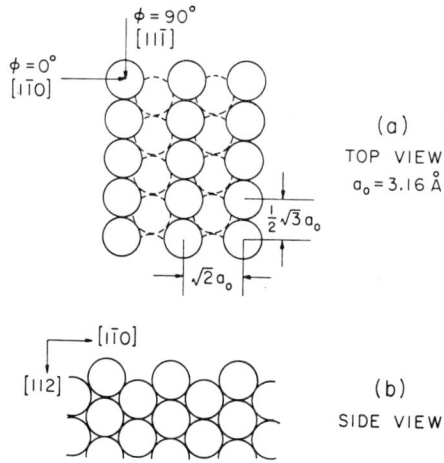

FIG. 27. Surface structure of W(112), schematically (Tendulkar and Stickney, 1971).

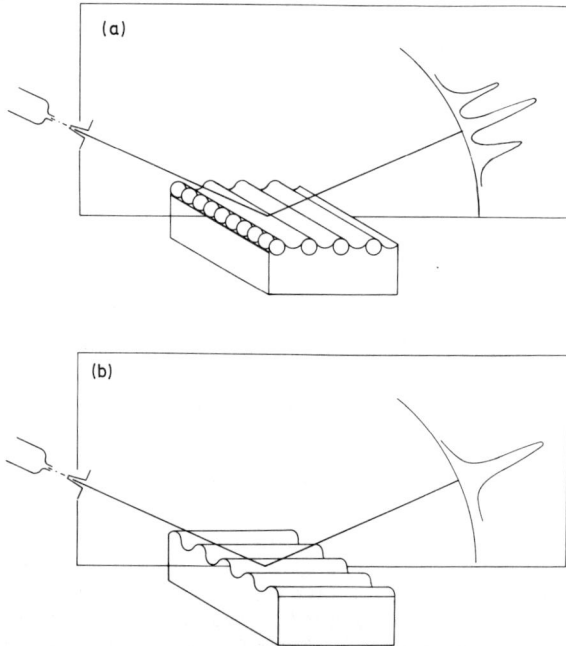

FIG. 28. Scattering geometry for diffraction from a surface with one-dimensional structure. (The angular distributions of Figs 29 and 30 are taken with the W(112) surface in position (a).)

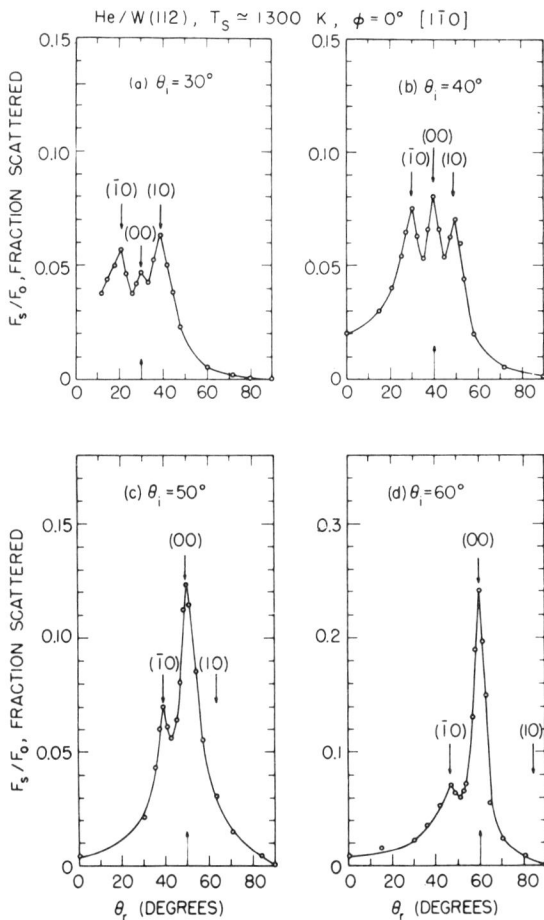

FIG. 29. Diffraction of He from W(112) for four angles of incidence. Beam conditions: He nozzle beam at $E_i \approx 58$ meV. Arrows indicate the expected positions of the diffraction peaks (Tendulkar and Stickney, 1971).

4.4. Diffraction from metals

While alkali halides readily show diffraction, many attempts that have been made to find diffraction from metal surfaces have remained without result. This has been a puzzle for a long time. It was quite recently that atomic diffraction has been observed on W(112) (Tendulkar and Stickney (1971); Stoll and Merrill (1973)). The reason becomes clear if we look at the (112) surface structure of W (Fig. 27).

This surface consists of rows of close-packed atoms at an inter-nuclear distance of 2.74 Å. The rows are separated from each other by a distance of 4.47 Å in the $\langle 110 \rangle$ direction. Diffraction has only been

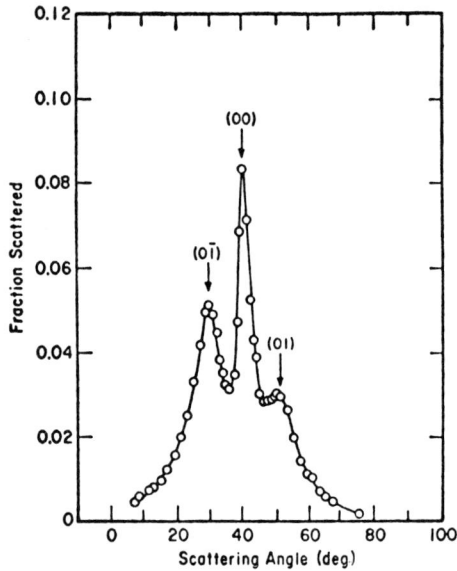

FIG. 30.   Diffraction of He from W(112).   Thermal He beam at T_g = 300 K.   (Stoll and Merrill, 1973.)

observed while the $\langle 110 \rangle$ direction was in the plane of incidence. No
structure effect was seen when the crystal was rotated round 90°, so that
the beam was incident along the rows. This is illustrated in Fig. 28. The
diffraction pattern for several angles of incidence is shown in Fig. 29.
(The distributions with the crystal rotated   round 90° are not shown here.)
The surface temperature was kept at T_S = 1300 K, because otherwise the
surface became contaminated. The background pressure was 10^{-9} Torr.
The arrows indicate the calculated position of the diffracted peaks. Compared to the alkali halides, the diffracted intensity is surprisingly large
at that surface temperature, but an estimate of the DWF with
$D \approx 5.8$ meV = $E_i/10$ (estimated) and $T_D \approx 270$ K gives for θ_i = 60°:

$$I_{el}/I_0 = e^{-2W} \approx 0.37$$

which compares well with the experimental findings.

    This experiment was repeated by Stoll and Merrill (1973), who con-
trolled the surface by LEED and AES. The results (Fig. 30, Maxwellian
He beam) essentially repeat the results of Tendulkar and Stickney, showing
that the diffraction is not due to an overlayer of tungsten carbide.

    The conclusion of these experiments is that this surface shows very
little structure if we proceed along a direction where the atoms are close
packed, and only appreciable spacing between the atoms leads to a
structure effect. From this we can conclude that the electron wave
function smears out the interaction potential along a distance of the order
of 2.7 Å, whereas the potential reproduces the structure of the lattice
only if the distance is of the order of 4.5 Å.

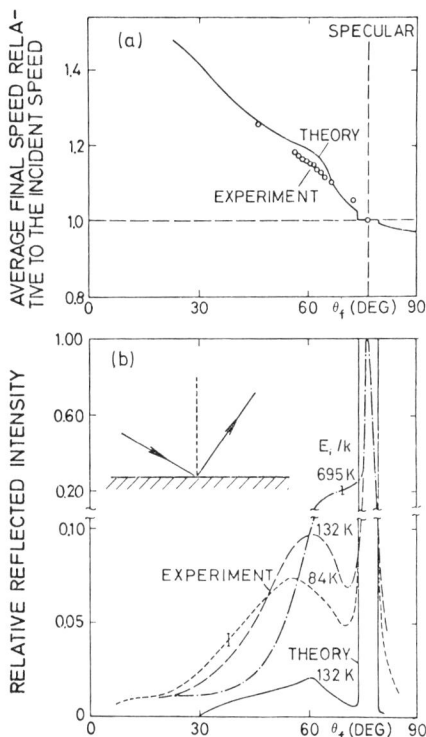

FIG. 31.   Inelastic scattering of He from Ag(111). Experimental results of Subbarao and Miller (1972) compared with calculations of Goodman (1972). Note the broken scale.

4.5. Inelastic scattering

We consider the interaction of a gas atom with a single phonon:

$$E_f - E_i = \pm \hbar\omega \tag{4.27}$$

$$\vec{K}_F - \vec{K}_i = \vec{G} \pm \vec{Q} \tag{4.28}$$

To measure a phonon spectrum $\omega(\vec{Q})$, which may be called a surface phonon spectrum, as opposed to $\omega(\vec{q})$, we have to measure two quantities: $\vec{K}_f$ (which is easy) and E_f (which is difficult). Even then the assumption $q_z = 0$ must be made, otherwise an ambiguity in the interpretation of the experimental results would arise.

There are three experiments that show indications for single-phonon events. The interpretation is not completely clear, however. Nevertheless, the experimental results are briefly discussed, together with the attempts that have been made to interpret them in terms of single-phonon scattering.

The first experiment was done by Subbarao and Miller (1969, 1972). A helium beam with narrow velocity distribution was scattered from

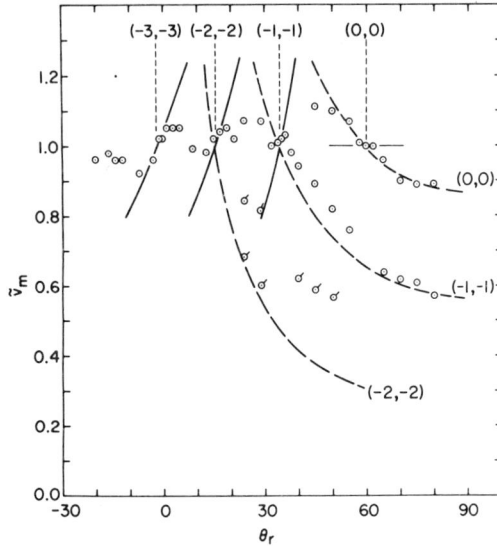

FIG. 32.   Time-of-flight analysis for He scattered from LiF(001) in the $\langle 100 \rangle$ azimuth (Fisher and Bledsoe, 1972).

epitaxially grown Ag(111). The result is shown in Fig.31. The surface temperature was T_S = 550 K (where Ag grows epitaxially). This is considerably greater than the Debye temperature of Ag, which is 255 K for the bulk and 155 K for the surface. Therefore, many-phonon processes should predominate. Three beam energies have been used: E_i = 60 meV, 11.4 meV and 7.8 meV (E_i/k_B = 695 K, 132 K and 48 K). The angular distributions of the scattered flux (Fig.31(b)) show a rather sharp specular peak. It is normalized to unity because the incident flux was not measured. The FWHM of that peak is about equal to the incoming beam. The specular peak is accompanied by an inelastic satellite which is broad and shifted towards the normal. The shift increases by decreasing E_i. The lobe cannot be assigned to a diffraction process, or to another effect of surface structure.

   This experiment was analysed by Goodman (1972) by an inelastic quantum theory based upon CCGM. Goodman used a simplified model of the solid with the following assumptions:

(a) Transfer of tangential momentum is neglected. The surface is effectively flat ($\vec{G} = \vec{Q} = \vec{0}$).

(b) The solid can be represented by an elastic Debye-type bulk continuum model.

(c) The interaction potential is a step potential without an attractive part (as with the hard-cube model).

(d) Only zero-phonon or one-phonon processes occur.

With these assumptions the qualitative feature of the experiment is reproduced (see Fig.31). In the theoretical curve (shown for E_i/k_B = 132 K)

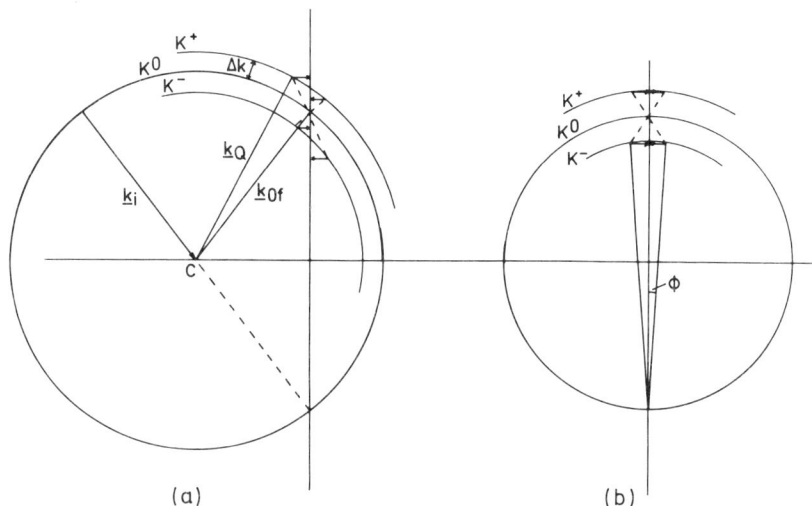

FIG. 33. Ewald construction for phonon scattering. Contribution of single-surface phonon modes is considered (a) in the plane of incidence, (b) perpendicular to this plane.

a detector resolution of 6° has been assumed, whereas the experimental resolution was ~3°. Even then, the quantitative agreement is bad. The inelastic intensity is at least one order of magnitude below the experimental value.

The same experiment was analysed by Beeby (1972), who showed that the ratio inelastic intensity : elastic intensity becomes closer to the experimental value if an attractive potential between He and Ag(111) (the exact value of which is unknown) is introduced. Beeby claims, however, that, under the given experimental conditions, multiphonon processes should preponderate (and therefore the inelastic peak becomes smaller again).

In the same experiment an energy analysis of the scattered atoms has been made by time-of-flight techniques. The experimental points in Fig.31(a) show that the energy is unchanged in the specular peak (as it has to be if the peak is truly elastic). The final energy increases with decreasing θ_f, indicating that the atoms gain energy by phonon annihilation. The theoretical curve is in rather good agreement with the experiment. Beeby's calculation for the same case (not shown here) gives similar agreement for the final speed data.

Another experiment was performed by Fisher and Bledsoe (1972), who measured the final speed of helium atoms in the vicinity of the diffraction peaks shown in Fig.23. The final speed data are shown in Fig.23(b) and (c). The plot in Fig.23(a) gives the speed distribution full width at half maximum (FWHM). The data are normalized by a factor which is chosen so that $\Delta \tilde{v}_r = 1$ belongs to a Maxwellian velocity distribution with the temperature of the surface (i.e. a completely thermalized beam).

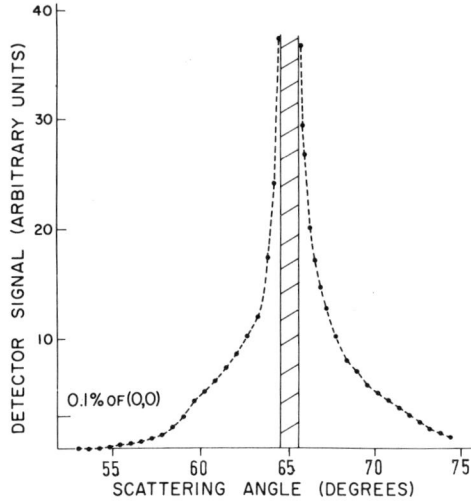

FIG. 34.   Inelastic scattering of He from LiF(001). Scan in the vicinity of the (00) peak, $\langle 100 \rangle$ azimuth. Helium nozzle beam at $E_i \approx 58$ meV; $T_S \approx 150$ K (Williams, 1971b).

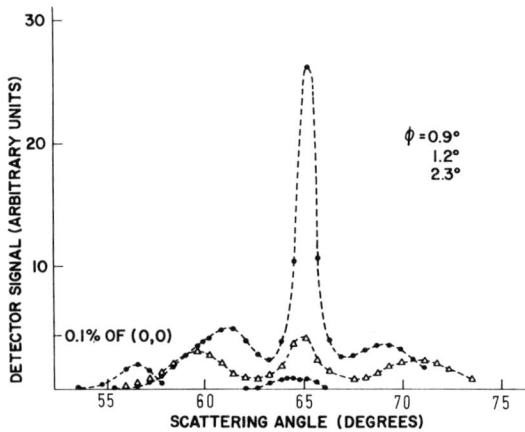

FIG. 35.   Inelastic scattering of He from LiF(001). Scan in the vicinity of the (00) peak at small out-of-plane angles (Williams, 1971b).

The velocity distribution showed up to three maxima depending on the scattering angle. These are the points shown in Fig.32. The dashed curves are theoretical curves calculated for the case that the tangential momentum of the gas atom is only changed by a $\vec{G}$ vector (giving rise to the diffraction peaks), but not by a $\vec{Q}$ vector. This is identical with the assumption $\vec{Q} \equiv 0$ for all $\vec{Q}$; only normal modes of lattice vibration contribute to the collision process. This assumption was made for the interpretation of the He-Ag(111) experiment, too. The solid lines indicate the positions that each diffraction peak would have if the beam velocity

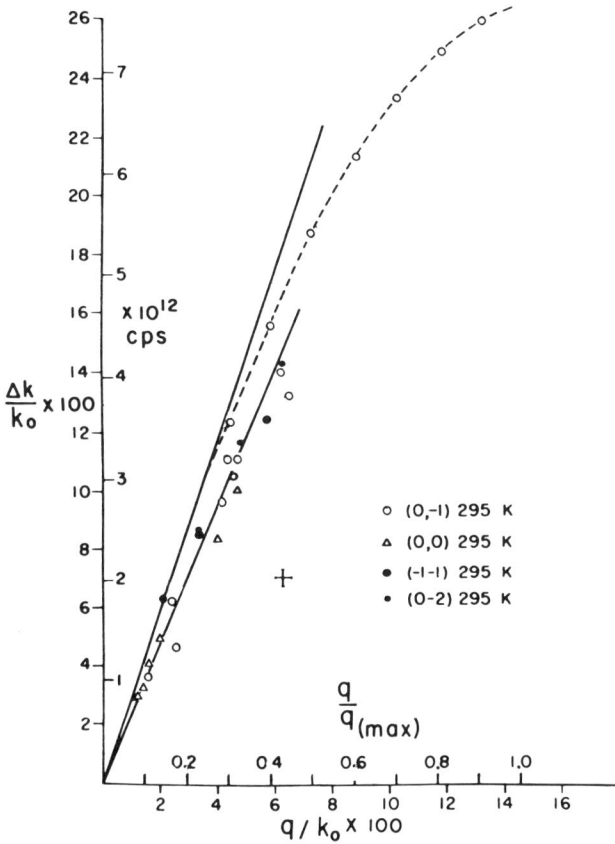

FIG. 36.   Phonon dispersion relation for surface phonons of the Rayleigh mode (Williams, 1971b).

was not equal to the nominal velocity. They are representative, therefore, of the velocity distribution of the incoming beam. The measurements show that:

(a) One-phonon processes are predominant;
(b) Phonon creation is more probable than phonon annihilation;
(c) Energy exchange occurs with little change of tangential momentum.

A complementary experiment was carried out by Williams (1971b). In this experiment no energy analysis of the scattered atoms was possible. However, the angular resolution was very good (FWHM of beam + detector was 0.6°) and out-of-plane measurements could be made. So Williams could resolve relatively small phonon vectors $\vec{Q}$ by out-of-plane measurements. The situation is illustrated by Fig.33.

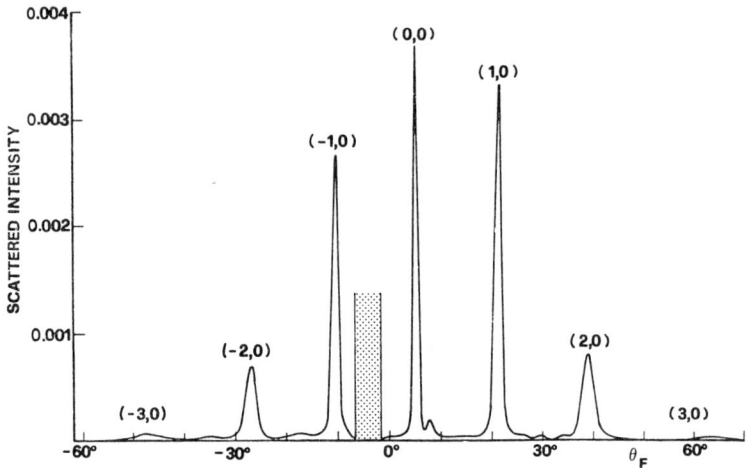

FIG. 37.   Excitation of rotational transitions in the scattering of H_2 from LiF(001). T_S = 80 K, $\langle 110 \rangle$ azimuth. (Boato et al., 1974.)

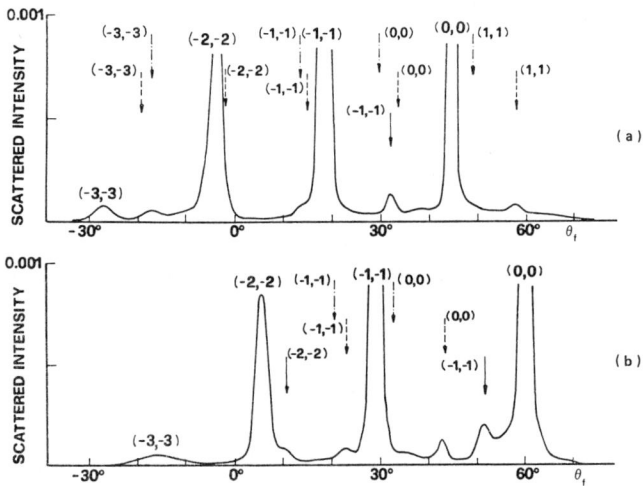

FIG. 38.   As Fig. 37; $\langle 100 \rangle$ azimuth.

From the law of conservation of energy we get $|k_f|$ by Eq.(4.27):

$$\frac{\hbar^2 k_f^2}{2m} = \frac{\hbar^2 k_i^2}{2m} \pm \hbar\omega$$

The sphere representative for phonon annihilation and creation is denoted by $|k^+|$, $|k^-|$, respectively.

To find the angular position of an atom scattered by a phonon (and some $\vec{G}$ vector) we must restrict our discussion to surface phonons with

$q_z = 0$. This seems to contradict the findings of Fisher and Bledsoe, who could explain their data by neglecting $\vec{Q}$, assuming bulk phonons only. But we shall see that the $\vec{Q}$ vectors are small indeed and that their contribution might have been averaged out by the rather poor angular resolution of Fisher and Bledsoe.

The surface phonon modes with the greatest $\vec{Q}$ are the Rayleigh modes, and we shall restrict our discussion to these modes for simplicity. So the parallel component of a scattered atom is given by

$$\vec{K}_f + \vec{Q} = \vec{K}_i + \vec{G} \qquad\qquad\qquad (4.29)$$

where a single interaction with a surface phonon $\vec{Q}$ is assumed. In the Ewald construction Eq.(4.29) gives us the condition that $\vec{K}_f + \vec{Q}$ ends on a rod, whereas k_f must end on a sphere. This situation is drawn in Fig.33, both in the plane of incidence and perpendicular to this plane. The largest Q vectors therefore give the largest angular separation from the diffracted beam.

The experiments have been made with He on LiF. Similar experiments with Ne on LiF showed diffraction and some structure, but could not be analysed in terms of one-phonon events (Williams (1971a)).

The scattered flux density is plotted versus θ_r. Angle of incidence and surface temperature were fixed: $\theta_i = 65°$; $T_{SF} = 150$ K. In Fig.34 an in-plane distribution is shown in the vicinity of the specular peak. An inelastic shoulder appears similar to that for H-LiF (Fig.24), but smaller (the intensity in the (00) peak was about 3000 detector units). In Fig.35 an out-of-plane distribution is shown in the vicinity of this peak. Structure appears in the inelastic shoulder. The maxima separate with increasing out-of-plane angle. This separation is (following Williams) due to the interaction with Rayleigh modes, as described by Eq.(4.29).

From the out-of-plane measurements (which are not all shown), Williams extracted a dispersion relation for the Rayleigh modes. This is shown in Fig.36. The experimental points are taken from the measured inelastic streaks using Eqs (4.27) and (4.29). The estimated error is indicated. The upper solid line is calculated from the bulk transverse sound velocity. The dashed line is a dispersion relation for bulk phonons with greater q measured by neutron scattering. The lower solid line represents the Rayleigh surface wave velocity calculated in the $\langle 100 \rangle$ direction. (References are given in Williams' paper.) The good agreement of the data with the Rayleigh mode dispersion relation supports the conclusion that the assumption — no bulk modes contribute to the scattering at large out-of-plane angles — was correct.

Another type of inelastic scattering occurs if not the solid but the gas particle is excited (or de-excited) by the collision. If this is a molecule, such a type of inelastic collision can occur at thermal energy. Discrete transitions between rotational levels in H_2 have been observed very recently by Boato et al. (1974). Their results are shown in Fig.37. An incident H_2 beam with narrow velocity distribution was diffracted from LiF(001) at $\theta_i = 5°$ and $\gamma = 0°$. The surface was kept at $T_S = 80$ K, thus preventing contamination of the surface by the beam. Diffraction orders up to (2,0) are clearly resolved and appear at the expected positions. Between these peaks there are very small additional peaks. This is better demonstrated in Fig.38, where θ_i was at 45° and 60°, and $\gamma = 45°$ ($\langle 100 \rangle$

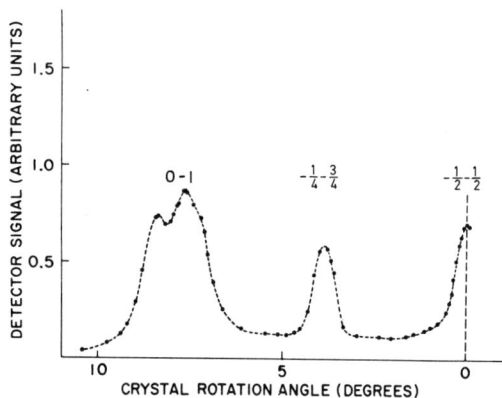

FIG. 39. Diffraction of He from an adsorbed layer of C_2H_5OH on LiF(001). The additional peaks are comparable in height to the original (0,-1) peak of LiF (Mason and Williams, 1972).

direction in the plane of incidence). The scale factor is enlarged, so that the additional peaks become more visible.

The arrows in Fig.38 indicate expected rotational transitions between the states $0 \to 2$, $2 \to 0$ and $3 \to 1$. Their position is calculated by assuming that

$$\frac{\hbar^2 k_f^2}{2m} = \frac{\hbar^2 k_i^2}{2m} \pm E_{rot} \qquad (4.30)$$

and

$$\vec{K}_f = \vec{K}_i + \vec{G} \qquad (4.31)$$

the tangential wave vector being only changed by diffraction but not by the rotational transition. E_{rot} is given for neighbouring states in $o-H_2$ or $p-H_2$ by

$$E_{rot} = 2(2\ell - 1) E_r \quad (\ell \geq 2) \qquad (4.32)$$

with $E_r \approx 7.3$ meV.

4.6. Study of adsorbed layers

Atomic scattering is very sensitive to adsorbed layers, probably much more so than LEED. In early surface scattering experiments a $\cos\theta_f$ distribution was almost invariably observed. When the experiments were repeated with a clean surface, the cosine distribution disappeared and the scattering pattern became visible. So cosine distribution was regarded as an indication that the surface was contaminated.

Hence the study of the rate at which an adsorption layer is built up is a very simple experiment. One uses a beam that gives a narrow scattering distribution, e.g. a He or H beam. Then the peak height is monitored while the adsorbing gas is allowed to strike the surface.

• $(1-1)$ • $(\frac{1}{2}-\frac{1}{2})$ ⊙ (00)

• $(\frac{1}{4}-\frac{3}{4})$

$(-\frac{1}{4}-\frac{3}{4})$

$(\frac{1}{4}-\frac{5}{4})$•   •   • $(-\frac{1}{2}-\frac{1}{2})$

• $(-\frac{1}{4}-\frac{5}{4})$

• $(-1-1)$

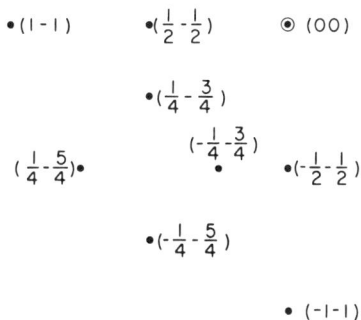

FIG. 40. Reciprocal lattice representing the observed diffraction peaks from the C_2H_5OH overlayer on LiF(001) (Mason and Williams, 1972).

The adsorption of C_2H_2 on Pt(111) has been studied by Smith and Merrill (1970). The surface has been monitored by LEED and AES. A similar study of CO and C_2H_2 on Pt(100) was performed by West and Somorjai (1971).

If the adatoms form a regular array on the surface, they themselves can act as diffraction grid. The surface Debye temperature is then given by the mass of the adatom and the bond strength. So chemisorption of light adatoms can result in rather high Debye temperatures. The (3×5) overlayer of tungsten carbide on W(110) is estimated to have $T_D \approx 1200$ K. Diffraction of D_2 and He on this overlayer has been reported by Weinberg and Merrill (1972).

Another study of diffraction from an ordered overlayer was performed by Mason and Williams (1972). They investigated adsorption of a number of molecules (CCl_4, C_6H_6, CH_3OH, C_2H_5OH and H_2O) on LiF (001) around $T_S = 150$ K. All the molecules under study reduced the diffracted intensity considerably. There was no means, however, of relating this to a quantitative degree of adsorption (in terms of atoms adsorbed per cm^2). But while all the other molecules formed an unordered overlayer, C_2H_5OH formed a regular array in a narrow temperature range 133 K $\lesssim T_S \lesssim 153$ K. This array was studied in great detail.

A number of new diffraction peaks appeared, some of them shown in Fig. 39. The intensity of both the new peaks and the original LiF peaks, which were still present, was smaller by about two orders of magnitude than the diffraction peaks of the clean surface. The reciprocal lattice representing the observed diffractive peaks is shown in Fig. 40. A possible arrangement for the adsorbed molecules consistent with the reciprocal lattice is shown in Fig. 41. The square arrangement is ruled out by the absence of a peak at the $(0,-1)$ LiF location.

The half-order peaks $(-\frac{1}{2},-\frac{1}{2})$ suggest an arrangement of the adsorbed molecules having a separation between rows of scattering centres which is twice that of LiF. Thus only a face-centred rectangular array is consistent with the reciprocal lattice. The authors suggest that the OH group is bound to the ionic surface and the methyl group is removed from the surface, this leading to the rather close packing shown in Fig. 40.

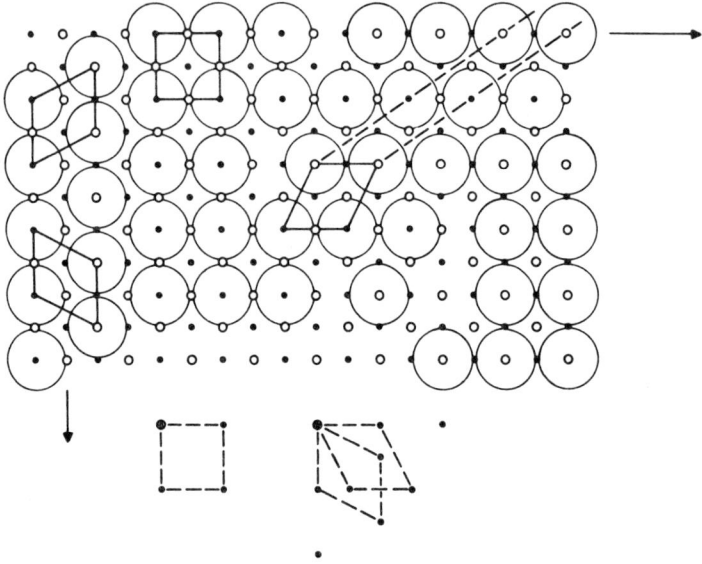

FIG. 41. Possible arrangement of C_2H_5OH on LiF(001). The large circles represent the approximate molecular diameter of the methyl group (Mason and Williams, 1972).

4.7. Surface Debye temperature

The atoms in the topmost layer of a solid are bound to a smaller number of neighbours than the bulk atoms are. So the bond strength (and therefore the Debye cut-off frequency and the Debye temperature) is smaller. As the surface atoms are in thermal equilibrium with the solid, they compensate the smaller bond strength by an increased amplitude of their thermal motion.

The Debye temperature of many surfaces has been studied by LEED. It should be mentioned, however, that electrons, as they penetrate into deeper layers of the solid, do not only probe the mean square displacement of the surface, but also that of the bulk. Therefore the surface Debye temperature is studied with electrons of very low energy ($E_i \lesssim 20$ eV) and the assumption is made that these electrons do not penetrate.

If one wishes to study insulator surfaces rather than metals, then problems with the static electrification may arise, and some surfaces have a tendency to dissociate under electron bombardment. So it is reasonable to study the surface Debye temperature of insulator surfaces with atomic scattering rather than with LEED.

By recalling the expression for the DWF:

$$I_{sp} = K \cdot I_i \, e^{-2W}$$

$$2W = 24\mu \, \frac{E_i}{k_B T_D^2} \left[\cos^2 \theta_i + \frac{D}{E_i} \right] T_S \qquad (4.33)$$

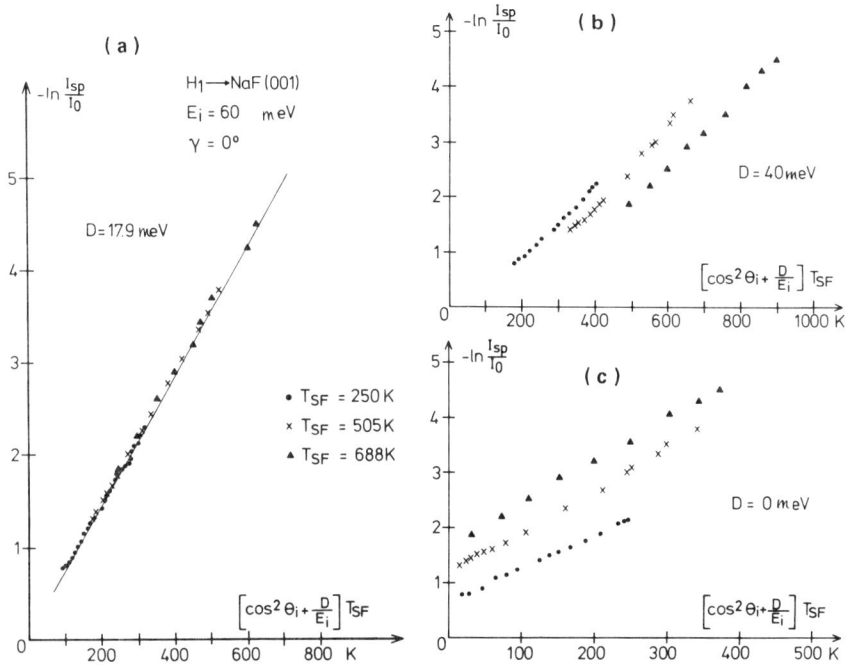

FIG. 42.   Surface Debye temperature of NaF studied by diffraction of atomic hydrogen.   A test for the role of the attractive well is made in (b) and (c).   (Wilsch et al., 1974.)

we find that we have to measure the specular intensity and K as function of any of the parameters E_i, θ_i or T_S, in order to get T_D.

A study of the surface Debye temperature by means of atomic beam scattering was first performed by Hoinkes et al. (1972b). The method is briefly described as follows. With the angular distribution at given θ_i, E_i, T_S, the elastic intensity is separated from the inelastic intensity (the 'shoulder' in Fig.24). This is repeated for each T_S and a representative set of angles of incidence. (In the case of H on LiF (and on NaF) the procedure is greatly simplified by the fact that almost all intensity is specular.) Now the logarithm of I_{sp}/I_i is plotted versus $(\cos^2 \theta_i + D/E_i) \cdot T_S$. The experimental points are expected to lie on a straight line the slope of which gives the Debye temperature.

As an example, a more recent measurement will be shown (Wilsch et al. (1974)). The surface Debye temperature of NaF was studied by using atomic beams of H and D, both Maxwellian and nearly monochromatic. The result for an H beam with E_i = 60 meV is shown in Fig.42(a). From the slope we get for the Debye temperature

NaF: T_D = 370 K ± 10%

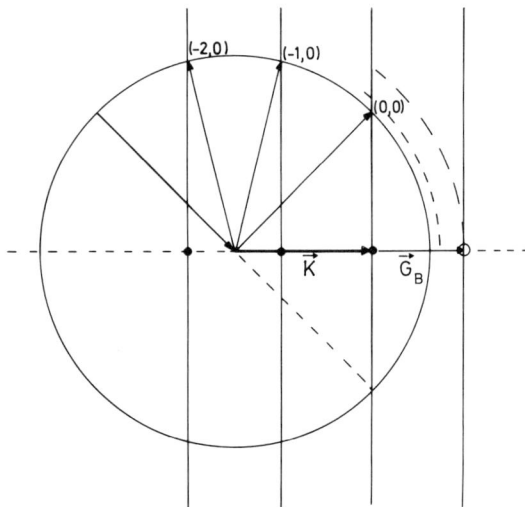

FIG. 43. Ewald construction in the plane of incidence. Dashed lines represent the energy of bound states in the interaction potential.

Figure 42(b) and (c) shows a test which was made to discover whether or not the attractive well plays a role in the DWF. In (b) an acceleration of the incoming atom by an (arbitrary) well depth of 40 meV was assumed. In (c) the acceleration was neglected (D = 0). In both cases the experimental points do not fall so well on a straight line as they do with D = 17.9 meV. (For the determination of D see the next section.) This lends support to the assumption (that we previously made) that the atoms are first accelerated before they notice the structure of the surface as it is disturbed by the thermal motion of the solid atoms.

Here again, there is a discrepancy in our conception of the scattering process. On one hand: a probing particle which must be accelerated before it notices the repulsive potential. On the other hand: an incident plane wave, coherent over many lattice spacings and interfering at great distance from the surface.

4.8. Resonant transition into bound states and the interaction potential

In this section we again consider the interaction potential. It was previously considered when the summation over pairwise potentials was discussed. Then we said that the pairwise parameters (and the whole method) is not reliable enough to calculate $V(\vec{r})$. Further experimental results are needed, either in the form of rainbow angles or as a precisely known 'master' potential from which we can extrapolate to other gas-solid systems by use of pairwise potential parameters fitted to the 'master' potential.

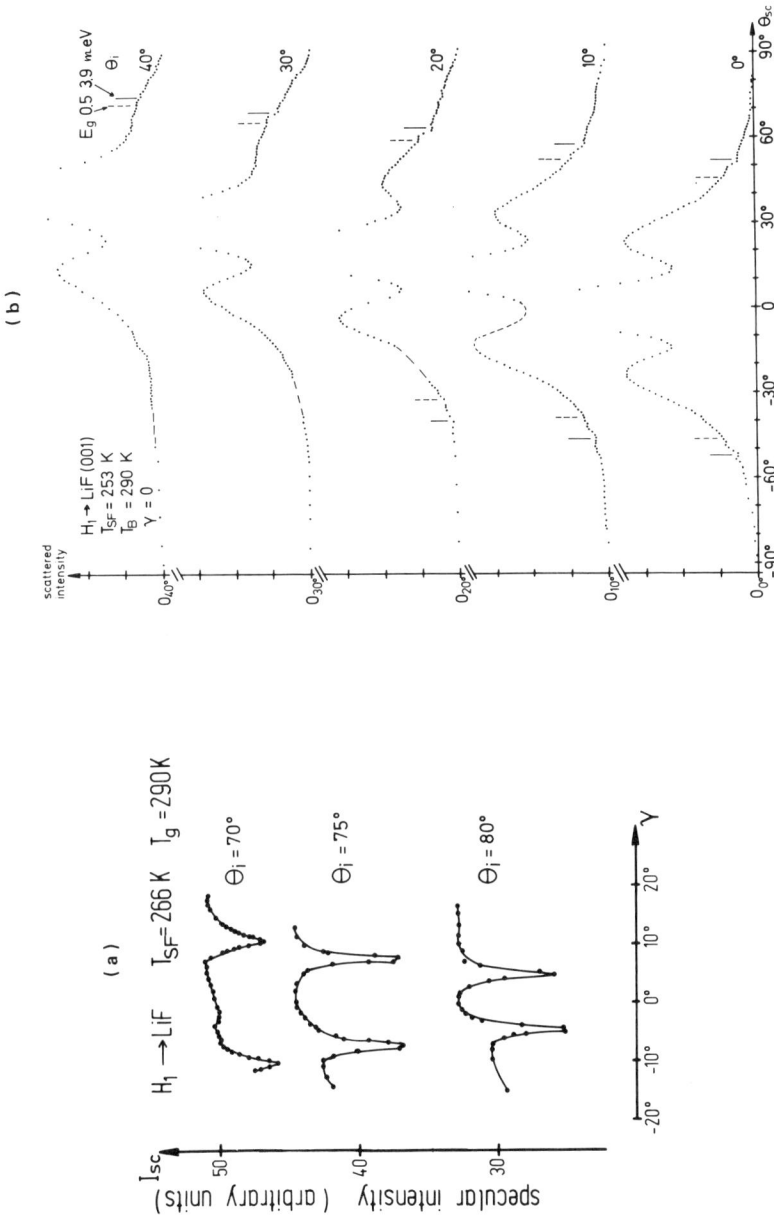

FIG. 44. Resonant transition of H into bound states on LiF(001) (selective adsorption), as indicated by minima in the angular distribution (b). Similar minima occur in the specular peak if the crystal is rotated round γ (a). (Results from Hoinkes et al., 1972a.)

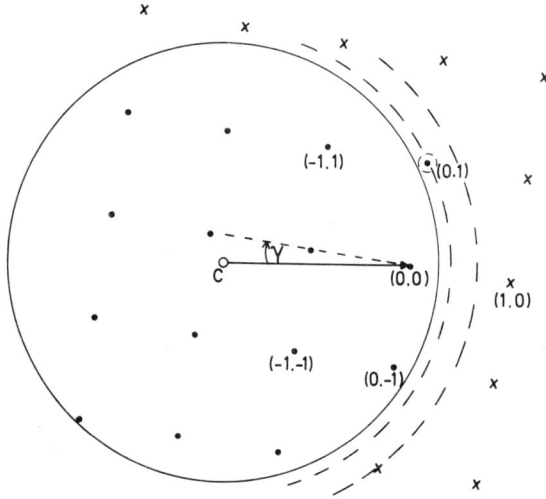

FIG. 45. Ewald construction for the transition into bound states. Surface plane is in the paper plane.

We shall now discuss resonant transitions into bound states (also denoted by selective adsorption) and shall see how this method can give us the energies of bound states of the gas-atom wave function in the inter-action potential. The conditions for resonant transitions into bound states are given by Eq.(4.24), which reads:

$$\frac{\hbar^2 k_i^2}{2m_g} = E_n + \frac{\hbar^2}{2m_g} (\vec{K} + \vec{G})^2 \tag{4.34}$$

where E_n are the eigenvalues of the Schrödinger equation (4.14) for $k_{\vec{G}z}^2 < 0$. We use the Ewald construction to demonstrate the situation (see Fig.43).

Elastic scattering is restricted to final wave vectors, which end on the sphere (the circle in Fig.43 is a cross-section through the centre of the elastic sphere). Phonon events will be neglected. Then no possibility exists for an atom to go into a final state associated with the reciprocal lattice vector $\vec{G}_B$ (except for a time which is short enough for the uncertainty relation to allow this transition). An extra energy ΔE would be necessary, so that

$$\frac{\hbar^2 k_i^2}{2m} + \Delta E = \frac{\hbar^2}{2m} (\vec{K} + \vec{G}_B)^2$$

This extra energy, which we need to fulfil the energy conservation law, is indicated by the dashed circle in Fig.43. If now the experimental condition is chosen so that ΔE equals a bound-state energy: $\Delta E = - E_n > 0$, this transition becomes possible.

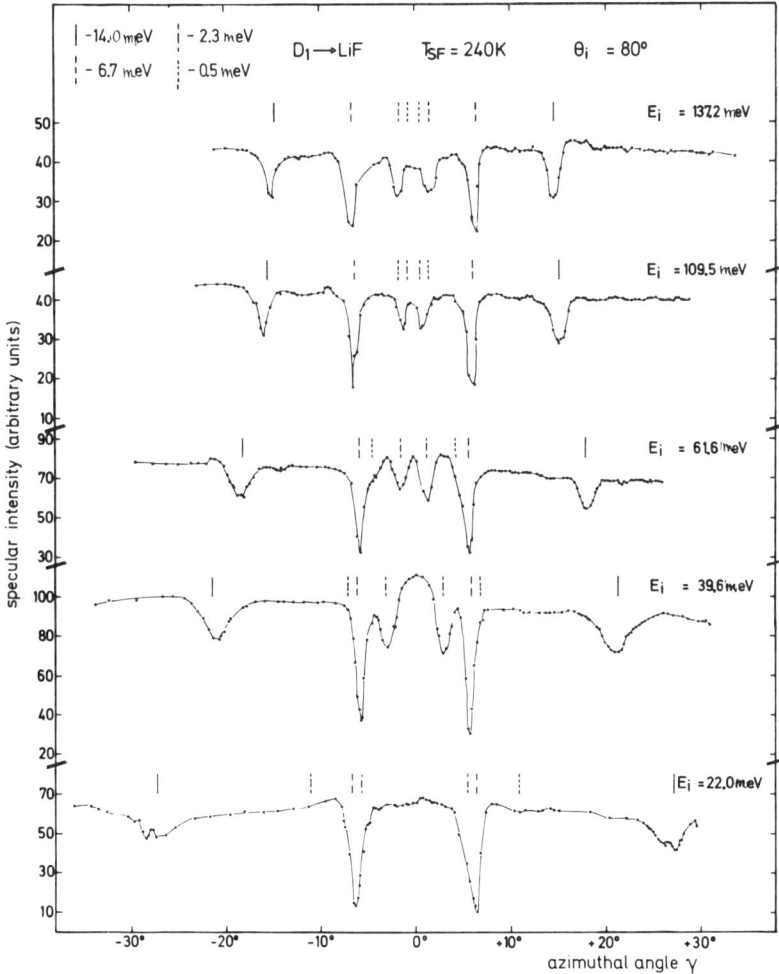

FIG. 46.   Resonant transitions of D into bound states on LiF(001) (Frank, 1973).

It has been shown by experiment that the transition probability can
be rather large (in excess of 50%) and that the transition results in a
minimum in all elastic channels (including the specular beam). The reason
may be (this is speculative) that the atoms, once in a bound state, become
incoherent before they can return into a channel associated with a diffracted
beam (this should be only the (00) beam, according to CCGM).

The first observation of selective adsorption in the system H-LiF(001)
is shown in Fig.44(b). Here the angular distribution is shown with
θ_i = 0°, 10°, 20°, 30° and 40°, respectively, and γ = 0. A Maxwellian beam
was used, so that the diffraction order (0,1) which is in plane at γ = 0°,
is spread out according to the velocity distribution. For the k_i, which fits
the condition (4.34), transition into the bound state E_n is possible, giving

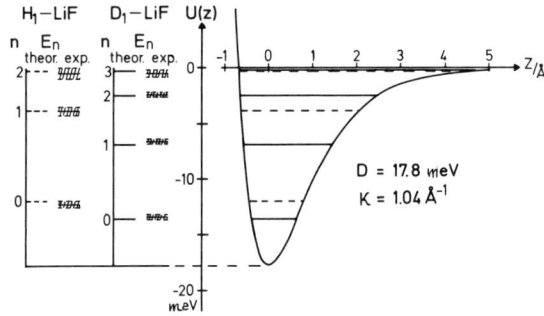

FIG. 47.   Bound-state energies of H and D on LiF(001).   The parameters of a Morse potential are chosen
to give best agreement with the experiment.

rise to a decrease in intensity in all elastic peaks associated with k_i.
Under these experimental conditions this can be seen only in the (0,1)
diffraction order as opposed to (0,0), as all possible k_i are represented
in the specular peak. Transition into two bound states is possible, indicated
in the distribution by two faint minima, which shift with angle of incidence,
as predicted by the theory.

Another method for finding resonant transitions is illustrated in
Fig. 45. Here the reciprocal lattice is seen from the top. The solid circle
is again the intersection of the paper plane with the elastic sphere. The
dashed circles give the energy of bound states. If we now rotate the
crystal round its z axis (angle γ), we rotate the reciprocal lattice as well
and it happens at certain angles γ that a reciprocal lattice vector coincides
with a dashed circle. So a transition into this bound state becomes
possible, giving rise again to a minimum in the diffracted beams. This
transition can now also be seen in the specular peak. If we monitor the
specular beam while we rotate the crystal round angle γ (detector position and θ_i being fixed), transitions are indicated by pronounced minima.
This can be seen in Fig. 44(a). The bound state, to which this transition
occurs, has an energy of -12.3 ± 0.3 meV.

Later experiments have been done by Frank (1973) with a velocity
selected beam. By that means the transitions become sharper and more
bound states can be found in the γ distribution alone. This is shown in
Fig. 46 for atomic deuterium, where the beam energy was varied at fixed
angle of incidence. (The energy level at E_n = 0.5 meV is known from other
experiments.) D has a mass twice that of H and therefore it has other
bound-state energies (but in the same potential, as the electron shell is
the same).

With these data the bound-state energies result as shown in Fig. 47.
The experimental values are given on the left-hand side together with the
estimated errors. For comparison, a Morse potential has been drawn,
where the bound-state energies are given by:

$$E_n = - \left(\frac{\sqrt{2mD}}{\alpha \hbar} - n - \frac{1}{2} \right)^2 \frac{\alpha^2 \hbar^2}{2m} \qquad (4.35)$$

The two parameters D and α were chosen to give the best agreement with
the experimental results.

BIBLIOGRAPHY

ADDISON, W.E., Structural Principles in Inorganic Compounds, Longmans, London (1961).

BAYH, W., private communication (1973).

BEDER, E.C., Adv. At. Mol. Phys. 3 (1967) 205.

BEEBY, J.L., J. Phys. C 4 (1971) 1.359.

BEEBY, J.L., J. Phys. C 5 (1972) 3438.

BOATO, G., CANTINI, P., MATTERA, L., Proc. 2nd Int. Conf. on Solid Surfaces, Kyoto (1974) 553.

CABRERA, N., CELLI, V., GOODMAN, F.O., MANSON, R., Surf. Sci. 19 (1970) 67.

DALGARNO, A., DAVISON, W.D., Adv. At. Mol. Phys. 2 (1966) 1.

DEVONSHIRE, A.F., Proc. R. Soc. (London) Ser. A 156 (1936a) 37; 158 (1936b) 269.

ESTERMANN, I., FRISCH, R., STERN, O., Z. Phys. 73 (1931) 348; 61 (1930) 95; 84 (1933) 430, 443. See also: STERN, O., Naturwissenschaften 17 (1929) 391.

FISHER, W.S., BLEDSOE, J.R., J. Vac. Sci. Technol. 9 (1972) 814.

FRANK, H., Master Thesis, Erlangen (1973), unpublished. See also: FINZEL, H., FRANK, H., HOINKES, H., LUSCHKA, M., NAHR, H., WILSCH, H., WONKA, U., Surf. Sci. 49 (1975) 577.

GARIBALDI, U., LEVI, A.C., SPADACINI, R., TOMMEI, G.E., Proc. 2nd Int. Conf. on Solid Surfaces, Kyoto (1974) 549.

GILLERLAIN, J.D., FISHER, S.S., Abstracts 8th Int. Symp. Rarefied Gas Dynamics, Stanford (1971) 605.

GOODMAN, F.O., J. Chem. Phys. 53 (1970) 2281.

GOODMAN, F.O., Surf. Sci. 24 (1971a) 667; 26 (1971b) 327.

GOODMAN, F.O., Surf. Sci. 30 (1972) 1.

HAYS, W.J., RODGERS, W.E., KNUTH, E.L., J. Chem. Phys. 56 (1972) 1652.

HOINKES, H., NAHR, H., WILSCH, H., Surf. Sci. 30 (1972a) 363; 33 (1972b) 516.

HOINKES, H., NAHR, H., WILSCH, H., J. Phys. C 5 (1972c) 143.

HOINKES, H., NAHR, H., WILSCH, H., J. Chem. Phys. 58 (1973) 3931.

LENNARD-JONES, J.E., DEVONSHIRE, A.F., Nature 137 (1936) 1069.

LENNARD-JONES, J.E., DEVONSHIRE, A.F., Proc. R. Soc. (London) Ser. A 158 (1937a) 242; 158 (1937b) 253.

LOGAN, R.M., KECK, J.C., J. Chem. Phys. 49 (1968) 860.

LOGAN, R.M., STICKNEY, R.E., J. Chem. Phys. 44 (1966) 195.

LORENZEN, J., RAFF, L.M., J. Chem. Phys. 54 (1971) 674.

MASON, B.F., WILLIAMS, B.R., J. Chem. Phys. 56 (1972) 1895.

MAVROYANNIS, C., STEPHEN, M.J., Mol. Phys. 5 (1962) 629.

McCLURE, J.D., J. Chem. Phys. 52 (1970) 2712.

McCLURE, J.D., J. Chem. Phys. 57 (1972a) 2810; 57 (1972b) 2823.

McCLURE, J.D., DOYLE, J., WU, Y., in Proc. 6th Int. Symp. Rarefied Gas Dynamics, Cambridge, Mass., 1968 (TRILLING, L., WACHMAN, H.Y., Eds) in Advances in Applied Mechanics, Suppl. 5, Academic Press, New York (1969) 1191.

O'KEEFE, D.R., PALMER, R.L., SMITH, J.N., Jr., J. Chem. Phys. 55 (1971) 4572.

OMAN, R.A., J. Chem. Phys. 48 (1968) 3919.

SMITH, D.L., MERRILL, R.P., J. Chem. Phys. 52 (1970) 5861.

SMITH, J.N., Jr., Surf. Sci. 34 (1973) 613.

SMITH, J.N., Jr., O'KEEFE, D.R., PALMER, R.L., J. Chem. Phys. 52 (1970) 315.

SMITH, J.N., Jr., PALMER, R.L., J. Chem. Phys. 56 (1972) 13.

STICKNEY, R.E., Adv. At. Mol. Phys. 3 (1967) 143.

STOLL, A.G., MERRILL, R.P., Surf. Sci. 40 (1973) 405.

SUBBARAO, R.B., MILLER, D.R., J. Chem. Phys. 51 (1969) 4679.

SUBBARAO, R.B., MILLER, D.R., J. Vac. Sci. Technol. 9 (1972) 808.

TENDULKAR, D.V., STICKNEY, R.E., Surf. Sci. 27 (1971) 516.

TOENNIES, J.P., J. Appl. Phys. 3 (1974) 91.

TSUCHIDA, A., Surf. Sci. 14 (1969) 375.

WEINBERG, W.H., MERRILL, R.P., J. Chem. Phys. 56 (1972) 2893.

WEST, L.A., SOMORJAI, G.A., J. Chem. Phys. 54 (1971) 2864.

WILLIAMS, B.R., J. Chem. Phys. 55 (1971a) 1315; 55 (1971b) 3220.

WILSCH, H., FINZEL, H., FRANK, H., HOINKES, H., NAHR, H., Proc. 2nd Int. Conf. on Solid
Surfaces, Kyoto (1974) 567.

YAMAMOTO, S., STICKNEY, R.E., J. Chem. Phys. 53 (1970) 1594.

ZAHL, H.A., ELLET, A., Phys. Rev. 38 (1931) 977.

CHEMISORPTION
A brief survey*

R. GOMER
James Franck Institute
  and Department of Chemistry,
University of Chicago,
Chicago, Ill.,
United States of America

Abstract

CHEMISORPTION: A BRIEF SURVEY.
  A very brief general survey is given of chemisorption and the principal areas of current interest in this field.

1. SUMMARY OF SALIENT FACTS

Chemisorption is defined as the adsorption of atoms or molecules on surfaces (metallic in the cases to be discussed) with binding energies in excess of ~ 0.5 eV, and involves electron transfer or sharing, as distinct from physisorption which results mainly from dispersion forces. The distinctions become hazy in borderline cases: there is a possibility that inert gas adsorption on clean metals involves charge transfer bonding, for instance. In most cases of chemisorption $E_b = 1 - 5$ eV.

Adsorption seems to be strongest on transition metals with their relatively localized, directional d-orbitals. Binding energies are usually higher on bcc than on comparable fcc metals. Adsorption energies and other properties vary with substrate crystallography within a given system. Almost all elements are adsorbed on metals: Thus H, O, C, N, S and metal atoms are strongly adsorbed. Since chemisorption involves bonding to the substrate, it is not surprising that 'saturated' molecules, i.e. those having only single bonds, like H_2 or saturated hydrocarbons, are strongly adsorbed only on dissociation, i.e. as H atoms or hydrocarbon fragments. Like H, O is strongly adsorbed only as atoms. In the case of N_2, strong adsorption also involves dissociation, but there is some evidence of intermediate adsorption in the molecular form. CO seems to be the only well established case of a diatomic molecule chemisorbed non-dissociatively. As we shall see later, however, there is a possibility that in its most tightly bound configurations CO is effectively dissociated.

Since many adsorbates are only chemisorbed dissociatively it may happen that the heat of adsorption relative to the molecular state is negative. This does not imply that the energy of adsorption relative to the gaseous atom is negative; only that it is less than half the heat of dissociation of the (diatomic) molecule.

---

* The complete text of the lectures will appear in Solid State Physics, Vol.XXX (1975).

Chemisorption generally involves a certain amount of electron transfer to or from the adsorbate particle. If the adsorbate has a small ionization potential, as in the case of alkali metals, electron transfer is from the adsorbate, in some cases leading to completely ionic adsorption. For adsorbates with high ionization potential, e.g. O, H, CO, there is some electron transfer to the adsorbate; usually this is of the order of 0.1 e or less.

Within a given system, distinct binding modes can exist on a given crystal plane. These may be characterized by different geometry, dipole moment, electronic structure, binding energy and cross-sections for electron-induced desorption. In a sense these states can be considered the analogues of different chemical isomers. Although there is clear evidence for distinct binding states in some cases, in others, e.g. the high-temperature states of CO on bcc metals, different binding states may be simulated by the coverage dependence of the binding energy and of other properties of what is essentially one type of adsorption.

In general the greatest variety of binding modes is observed on bcc substrates. This is probably due to the atomically more open structure of bcc surfaces, which permits a larger variety of configurations to occur.

Even adsorbates with very high binding energies become mobile at relatively low temperatures, indicating that the potential barrier for moving from site to site is much less than the energy of adsorption. It turns out that activation energies of surface diffusion for monatomic adsorbates vary from 10% to 25% of the adsorption energy, and that this fraction increases with the atomic roughness of the substrate. For polyatomic adsorbates the diffusion energy may be higher and become comparable to the binding energy.

In almost all known cases of adsorption on clean metals the activation energy of adsorption is extremely, almost negligibly, small except for the dissociative adsorption of CO, where the activation energy is so high that non-dissociative adsorption occurs, at least at moderate temperature.

The sticking coefficient is defined as the probability that an impinging molecule will be adsorbed. Obviously this definition is meaningful without modification only under conditions of nearly infinite surface lifetime of the chemisorbed species. For non-dissociative adsorption (i.e. CO or metal atoms) sticking coefficients are very high: between unity and 0.5 even at high substrate and gas temperatures. For dissociative adsorption there is a marked decrease, and sticking coefficients can vary from 0.3 for O_2 on W to 10^{-3} or less for N_2 on W. As might be expected, sticking coefficients are smallest on smooth, close-packed planes.

2. ADSORPTION, ABSORPTION, RECONSTRUCTION

In some cases adsorption of atoms or molecules A on a given metal M is the only interaction between these partners which leads to a decrease in free energy. This occurs when bulk compounds A_nM_m do not exist. Where bulk compounds exist (e.g. oxides) adsorption is thermodynamically a precursor to bulk compound formation, at least at finite pressure. In such cases interaction may still stop at the monolayer (i.e. at the adsorption) stage if the activation energy of absorption, or bulk compound formation, is too high.

Finally, it may happen that, even for monolayer amounts of adsorbate, the thermodynamically stablest structures consist not of adsorption on top of the unperturbed surface, but of a re-arrangement involving one or more layers of substrate atoms. This phenomenon is usually referred to as surface reconstruction. There is evidence that reconstruction occurs for O adsorption at sufficiently high temperatures and/or pressures on most metals. This suggests that kinetic as well as thermodynamic considerations determine whether reconstruction will occur in a given case.

3. COMPARISON OF CHEMISORPTION AND 'ORDINARY' CHEMICAL BONDING

While chemisorption does not differ in principle from bonding in other situations, the interaction between an adsorbate atom and a solid is an $N+1$ atom problem even if many-body effects are ignored, and this leads to broadening of what would be molecular orbitals in small molecules. Nevertheless many concepts useful in chemistry, e.g. the use of atomic orbitals for the construction of molecular orbitals, can be extended to chemisorption. Since we do not wish to solve the $N + 1$ atom problem ab initio, we shall wish to take over as much of the solutions of the N atom, i.e. substrate problem, as possible. The principal difficulty turns out to be that there is as yet relatively little information on some of the quantities we would most like to know, for instance surface densities of state of metal orbitals, Wannier, or atomic, projected onto Bloch states.

Although we naturally hope to extrapolate from 'ordinary' chemistry, a word of caution is in order. We know from the chemistry of linearly connected atoms that bond lengths and angles in such compounds are remarkably constant for given atoms and bond types. In adsorption the situation may be different. Solids represent a nearly fixed matrix, on which the adsorbate must arrange itself as best it may. This could give rise to geometric configurations, with bond distances and angles unlikely to be seen in ordinary molecules where atoms are much freer to move to optimal positions. As yet, adsorbate-substrate configurations are known only in a small number of simple cases, so that the above serves more as warning than confirmed fact.

4. TOPICS OF CURRENT INTEREST

It will be useful to list some of the topics now actively under study, together with the principal methods of attack. Somewhat arbitrarily, this list is divided into static and dynamic phenomena. It is intended merely to make a compilation, for the record so to speak, without comment.

(a) Statics

Geometry (LEED; field ion microscopy).

Number density (LEED; sticking coefficient and effusion techniques; thermal desorption).

Characterization of different adsorption states (thermal and electron impact desorption; work function measurements; photoelectron spectroscopies (u.v. and X-ray); field emission spectroscopy; work function measurements).

Binding energies (thermal desorption; calorimetry).

Adsorbate charge (work function measurements: field emission, Kelvin diode, retardation methods, combined with number densities; X-ray electron spectroscopy).

Electronic structure (photoelectron spectroscopy; field emission spectroscopy; ion neutralization spectroscopy).

Quantum-mechanical description of binding (generalizations of LCAO-MO method, in particular the Anderson model; generalizations of valence bond method; dielectric response theory).

(b) Dynamics

Rates of adsorption and sticking coefficients (pressure changes in adsorption; reflection methods).

Surface diffusion (field emission and field ion microscopy).

Kinetics of binding state interconversion (electron impact desorption, thermal desorption; photoelectron and field emission spectroscopy).

Kinetics of desorption (thermal and electron impact desorption).

Kinetics and mechanism of electron impact desorption (electron impact desorption).

5. WHERE DO WE STAND?

We conclude this brief discussion by a very short assessment of the present status of chemisorption research.

At present, experimental information on the existence and many properties of adsorption states on single-crystal planes of metals is beginning to accumulate. This includes such quantities as number density, dipole moment, electron impact desorption cross-sections, binding energy, and electronic structure. We are most ignorant about the geometry of substrate-adsorbate configurations and their correlation with binding modes. The reason for this is that we are trying to probe monolayers and, in any scattering or diffraction experiment, must ensure adequate cross-section, e.g. by using slow electrons. This leads ipso facto to multiple scattering, and hence invalidates in LEED the simple kinematic approach valid for X-ray diffraction where multiple scattering can be neglected. Consequently we are just beginning to obtain sorely needed structural information.

On the theoretical side, a reasonable start at understanding the fundamentals of chemisorption has been made, although many details are far from settled. It is my own view that extensions and refinements of LCAO-MO methods, possibly pushing beyond the Hartree-Fock approximation, will prove both feasible and adequate for a reasonably quantitative description of at least the statics of chemisorption. This view is by no means universally accepted, and some workers feel that correlation must be treated in a more fundamental way. Even if my optimistic view is accepted, we are still very much in the model-Hamiltonian stage, and it will be some time before we shall be able to calculate correctly, let alone predict, say the energy differences between two different binding modes of H on a (100) tungsten surface.

Part IV

BIOLOGICAL SURFACES

BIOLOGICAL SURFACES
(Biological membranes)

J. SCHNAKENBERG
Institut für Theoretische Physik,
Rheinish-Westfälische Technische Hochschule,
Aachen, Federal Republic of Germany

Abstract

BIOLOGICAL SURFACES (BIOLOGICAL MEMBRANES).
'Biological surfaces' are interpreted as 'biological membranes', which represent the wall elements in the structure of living organisms. Every cell and likewise the subunits of the cell are surrounded by a membrane. Beyond their structural role as wall elements, biological membranes receive their essential biological significance from their functional role as the contact elements of the cell and its subunits with the surroundings. Thus, transport of both neutral molecules and ions across membranes is one of the main subjects of membrane physiology. A further important phenomenon in some biological membranes is that of electrical excitation. In the case of the membrane of nerve cells, this phenomenon is the basic process for the propagation of information along the nerve cell. This paper reports briefly on some of the contributions which can be expected from theoretical physics to support the functional analysis of biological membranes. A brief review is given of irreversible thermodynamics as applied to transport across membranes. The series of the further sections may be characterized by an increasing amount of detailed assumptions on the molecular level. Phenomenological descriptions are presented of membrane transport which exceed the black box theory of thermodynamics. A physical picture is given of the phenomenon of membrane excitation upon which three models, given in the paper, are based. Whereas the pore-blocking model in its very essence is a rather general description of the phenomenon, the dipole chain model interprets ionic transport across membranes as diffusion along protein chains and, finally, Adam's model considers excitation as a phase transition in the membrane plane.

1. INTRODUCTION

Interpreting a 'surface' as a two-dimensional boundary between different solid phases, biological sciences certainly offer a lot of 'biological surfaces' since a multiphase composition is crucial for the existence and function of living organisms. The usual designation for a biological surface in this sense is 'biological membrane', and this will be used henceforth. The significance of biological membranes lies in the fact that they belong to the surprisingly small number of what one might call the basic principles of structure, organization and function of living organisms. Among these principles is that of the cell with its role as structural unit exhibiting the same structural scheme for all different kinds of cells which are needed in a complex organism. Other principles are that of coding and storing genetic information in the DNA or that of the proteins which act as the biopolymers.

Biological membranes are first of all the 'wall elements' of the living nature. Every cell is surrounded by a membrane and thus receives its individuality from the membrane. The same applies to the subunits of the cell, as for example to the nucleus or the mitochondria (the power supply of the cell). Being a wall, however, is only one aspect of biological membranes and certainly not the most important since a complete wall round a cell would make the cell senseless for the rest of the organism. The biologically much more relevant function of membranes is that of a contact element, e.g.

between the cell and its exterior. The possibility of contact has to be provided by the membrane for matter, both neutral molecules and ions, for
energy, and even for information.

The material contact, i.e. the permeability of the membrane for certain
molecules or ions, has to be highly selective since otherwise the system
would approach the thermodynamic equilibrium, which means just 'death'
to the living individuum. This selectivity, however, is not a constant in
time. The membrane must be able to change the permeabilities due to the
very situation of the whole system. The information of such situations
must proceed to the individual membrane in some appropriate way.

A particularly important and interesting phenomenon of material transport across biological membranes is that of 'active transport'. By this term,
one understands the capability of some cells to take up, for example,
potassium ions from the exterior although the interior potassium concentration is already larger than that of the exterior by a factor of 10 or 100.
Thermodynamics of irreversible processes tells us that such a transport
process is possible only at the expense of some other degradation or dis-
sipation process. Indeed, active transport is performed in biological
membranes by coupling the material flux to some appropriate source of free
energy, mostly an enzymatic chemical reaction. We shall return briefly to
this phenomenon in the next section.

A further phenomenon in a special group of biological membranes is that
of membrane excitation. It is observed that this phenomenon represents a
particular challenge to physicists since many properties of excitable
membranes suggest analogies with excitation processes, e.g. in solid-state
physics or in physical systems with phase transitions. To give a very rough
preliminary explanation for membrane excitation, we might look at an
excitable membrane as a bi-stable system, the transitions between the two
stable states being related to material transport or at least partial transport.
The biological significance of membrane excitation, however, is not material
transport across but information transport along the membrane plane of
nerve cells in that a single excitation process represents a unit of information
which is to proceed along the spatially extended nerve cell. A great deal of
the model studies to be discussed in the final sections will be devoted to
membrane excitation.

Let us stop the presentation of biological membrane phenomena at this
point. Our list of phenomena is still very far from being complete and only
some of those phenomena have been selected which attract special interest
from a physical point of view. The question we shall raise now, and leave
without a final answer, is whether a meaningful contribution of methods and
experiences from the physical and chemical science of surfaces can be
expected in order to make progress in the analysis of structure and function
of biological membranes.

It is nowadays generally accepted that classical experimental physical
and chemical methods such as light and electron microscopy, X-ray diffraction, EPR, NMR, radiochemistry and others are essential techniques in
analysing the structure of biological membranes. An enormous amount of
detailed information is obtained by the application of such techniques and yet
there is no complete agreement on the basic structure of biological membranes.
The most commonly accepted structural model is still the so-called unit
membrane model of Danielli and Davson [1] set up in 1935. This structural
model is roughly sketched in Fig.1. It describes the biological membrane

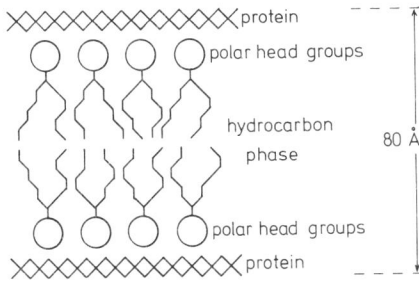

FIG.1.   Unit membrane model.

as a double layer of lipid molecules which form a hydrophobic interior
membrane phase with their twofold hydrocarbon chains and hydrophobic
exterior boundaries with their polar head groups. The hydrocarbon interior
may be considered almost as a fluid with a dielectric constant of order of
magnitude 3 to 5, thus representing a highly insulating barrier for ions under
the influence of a gradient of electrochemical potential across the membrane.
The polar boundaries are covered by particular membrane proteins which are
assumed to be responsible for special properties of distinguished membranes.

In contrast to the above experimental techniques, the application of
theoretical physical methods and theories is far from being generally accepted.
Many biologists and physiologists vigorously refuse any kind of theories and
models as long as there is no complete and certain knowledge of the molecular structure, and such objections should be taken seriously. On the other
hand, physics is able to provide a theory for completely unknown systems,
namely the phenomenological theory of irreversible thermodynamics. As
everybody would believe that all phenomena in nature, including those in
living organisms, satisfy the first and second law of thermodynamics, the
phenomenological theory of irreversible thermodynamics will certainly
apply also to biological membranes as a whole. In recent years, there has
been considerable progress in this theory, particularly for systems far
from thermodynamic equilibrium like those in living organisms. A detailed
presentation of the recent state of the theory is given in the book of Glansdorff
and Prigogine [2].

Being a phenomenological theory, however, irreversible thermodynamics
can provide only limited results and predictions which will play the role of
conditions for all observable phenomena occurring in biological systems.
A brief review of the basic statements of irreversible thermodynamics as
applied to transport processes across membranes is given in Section 2.

Phenomenological methods like that of the Nernst-Planck equations which
go beyond irreversible thermodynamics were already developed decades ago
and are still in use. New methods are currently suggested, the most
promising one being perhaps 'network thermodynamics' by Oster et al. [3].
A brief report on both these methods is given in Section 3.

Still bearing in mind the serious objections against model studies in biophysics, the remaining part of the paper is devoted to models for membrane
excitation. A brief summary of some basic experimental findings in the case

of nervous excitation is presented in Section 4, where we also try to derive a
possible excitation mechanism which is realized by three different models in
Sections 5, 6 and 7. Our models can be characterized in this sequence by an
increasing amount of co-operativity, the last one being the Adam model,
which interprets membrane excitation as a phase transition. To conclude
our introduction, let us agree that we shall consider the models as an
attempt to 'offer' nature simple mechanisms for membrane excitation.
Whether nature really makes use of the proposed mechanisms can only be
decided by comparison with experimental results. As this comparison will
always be incomplete we must be aware of the possibility that a model is not
a molecular picture of what happens in the living organism but only a simulation of the process. From such a simulation the most promising next
experimental step may then be derived the results of which will, hopefully,
tell us whether our theoretical ideas are still on the right path.

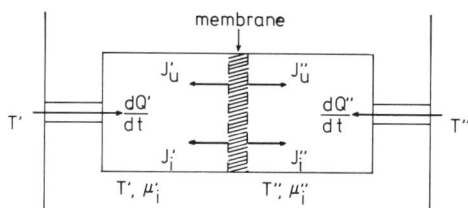

FIG.2.   Irreversible processes across membranes.

2.   IRREVERSIBLE THERMODYNAMICS OF MEMBRANES

In this section some of the basic results of irreversible thermodynamics
as applied to membranes are presented. The presentation follows that of
Sauer's article in the Handbook of Physiology [4], which is recommended for
further details. For a general introduction to the theory, the reader is
referred to the books of de Groot and Mazur [5] and Katchalsky and Curran [6].
We shall treat the membrane as a black box with only the assumption
that it is in a steady state, i.e. its extensive and intensive parameters do
not vary with time on a short time or microscopic scale. Macroscopic time
variations are included. To define the boundary conditions for the membrane,
let us fix it between two equilibrium systems ' and ", which need not be in
equilibrium with respect to each other. Each of the equilibrium systems is
characterized by its values for temperature T, pressure p and chemical
potential μ_i of the components i = 1, 2,...,N. The system as a whole is assumed
to be closed. The subsystems ' and " may exchange heat dQ'/dt and dQ''/dt,
respectively, with different external heat baths in a reversible way and heat
and matter with each other by irreversible processes across the membrane.
This situation is depicted in Fig.2.
For simplicity let us assume that the position of the membrane is fixed
and that there is no variation of the volumes V' and V" of the subsystems.

Applying the first law of thermodynamics to each of the subsystems yields

$$\frac{dU'}{dt} = \frac{dQ'}{dt} + J'_u \quad , \quad \frac{dU''}{dt} = \frac{dQ''}{dt} + J''_u \tag{2.1}$$

where U', U'' are the internal energies and J'_u and J''_u denote the exchange of energy other than the reversible exchange with the heat baths. From (2.1) we obtain the first law of the whole system:

$$\frac{dU}{dt} = \frac{d}{dt}(U' + U'') = \frac{dQ'}{dt} + \frac{dQ''}{dt} + J'_u + J''_u \tag{2.2}$$

If no external work is done to the system (including e.g. mechanical, electrical, gravitational), we have

$$J'_u + J''_u = 0 \tag{2.3}$$

since the membrane as part of the whole system is assumed to be in a steady state.

The second law of thermodynamics is formulated as Gibbs relations for the subsystems:

$$\frac{dS'}{dt} = \frac{1}{T'}\frac{dU'}{dt} - \sum_i \frac{\mu'_i}{T'} J'_i$$

$$\frac{dS''}{dt} = \frac{1}{T''}\frac{dU''}{dt} - \sum_i \frac{\mu''_i}{T''} J''_i \tag{2.4}$$

where S' and S'' are the entropies and J'_i, J''_i denote the mutual exchange of matter of component i satisfying

$$J'_i + J''_i = 0 \tag{2.5}$$

since the system as a whole was assumed to be closed and the membrane is in a steady state.

Let us now calculate the time variations of the entropy S of the whole system. As the entropy is an extensive quantity and thus additive, and the membrane does not contribute to time variations, we obtain

$$\frac{dS}{dt} = \frac{dS'}{dt} + \frac{dS''}{dt} \tag{2.6}$$

Combining Eqs (2.1), (2.3) - (2.5) and (2.6), we immediately get

$$\frac{dS}{dt} = \frac{d_e S}{dt} + \frac{d_i S}{dt} \tag{2.7}$$

where $d_e S/dt$ denotes the reversible external change of entropy defined as

$$\frac{d_e S}{dt} = \frac{1}{T'} \frac{dQ'}{dt} + \frac{1}{T''} \frac{dQ''}{dt}$$

(2.8)

and $d_i S/dt$, the irreversible contribution to the entropy change, or the so-called entropy production, given as

$$\frac{d_i S}{dt} = \Delta \frac{1}{T} J_u' + \sum_i \Delta \frac{\mu_i}{T} J_i''$$

(2.9)

with the abbreviations

$$\Delta \frac{1}{T} = \frac{1}{T'} - \frac{1}{T''} \quad , \quad \Delta \frac{\mu_i}{T} = \frac{\mu_i'}{T'} - \frac{\mu_i''}{T''}$$

The extension of the second law of thermodynamics to the system as a whole implies that $d_i S/dt$ is non-negative and vanishes if and only if the subsystems are in equilibrium with each other. Furthermore, we see that $d_i S/dt$ in (2.9) has the form of a bilinear expression:

$$\frac{d_i S}{dt} = \sum_\alpha X_\alpha J_\alpha \geq 0$$

(2.10)

of so-called generalized thermodynamic forces X_α and fluxes J_α. In our example, the X_α and J_α are given as

$$X_\alpha: \quad \Delta \frac{1}{T} \quad \Delta \frac{\mu_i}{T}$$

$$J_\alpha: \quad J_u' \quad\quad J_i''$$

(2.11)

The X_α are differences (in continuous systems: gradients) of intensive parameters and the J_α are time derivatives (in continuous systems: flux densities) of extensive parameters.

If in addition to the differences in temperature and chemical potentials a difference of the electrical potential Φ is present across the membrane, Eq.(2.3) has to be replaced by

$$J_u' + J_u'' = \sum_i J_i'' e_i \Delta \Phi$$

(2.12)

where e_i is the electric charge per mole of component i. Repeating the same calculation which led us to Eq.(2.9), we now obtain

$$\frac{d_i S}{dt} = \Delta \frac{1}{T} \bar{J}_u + \sum_i \left(\Delta \frac{\mu_i}{T} + e_i \Delta \frac{\Phi}{T} \right) J_i'' \geq 0$$

(2.13)

where $\overline{T}$ and $\overline{J}_u$ are averaged temperatures and heat fluxes defined by

$$\frac{1}{\overline{T}} = \frac{1}{2}\left(\frac{1}{T'} + \frac{1}{T''}\right) \quad, \quad \overline{J}_u = \frac{1}{2}\left(J_u' - J_u''\right)$$

For applications to biological membranes, we can assume vanishing temperature differences such that Eq.(2.13) reduces to

$$\frac{d_i S}{dt} = \frac{1}{T}\sum_i \Delta\eta_i\, J_i'' \tag{2.14}$$

where η_i denotes the electrochemical potential:

$$\eta_i = \mu_i + e_i\Phi \tag{2.15}$$

For dilute electrolytes, μ_i can be considered as independent of Φ and,
approximately,

$$\mu_i = \mu_i^0(T,p) + RT\log x_i \tag{2.16}$$

where x_i is the molar fraction of component i.

If, in addition to the situation described so far, chemical reactions can
occur somewhere in our system (including the membrane interior), Eq.(2.5)
has to be replaced by

$$J_i' + J_i'' = \sum_\rho \nu_{i\rho}\, W_\rho \tag{2.17}$$

where $\nu_{i\rho}$ is the stoichiometric coefficient of component i in the reaction ρ
with a molar rate W_ρ. The result for $d_i S/dt$ with vanishing temperature
difference, but including the difference of electrical potential, now reads:

$$\frac{d_i S}{dt} = \frac{1}{T}\sum_i \Delta\eta_i\, J_i'' + \frac{1}{T}\sum_\rho A_\rho'\, W_\rho \geq 0 \tag{2.18}$$

where A_ρ' is the affinity of reaction ρ expressed in terms of the variables of
the ' subsystem:

$$A_\rho' = -\sum_i \mu_i' \nu_{i\rho} \tag{2.19}$$

Equation (2.18) could easily be rewritten in a more symmetric form but this
will not be necessary for our purposes.

The generalized thermodynamic fluxes J_α can be considered as functions
of all forces X_α or vice versa. The fluxes have been constructed such that
they vanish in the thermodynamic equilibrium, i.e. if all forces vanish.

A Taylor expansion of the fluxes with respect to the forces will start with the linear terms. The so-called linear theory of irreversible thermodynamics truncates the Taylor expansion after the linear terms such that

$$J_\alpha = \sum_\beta L_{\alpha\beta} X_\beta \qquad (2.20)$$

The so-called phenomenological coefficients $L_{\alpha\beta}$ satisfy Onsager's reciprocity relations:

$$L_{\alpha\beta} = L_{\beta\alpha} \qquad (2.21)$$

Sauer [4] has shown that the reciprocity relations remain valid if Eq.(2.20) is considered as an exact pseudolinear relation.

For applications to biological membranes, the linear theory is in most cases a very poor approximation since it is an intrinsic characteristic of life to be far from equilibrium. Nevertheless, let us discuss the possibility of active transport in the linear theory. As mentioned in the introduction, active transport is a coupling phenomenon between a material flux J and a chemical reaction with affinity A. In the linear theory, this coupling is written as

$$J = \frac{1}{T}\left(L_M \Delta\eta + L_{MR} A\right)$$

$$W = \frac{1}{T}\left(L_{RM} \Delta\eta + L_R A\right) \qquad (2.22)$$

where W is the molar reaction rate and $\Delta\eta$ the difference of the electro-chemical potential of the component to be transported actively. The conditions for active transport, i.e. J and $\Delta\eta$ with opposite directions, are easily derived from (2.22). It should be emphasized that the actively transported component need not take part in the reaction. The phenomenological coefficients $L_{MR} = L_{RM}$ couple a flux J with vector character to a force A which is a scalar, or, vice versa, a scalar flux W to a vector force $\Delta\eta$. In our reduced system with a plane membrane and one-dimensional vectors perpendicular to the membrane, this means that the coupled quantities J, A or W, $\Delta\eta$ have different parity with respect to an inversion between the left and the right side of the membrane. As a consequence, $L_{MR} = L_{RM} = 0$ unless the system, i.e. the membrane, exhibits an intrinsic asymmetry with respect to a left-right inversion in its internal structure. This necessary condition for active transport in the linear theory is a particular consequence of Curie's principle.

3.   PHENOMENOLOGICAL MEMBRANE MODELS

This section contains a brief report on two very different approaches to our membrane problem which go a little beyond the thermodynamic theory but nevertheless are phenomenological in that they do not assume any detailed concept on the molecular level.

3.1. Nernst-Planck theory

The first of these phenomenological approaches dates back to the last century and was developed by Nernst [7, 8] and Planck [9, 10] from 1888 until 1890 for application to diffusion problems in electrolytes and other systems. In the Nernst-Planck approach, the membrane is assumed to consist of porous material which possibly carries positive or negative fixed charges. The permeating molecules and ions are assumed to pass the pores or channels of the membrane material by a diffusion-like motion under the influence of concentration gradients and both applied external and fluctuating internal electrical fields due to fixed charges. For the density of the molar flux J_i of component i, an extended Fick's law ansatz is made:

$$J_i = -D_i \left(\frac{dc_i(x))}{dx} + \frac{e_i}{RT} c_i(x) \frac{d\Phi(x)}{dx} \right) \qquad (3.1)$$

where x is the co-ordinate perpendicular to the membrane plane; $c_i(x)$ is the molar concentration of component i; and D_i is the diffusion constant for component i. The constants inside the brackets in (3.1) are chosen such that for equilibrium $J_i \equiv 0$ we have

$$\frac{dc_i(x)}{dx} + \frac{e_i}{RT} c_i(x) \frac{d\Phi(x)}{dx} = 0$$

and upon integration

$$\log c_i(x) + \frac{e_i}{RT} \Phi(x) = \text{const} = f_i(T,p) \qquad (3.2)$$

Equation (3.2) means that the electrochemical potential η_i of Eqs (2.15), (2.16) is independent of x.

The Nernst-Planck approach is completed by the condition of electro-neutrality which reads:

$$\sum_i e_i c_i(x) + \rho(x) = 0 \qquad (3.3)$$

where $\rho(x)$ is the density of fixed charges. The condition of electroneutrality introduces the possibility of coupling between different fluxes into the theory. This kind of coupling, however, is rather indirect and one cannot hope to obtain thereby a satisfactory description of coupling phenomena which are typical for biological membranes like active transport or excitability. Nevertheless, the Nernst-Planck theory gives insight into a lot of less complicated transport phenomena. As an exercise, the reader may easily verify that a special solution of Eq.(3.1) for $E = -d\Phi(x)/dx = \text{const}$ is given by

$$J_i = \frac{D_i\, w_i\, \overline{c}_i}{a} \; \frac{\exp\left(\dfrac{1}{2}\dfrac{\Delta\eta}{RT}\right) - \exp\left(-\dfrac{1}{2}\dfrac{\Delta\eta}{RT}\right)}{F(E)}$$

$$\overline{c}_i = \left(c_i'\, c_i''\right)^{1/2}$$

$$F(E) = \left(\frac{1}{2}\frac{e_i E}{RT} a\right)^{-1} \sinh\left(\frac{1}{2}\frac{e_i E}{RT} a\right)$$

(3.4)

$\Delta\eta_i$ is the difference of chemical potential across the membrane; a is the membrane thickness; c_i', c_i'' are the concentrations left and right of the membrane; and w_i is a constant related to the difference of the standard chemical potential of component i between the external solution and the membrane material.

The result in (3.4) already gives an example for a non-linear relationship between a flux J_i and its driving force $\Delta\eta_i$. For small forces, (3.4) reduces to

$$J_i = L_{ii}\, \frac{\Delta\eta_i}{T}$$

(3.5)

where

$$L_{ii} = \frac{D_i\, w_i\, \overline{c}_i}{aR}$$

(3.6)

Situations of more complexity than our exercise are very difficult to solve. Quite a variety of integrations of the Nernst-Planck equation involving very general solutions have been presented by Schlögl [11].

3.2. Network thermodynamics

Network thermodynamics as proposed by Oster et al. [3] starts from a point of view completely different from that of the Nernst-Planck theory.

The simplest phenomenological way to look at a membrane is a material capacitor C which is separated from the adjoining electrolytes by two material resistances R (not to be confused with the gas constant which will henceforth be denoted by R_G). The electrical analogue network is shown in Fig.3.

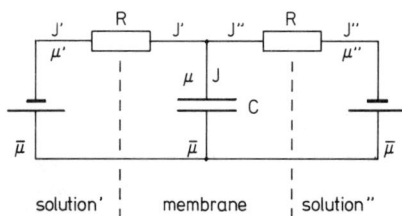

FIG.3.   Simple network model for membranes.

The model in Fig.3 describes transport of one component without interaction with any other component and without influence of an applied electric field. The batteries $\mu'-\bar{\mu}$ and $\mu''-\bar{\mu}$ represent the external solutions. The constitutive relations for the elements R and C are $(\bar{\mu} = 0)$

$$J' = \frac{1}{R}(\mu'-\mu) \quad , \quad J'' = \frac{1}{R}(\mu - \mu'')$$

$$J = \frac{dn}{dt} = \frac{dn}{d\mu}\frac{d\mu}{dt} = C\frac{d\mu}{dt}$$

(3.7)

where the material capacity C for an ideal fluid within the membrane is given as

$$C = \frac{dn}{d\mu} = \frac{d}{d\mu}V\exp\left(\frac{\mu}{R_GT}\right) = \frac{n}{R_GT}$$

(3.8)

n being the number of moles in the membrane and V the membrane volume. Combination of (3.7) with Kirchhoff's current law $J' = J + J''$ leads to the differential equation:

$$C\frac{d\mu}{dt} = \frac{1}{R}(\mu' + \mu'') - \frac{2}{R}\mu$$

(3.9)

For small variations of μ such that $C \simeq$ const, (3.9) can be solved to give

$$\mu = \tfrac{1}{2}(\mu' + \mu'') + A_0\exp\left(-\frac{t}{\tau}\right)$$

(3.10)

A_0 being an integration constant and τ the relaxation time defined as

$$\tau = \tfrac{1}{2}RC$$

(3.11)

In the steady state, $t \to \infty$,

$$\mu = \tfrac{1}{2}(\mu' + \mu'') \quad , \quad J' = J'' = \frac{1}{2R}\Delta\eta$$

(3.12)

we obtain a linear relationship between flux and force which can be compared with the linear result (3.5) of the Nernst-Planck theory to yield

$$R = \frac{aR_GT}{2Dc_e}$$

(3.13)

where $c_e = w\bar{c}$ is the effective membrane concentration. Inserting (3.8) and (3.13) into (3.11) and interpreting C as capacitance per unit area such that $V = a$ and $c_e = n/a$ leads to

$$2\frac{D}{2}\tau = \left(\frac{a}{2}\right)^2$$

(3.14)

which is nothing else but the classical Einstein relation for diffusion coefficient and relaxation time (the factor $1/2$ because of the symmetric arrangement of our system).

Before the network method is applied to more complex situations, a simplification of notation is achieved by introducing the so-called bond graph representation. In this representation, a junction or parallel arrangement satisfying Kirchhoff's voltage law is replaced by a 'one-junction'. As the zeros of the chemical and electrical potentials are arbitrary, we can also omit the ground terminals in our networks. With this notation, the simple network model of Fig.3 is translated into a graph which is shown in Fig.4. From Fig.4 we conclude that the zero-junctions satisfy Kirchhoff's current law, the voltages being equal for all branches, whereas the one-junctions satisfy Kirchhoff's voltage law, the fluxes being equal in the branches:

1-junctions: $\mu' = \mu_1 + \mu$, $\mu = \mu_2 + \mu''$

0-junction: $J' = J + J''$

$$(3.15)$$

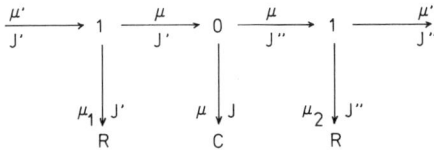

FIG.4. Bond graph representation of the simple membrane model.

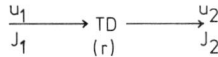

FIG.5. Transducer.

The network models can immediately be extended to describe multi-compartment models including several internal capacitances; however, the network language so far represented does not include any coupling phenomena. For this purpose, a transducer element TD is introduced, the symbol of which is shown in Fig.5. The action of the transducer is defined by its constitutive relations:

$$u_1 = \frac{1}{r} u_2 \ , \quad J_1 = r J_2 \qquad\qquad (3.16)$$

where r is the 'modulus' of the transducer.

In Fig.5 and Eq.(3.16), the symbol u has been used for the potential or force variable to indicate that the force may be any chemical or the electrical potential. From (3.16) it also follows that the transducer is a non-dissipative ideal element for we have $u_1 J_1 = u_2 J_2$.

The transducer enables us to introduce coupling into our network models. In Fig.6, coupling between material flux J and electrical flux j with forces μ and Φ, respectively, is shown. The modulus of the transducer in this case is e = electric charge per mole. Connecting a further material resistance R to the one-junction in the material channel directly leads to a linear relationship between the material flux J and the difference of the electrochemical potential $\Delta\eta$ as given in (3.5).

Our own investigations in Aachen on network thermodynamic modelling of membrane phenomena indicate that inclusion of transducers is still not sufficient to describe highly non-linear phenomena like excitation in a satisfactory way. For this purpose, parametric coupling must be introduced between the moduli of transducers in different material channels.

The essential advantages of network thermodynamics are the possibilities to:

(a)   Design phenomenological descriptions of systems without analysing them on a molecular level, i.e. to obtain 'minimal models' without assuming unnecessary details;

(b)   Apply a lot of mathematical theorems from graph theory to those models, particularly concerning statements on conditions for stability or instability.

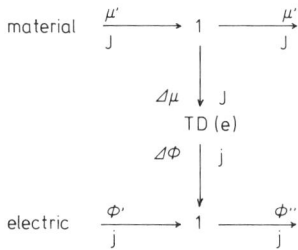

FIG.6.   Transducer coupling of material and electric flux.

4.   MEMBRANE EXCITATION

Membrane excitation is the elementary process by which nervous conduction is performed in living organisms. A local excitation of the nerve cell membrane propagating along the far-reaching extensions of the nerve cell, the so-called neurites, represents a unit of information to be transmitted. Although excitation is an all-or-nothing process, the nervous system is not a digital computer but a frequency-modulated analogue computer. The intensity of information or sensation to be transmitted is coded as the frequency of excitations passing a local region of the membrane.

In our models, interest will be focused on the local excitation process. The propagation mechanism is the subject of an extra theory and will not be discussed here.

The local record of the electrical membrane potential during nervous conduction, so-called firing of the nerve, shows a sequence of peaks with a rest potential of -60 mV (inside negative) and a peak potential of the order of magnitude of +50 mV (Fig.7). The transition from -60 mV to +50 mV is called the depolarization of the nerve cell. The frequency of peaks can be as high as 1000 Hz. Inspection of the ionic composition of the internal medium of the cell (axoplasm) and the external medium shows that rest and peak potentials are approximately given by the Nernst potentials of potassium and sodium:

$$\frac{RT}{e} \log \frac{[K^+]_e}{[K^+]_i} = U_K \simeq -60 \text{ mV} \tag{4.1}$$

$$\frac{RT}{e} \log \frac{[Na^+]_e}{[Na^+]_i} = U_{Na} \simeq +50 \text{ mV} \tag{4.2}$$

the subscripts e and i denoting exterior and interior (axoplasm) of the cell. From (4.1) and (4.2) we also conclude that the axoplasm contains much potassium and little sodium as compared with the external medium.

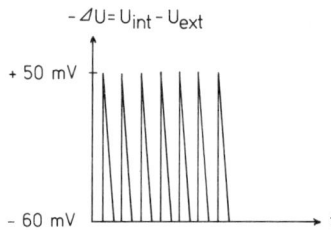

FIG. 7.   Sequence of excitations.

A possible interpretation of (4.1) and (4.2) says that the membrane is potassium-permeable but sodium-impermeable in the rest state and vice versa during excitation. The excitation process is then considered as a sudden influx of sodium due to an external change of the membrane potential. After the excitation, the system is restored to its ground state by a less rapid outflux of potassium.

There is, of course, considerable further experimental information which supports the hypothesis of permeability changes. Within the present framework, we do not want to go through all the experimental results obtained for nerve excitation. For further reading, the excellent review article of Hodgkin [12] is recommended. The aim of the remainder of this paper is to construct meaningful models for the increase of sodium permeability during the depolarization process. Let us describe this phenomenon in a more physical language. In the rest state with its potential difference of $-\Delta U = U_{int} - U_{ext} = -60 \text{ mV}$ we have an electric field of the order of magnitude of 10^5 V/cm directed towards the axoplasm, whereas during depolarization with $-\Delta U = +50$ mV the electric field changes its direction towards the exterior. The basic question now is: how does such a drastic decrease or reversal of electric field produce an increase of the sodium permeability?

The essential point of this section is to argue that the field-dependent sodium permeability can be understood on the basis of a flux-voltage characteristic for sodium which shows a voltage-controlled instability. Such a characteristic is shown in Fig.8. In this figure, the membrane potential is represented by a battery the characteristic of which, the 'load line', is assumed to be a straight line corresponding to a constant negative internal resistance. The load lines e, 0, r correspond to different values of the membrane potential or to the electric field, e being a low field line and r a high field line. The intersections of the membrane characteristic and the load line represent the steady-state working points of the system. We have neglected the electrical capacitance of the membrane parallel to resistance and battery since we shall restrict ourselves to the steady state only.

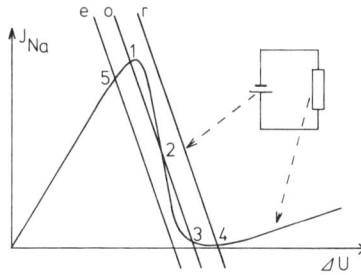

FIG.8.   Sodium flux instability.

For medium values of the electric field within a certain range (load line 0 in Fig.8), the system has three working points, 1,2,3, of which 2 is metastable and 1 and 3 are stable. For high (r) and low (e) fields, the system has only one working point (4 and 5); however, the high field working point 4 has a low sodium flux whereas the low field working point has a high sodium flux. Clearly, we now identify the rest state of the membrane with the high field case and the excited state with the low field case. Thus, within the framework of our rather simplified construction, we have reduced the problem of excitation to that of finding sodium conduction mechanisms exhibiting voltage-controlled instabilities.

At this point the influence of another cation, calcium, must be mentioned. In many excitable membranes, the excitation process crucially depends on the presence of calcium ions in the external medium. More precisely, the voltage ΔU_c, where the differential conductivity has its largest negative value, depends on the outside calcium concentration like

$$\Delta U_c = -\Gamma \log [Ca^{2+}]_e + const \tag{4.3}$$

where Γ ranges between 5 mV and 20 mV (per e-fold change of Ca^{2+}). This finding suggests that calcium has a blocking effect on the sodium permeability. It could be speculated that, at high fields, calcium ions are blocking the sodium transport channels whereas at low fields calcium is released from its blocking sites and thus gives way to sodium permeation. This idea is the basis for our first excitation model to be presented in the next section.

5.   PORE-BLOCKING MODEL

The basis of the pore-blocking model is the assumption of the existence of particular transport channels, 'pores', through which ions can permeate in order to pass across the membrane. The assumption of some appropriate transport facility for ions is inevitable since the 'bare' membrane, with its hydrophobic hydrocarbon interior, represents an impermeable barrier for ions.

The pores are considered as one-dimensional paths of stable ionic sites across the membrane and approximately perpendicular to the membrane plane. Ionic permeation across the membrane is then described as a series of elementary diffusion-like jumps between neighbouring sites. The stable sites and the barriers separating them can be imagined as corresponding to minima and maxima, respectively, of the profile of the chemical potential of the ions.

The critical point of the pore hypothesis is the fact that no experimental proof of their existence could yet be given. On the other hand, the theoretical model to be developed in this section, as well as other models on the basis of the pore hypothesis, do not critically depend on a detailed picture of 'pores' on a molecular level. The assumption of a diffusion-like motion of ions across membranes is certainly of sufficient flexibility to include pore structures or transport channels which are quite different from tubes or protein helices. A much less particular structure of a pore will be discussed in the next section. Here, we shall continue to use the term 'pore' in a generalized sense.

If there are no fixed charges in the membrane material bordering the pore, the electrostatic repulsion between cations within distances of less than 100 Å will prevent simultaneous occupation of one pore by more than one ion. If, on the other hand, the stable sites correspond to fixed charges, a complementary model describing ionic transport by 'hole propagation' would be more appropriate. We restrict ourselves here to the 'one-particle' version of the pore model.

To account for the calcium-sodium interaction in excitation phenomena, we assume that the pores under consideration can be entered by either a sodium or a calcium ion. If the pore contains the number of N-1 stable sites separated by N barriers, the following states for the pores are possible:

state 0 : empty pore

state A_i : occupation of site i by Na^+

state B_i : occupation of site i by Ca^{2+}

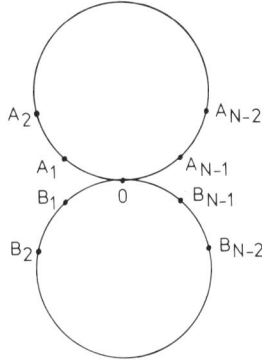

FIG.9.   Diagram of the pore-blocking model.

These states can be arranged to form a 'figure-8' diagram as shown in Fig.9. Each line or edge which connects two points or vertices in the diagram means a possible direct transition between the corresponding states.

To obtain a dynamic description of the model, a probability P_α is defined for each state $\alpha = 0$, A_i or B_i such that

$$\sum_\alpha P_\alpha(t) = P(0,t) + \sum_{i=1}^{N-1} \{P(A_i,t) + P(B_i,t)\} = 1 \qquad (5.1)$$

For the time evolution of the $P_\alpha = P_\alpha(t)$ a differential equation ansatz is made which reads:

$$\frac{dP_\alpha(t)}{dt} = \sum_\beta \{W(\alpha,\beta)\,P_\beta(t) - W(\beta,\alpha)\,P_\alpha(t)\} \qquad (5.2)$$

Equation (5.2) is the phenomenological or deterministic rate equation of the system. If one wants to include higher-order probabilities like fluctuations in the theory, a stochastic generalization of Eq.(5.2) is needed. The theory of such generalizations has been worked out by Hill and Plesner [13].

The $W(\alpha,\beta)$ in Eq.(5.2) correspond to the edges of the diagram in Fig.9 and represent the transition probabilities per time for a transition from state β to state α. All particular information on the transport system is involved in the $W(\alpha,\beta)$. A few conditions on the $W(\alpha,\beta)$ can immediately be formulated, e.g.:

$$W(A_1,0) \sim c_A' \qquad , \qquad W(A_{N-1},0) \sim c_A''$$
$$\qquad (5.3)$$
$$W(B_1,0) \sim c_B' \qquad , \qquad W(B_{N-1},0) \sim c_B''$$

where $c_A^!$, $c_A^"$ are the concentrations of sodium on the ' and the " side of the membrane and similarly $c_B^!$, $c_B^"$ for calcium. The numbering i of the sites has been chosen such that site i = 1 is adjoining the ' side of the membrane and site i = N-1 the " side. Moreover, if a voltage difference ΔU is applied to the membrane, we have

$$W(A_{i+1}, A_i) \sim e^{\Phi} \quad , \quad W(A_i, A_{i+1}) \sim e^{-\Phi}$$
$$W(B_{i+1}, B_i) \sim e^{2\Phi} \quad , \quad W(B_i, B_{i+1}) \sim e^{-2\Phi}$$

(5.4)

where

$$\Phi = \frac{e\Delta U}{2NRT}$$

(5.5)

and $\Delta U > 0$ corresponds to an electric field with direction from the ' to the " side of the membrane. The equilibrium condition for the system requires that

$$\log \prod_{i=0}^{N-1} \frac{W(A_{i+1}, A_i)}{W(A_i, A_{i+1})} = \frac{e\Delta U}{RT} + \log \frac{c_A^!}{c_A^"}$$

$$\log \prod_{i=0}^{N-1} \frac{W(B_{i+1}, B_i)}{W(B_i, B_{i+1})} = \frac{2e\Delta U}{RT} + \log \frac{c_B^!}{c_B^"}$$

(5.6)

where by definition

$$A_0 \equiv B_0 \equiv A_N \equiv B_N \equiv 0$$

(5.7)

A determination of the transition probabilities exceeding the above conditions requires particular assumptions on the model. To start with the simplest case, the remaining constants would be chosen equal to some constant which can be fitted to experimental curves. Such a version of the model, however, would not include any excitation phenomenon since both sodium and calcium could pass the pore without a negative differential conductivity.

To obtain pore-blocking by calcium we assume that one of the barriers along the pathway of sites, say between i = M-1 and i = M, is impermeable for calcium due to the presence of a calcium specific blocking site. For small electric fields, only a small fraction of pores will be occupied and thus blocked for sodium transport such that the electric current starts as a linear function of the voltage difference like an Ohmic resistance. For increasing electric fields, the fraction of blocked pores increases due to the twofold charge of calcium until finally all pores are closed and the electric current tends to zero exponentially with increasing voltage difference.

This expected behaviour of the model is confirmed by analytical and computer evaluations of the rate equation (5.2) for the steady state. Of particular interest is the calculation of the 'critical voltage difference' ΔU_c for which the differential conductivity has its largest negative value. In accordance with Eq.(4.3), the analytical calculations lead to

$$\Delta U_c \simeq - \frac{N}{2M} \frac{RT}{e} \log [Ca^{2+}]_e + const$$

(5.8)

Thus comparison with experimental results leads to the determination of the relative penetration depth M/N of calcium.

A full description of the model and its application to the excitation phenomenon of the surface membrane in frog-skin epithelium has been given by Heckmann et al. [14]. The analytical evaluation of Eqs (5.1) and (5.2) causes mathematical complications even in the steady state ($dP_\alpha/dt=0$) although in this case the problem is reduced to a linear equation for the P_α. The usual technique for solving a linear equation in terms of determinants and sub-determinants is useless for a discussion of the properties of the solution. In place of that, use is made of a theorem from the mathematical theory of graphs which allows the steady-state solutions and the steady-state fluxes to be expressed in terms of partial graphs of the diagram of Fig.9. The proof of this theorem dates back as far as to the beginning of network theory with Kirchhoff [15] and was independently rediscovered several times more than a hundred years later. A full description of the underlying theory is given by Hill [16] and Schnakenberg [17].

6.   ONSAGER'S DIPOLE CHAIN MODEL

The concept of 'pores' as ionic transport channels, as developed in Section 5, should not be taken as a molecular picture in the literal sense. Much less artificial structures than pores or tubes may act as pores by providing an appropriate transport mechanism which enables the ion to penetrate through the hydrocarbon membrane phase in the interior. From a physical point of view, this penetration problem is mainly the problem of screening the electrical charge of the ion against the non-polar hydrocarbon surrounding.

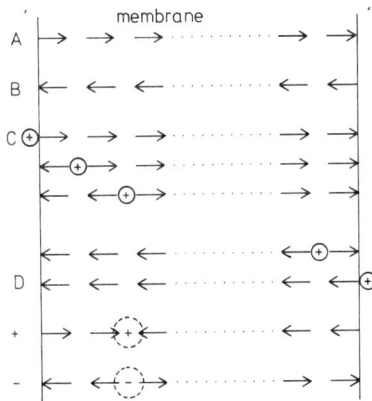

FIG.10.   Onsager's dipole chain model.

A strikingly simple but very convincing model which is based upon the idea of screening has been suggested by Onsager [18]: ionic transport along a chain of elementary molecular dipoles, which are capable of rotating about their centres. Such a chain could be formed by a single extended protein molecule, and possible candidates for the elementary molecular dipoles are hydroxyl and carboxyl groups in the side chains of amino acids.

Such a dipole chain has two low-energy configurations: all positive ends of the dipoles are directed towards the ' or the " side of the membrane (configurations A and B in Fig.10). Clearly, an applied electric field causes an energy difference between these configurations proportional to the field F.

Let the dipole chain be in its low-energy configurations with all positive ends towards the " side of the membrane (configuration A in Fig.10). A cation from the ' side of the system, say of type 1, can now cross the membrane along the chain by adjoining the negative end of the dipole nearest to the ' side, jumping over it and simultaneously reversing its direction and so forth, until it arrives at the " side of the membrane (configurations C to D in Fig.10). Having passed the chain, the ion has completely reversed the chain's direction compared with that at the beginning. Thus, no further ion of type 1 can pass the same chain in the same direction, and a steady ionic flux of type 1 will not be sustained unless the chain is capable of re-arranging itself appropriately to the original polarization. Such a re-arrangement could be accomplished by the passage of:

(a) An anion (type 2) in the same direction as cation 1;
(b) A cation (type 2) in the opposite direction;
(c) A plus or minus defect (configurations +/- in Fig.10) in any of the two possible directions.

In case (c), there is a net flux of electric charge, and the ionic flux is controlled by an applied field. This also holds for cases (a) and (b) if the absolute values of the valencies of ions 1 and 2 are different.

From the physical point of view, Onsager's model actually provides a mechanism for screening the ionic charge against the hydrophobic interior of the membrane. While an ion of charge $\nu_1 e$ (e is the elementary charge, ν_1 the valency) traverses the dipole chain from ' to ", the charge actually transferred is $\nu_1 e - 2q$ (q is the effective amount of charge at the ends of the elementary dipoles) rather than $\nu_1 e$. If the chain is re-arranged by the passage of a cation of charge $\nu_2 e$ from right to left or of a defect in any direction, the actual charge transfer in the second step is $-\nu_2 e + 2q$ or $2q$.

Since the permeability for anions across biological membranes is always very small, we omit the case (a). Combination of cases (b) and (c) leads to a model with a diagram shown in Fig.11. The 'channels' 1, 2, plus and minus in Fig.11 correspond to the passage of cations of type 1, 2 and to that of plus and minus defects.

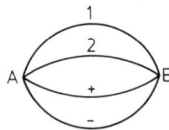

FIG.11.   Diagram of the dipole chain model.

To apply the model to membrane excitation, let us interpret cations 1 and 2 as calcium and sodium such that $\nu_1 = 2$, $\nu_2 = 1$. To start with, let us assume that the formation energy of the defects is so large that, at least for not too high fields, the channels plus and minus remain closed.

The quantitative evaluation of the model follows the rate equation theory given in Section 5. From Fig.11 an analytic solution can be obtained for the steady state. We restrict ourselves to a discussion of the results; a full description of the model and its evaluation is given by Schnakenberg [19].

The most interesting result of the model is again an unstable characteristic of the electrical current as a function of the applied voltage difference similar to that in Fig.8. Thus, Onsager's model gives a possible basis for describing membrane excitation. In contrast to the pore-blocking model, however, excitation is obtained in Onsager's model without assuming the existence of a calcium specific binding site. This can be understood by simple physical arguments. In the simplified version of our model without formation of defects, the complete transport cycle consists of two phases: $A \to B$ via channel 1; and $B \to A$ via channel 2. Each phase involves a change in electrostatic energy:

$$\Delta E_1(A \to B) = -e\Delta U(\nu_1 - 2\alpha)$$
$$\Delta E_2(B \to A) = e\Delta U(\nu_2 - 2\alpha) \tag{6.1}$$

where e is the elementary charge, ΔU is the applied voltage difference, ν_1 and ν_2 are the valencies and

$$\alpha = \frac{p}{e\ell} \tag{6.2}$$

In Eq.(6.2), p is the dipole moment of the elementary dipoles and ℓ their length, such that α is dimensionless. A realistic estimate of the order of magnitude of α will yield a value not only smaller than 1 but nearer to $\alpha \approx 0.1$. This means that

$$\Delta E_1(A \to B) < 0 \quad, \quad \Delta E_2(B \to A) > 0 \tag{6.3}$$

for $\Delta U > 0$. On the other hand, irreversible thermodynamics tells us that

$$-\{\Delta E_1(A \to B) + \Delta E_2(B \to A)\} = e\Delta U(\nu_1 - \nu_2) = T\Delta_i S \geq 0 \tag{6.4}$$

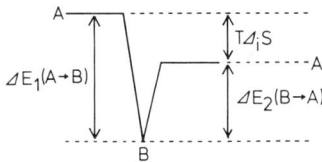

FIG.12.   Energy diagram of the dipole chain model.

where $\Delta_i S$ is the internal production of entropy for the complete cycle. In Fig.12, the energies are schematically represented as given by conditions (6.3) and (6.4). We see that the intermediate state B has the lowest energy and therefore acts as an 'energetic trap' for the whole process. As the depth of the trap increases with the voltage difference, we expect a vanishing electric current for $U \to +\infty$. On the other hand, at low values of ΔU the system will certainly behave like an Ohmic resistance such that in summary we expect a current-voltage characteristic similar to that in Fig.8 but without increasing part of the current at very high ΔU. Such an increase of the current will be produced by the formation of plus and minus defects at very high ΔU due to the large values of their formation energies compared with that of the calcium and sodium complexes in the chain. The graph theoretical evaluation of Fig.11 confirms our qualitative results.

Although Onsager's model is so attractive because of its very simple structure and the small number of additional assumptions made, one may still raise the standard objection which is often made to theoretical models: if the number of the adjustable parameters is large enough, every model can be forced to describe every experiment with sufficient accuracy. We feel, however, that in this case the situation is somewhat different. First, the number of parameters is not so large. Second, Onsager's model claims a molecular mechanism. This means that almost all parameters bear a defined physical meaning on a molecular scale. Therefore, it should be possible at least to narrow down the range in which the parameters can be adjusted by estimating their order of magnitude. The third point is that an attempt to adjust the model to experimental curves by making use of a feedback computer routine has shown us that the model is not very flexible. In fact, we have not been able to reproduce every kind of N-shaped characteristic that we wanted.

From these considerations we conclude that it will not be very difficult to put the whole model to a crucial test by comparing the predicted results with experimental findings which either have been obtained by other researchers or could readily be obtained in the future.

7.   MEMBRANE EXCITATION AS A PHASE TRANSITION

The pore-blocking model of Section 5 and Onsager's dipole chain model of Section 6 have been shown to yield membrane excitation and they have also been used to describe excitation in special biological membranes. Unfortunately, both the pore-blocking and the dipole chain model are not appropriate to describe excitation of the nerve membrane. One reason for this conclusion lies in the fact that, for example, the steady-state current-voltage characteristics of the pore-blocking and of the dipole chain model are too smooth compared with that of the nervous membrane. The latter shows instabilities with a large negative differential conductivity within a very narrow voltage region such that the characteristic in this region is almost a discontinuous step function rather than a continuous curve. Such a drastic change of the current within a narrow voltage interval suggests that membrane excitation in this case is much more like a phase transition than a continuous transition from sodium- to calcium-occupied pores.

The idea of membrane excitation as a phase transition has been worked out by Adam [20, 21]. In his model, the pores for ionic transport across the

membrane are reduced to 'active centres' which can bind either a divalent
or two monovalent cations. It would not cause any difficulties to replace
the active centres by extended pores like those in the pore-blocking model,
but qualitative changes of the results are not to be expected from such a
generalization.

In contrast to the sodium-calcium interaction hypothesis in the models
so far discussed here, Adam interprets excitation as an interaction of
potassium with calcium. Thus the active centres are occupied either by a
calcium ion, this being the ground state, or by two potassium ions in the
excited state. If n is the fraction of excited active centres, the rate equation
analogue to Eq.(5.2) now reads:

$$\frac{dn}{dt} = W_{10}(1-n) - W_{01}n \qquad\qquad (7.1)$$

where W_{10} and W_{01} are the transition probabilities per time for $0 \to 1$ and $1 \to 0$,
respectively, and 0 and 1 denote ground state and excited state, respectively.

The transition probabilities first of all contain the concentrations of
potassium c_1', c_1'' and calcium c_2', c_2'' on the ' and " side of the membrane.
Secondly, the voltage difference ΔU is included in the W_{10} and W_{01}, as described
in Eq.(5.4). Combining both these influences we obtain

$$W_{10} \sim \gamma_1 c_1'^2 e^{2\Phi} + \gamma_2 c_1' c_1'' + \gamma_3 c_1''^2 e^{-2\Phi} \qquad\qquad (7.2)$$

$$W_{01} \sim c_2' e^{2\Phi} \qquad\qquad (7.3)$$

where

$$\Phi = \frac{e\Delta U}{4kT} \quad (e = \text{elementary charge}) \qquad\qquad (7.4)$$

In (7.2) the γ_1, γ_2, γ_3 are relative weight factors, and in (7.3) use has been
made of the fact that axoplasm, the interior of the nerve cell, contains no
calcium: $c_2'' = 0$.

The essential point of Adam's model is the assumption that the active
centres are not independent of each other but that there are interaction
energies $w_0 > 0$ and $w_1 > 0$ for pairs of active centres such that the pair
receives an extra binding contribution by w_0 or w_1 if both partners are in the
ground state or in the excited state.

With this interaction hypothesis, the model becomes equivalent to a two-
dimensional Ising model. At first, one might think of applying the exact
solution for the two-dimensional Ising model as given by Onsager to the
nervous membrane. This is not possible since (a) the active centres do not
form a regular lattice, and (b) Onsager's exact solution is restricted to the
thermodynamic equilibrium whereas for application to nervous excitation
one is interested in situations far from equilibrium. This latter point
means that one has to formulate and solve a kinetic equation of the type of
(7.1), but Eq.(7.1) is certainly not an exact equation since a co-operative
model like the Ising model cannot be described by global variables like the
fraction n of excited pores. Such a global description is only an approximation, the so-called mean field theory or the Bragg-Williams approximation
which is often used in the theory of co-operative systems.

Applying this approximation to our model, we now express the co-operativity by introducing the pair interaction energies w_0 and w_1 into the transition probabilities W_{10} and W_{01} in the following way:

$$W_{10} \sim \exp\left\{-\frac{(1-n)w_0}{kT}\right\} \tag{7.5}$$

$$W_{01} \sim \exp\left\{-\frac{nw_1}{kT}\right\} \tag{7.6}$$

Inserting (7.2), (7.3), (7.5) and (7.6) into the kinetic equation (7.1), we obtain for the steady state $dn/dt = 0$:

$$\frac{n}{1-n}\exp\left\{-\frac{nw}{kT}\right\} = f(\Phi) \tag{7.7}$$

$$w = w_0 + w_1 > 0 \tag{7.8}$$

$$f(\Phi) \sim \gamma_1 \frac{c_1'^2}{c_2'} + \gamma_2 \frac{c_1' c_1''}{c_2'} e^{-2\Phi} + \gamma_3 \frac{c_1''^2}{c_2'} e^{-4\Phi} \tag{7.9}$$

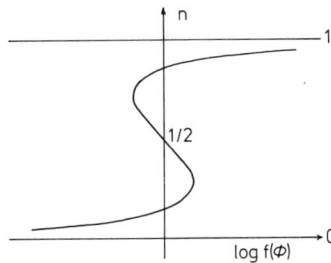

FIG.13.   Fraction of excited centres n for a phase transition in the membrane.

In Fig.13, n is shown as a function of $\log f(\Phi)$. If $w > 4kT$ as has been assumed in Fig.13, n becomes a multiple-valued function of $\log f(\Phi)$ and thus of Φ or ΔU. This situation is typical for the appearance of a phase transition in the mean field theory. Since $f(\Phi)$ is a monotonic function of Φ and thus of ΔU, and

$$f(\Phi) = \begin{cases} \gamma_1 \dfrac{c_1'^2}{c_2'} & \Phi \to +\infty \\ +\infty & \Phi \to -\infty \end{cases} \tag{7.10}$$

we see that, for high voltage differences ΔU corresponding to the rest state of the membrane, almost all active centres are in the ground state. For decreasing ΔU, i.e. for increasing depolarization, only a very small increase of the fraction n of excited centres is observed, but this part of the n curve becomes metastable as soon as ΔU reaches that interval where n is multiple-valued. For low values of ΔU, i.e. for depolarization of the membrane, almost all centres are excited. The stable transition from the n ≃ 0 to the n ≃ 1 part of the curve is obtained by the usual Maxwell construction for multiple-valued equations of the thermodynamic state.

If we introduce the co-operativity between active centres into a pore model for steady-state transport, the discontinuous transition from the ground state to the excited state leads to a vanishing ΔU interval for the negative differential conductivity, such that the current-voltage curve becomes a step function.

Adam has also applied his model to non-steady-state phenomena in the nervous membrane, and he obtained a surprisingly good agreement with experimental results. Recently, phase-transition phenomena in quite different membranes have also been successfully described on the basis of this model. The initial rejection of the idea of co-operativity by the physiologists may eventually be withdrawn in view of the success of the model.

REFERENCES

[1] DANIELLI, J.F., DAVSON, H., J.Cell.Comp.Physiol. 5 (1935) 495.
[2] GLANSDORFF, P., PRIGOGINE, I., Thermodynamic Theory of Structure, Stability and Fluctuations, Wiley, London, New York, Sydney, Toronto (1971).
[3] OSTER, G.F., PERELSON, A.S., KATCHALSKY, A., Network thermodynamics: dynamic modelling of biophysical systems, Q.Rev.Biophys. 6 (1973) 1.
[4] SAUER, F., in Handbook of Physiology, Section 8: Renal Physiology (ORLOFF, J., BERLINER, R.W., Eds), American Physiological Society, Washington,D.C. (1973).
[5] DE GROOT, S.R., MAZUR, P., Non-Equilibrium Thermodynamics, North-Holland, Amsterdam (1962).
[6] KATCHALSKY, A., CURRAN, P.F., Non-Equilibrium Thermodynamics in Biophysics, Harvard University Press, Cambridge (1967).
[7] NERNST, W., Z.Phys.Chem. 2 (1888) 613.
[8] NERNST, W., Z.Phys.Chem. 4 (1889) 129.
[9] PLANCK, M., Ann.Phys.Chem.NF 39 (1890) 161.
[10] PLANCK, M., Ann.Phys.Chem.NF 40 (1890) 561.
[11] SCHLÖGL, R., Stofftransport durch Membranen, Dietrich Steinkopf Verlag, Darmstadt (1964).
[12] HODGKIN, A.L., The Conduction of the Nervous Impulse: The Sherrington Lectures VII, Liverpool University Press (1967).
[13] HILL, T.L., PLESNER, I.W., J.Chem.Phys. 43 (1965) 267.
[14] HECKMANN, K., LINDEMANN, B., SCHNAKENBERG, J., Biophys.J. 12 (1972) 683.
[15] KIRCHHOFF, G., Poggendorffs Ann.Phys.Chem. 72 (1847) 495.
[16] HILL, T.L., J.Theor.Biol. 10 (1966) 442.
[17] SCHNAKENBERG, J. (to be published).
[18] ONSAGER, L., in Physical Principles of Biological Membranes (SNELL, F., WOLKEN, J., IVERSON, G.J., LAM, J., Eds), Gordon and Breach, New York (1970).
[19] SCHNAKENBERG, J., Biophys.J. 13 (1973) 143.
[20] ADAM, G., Z.Naturforsch. 23b (1968) 181.
[21] ADAM, G., in Physical Principles of Biological Membranes (SNELL, F., WOLKEN, J., IVERSON, G.J., LAM, J., Eds), Gordon and Breach, New York (1970).

Part V

APPLIED SURFACE SCIENCE

ELECTRODE REACTIONS AND CORROSION

R. PARSONS
Department of Physical Chemistry,
University of Bristol,
Bristol,
United Kingdom

Abstract

ELECTRODE REACTIONS AND CORROSION.
1. Introduction: general characteristics of electrode reactions. 2. How do we measure the kinetics of electrode reactions? 3. Consequences of charge transfer in kinetics. 4. Mass transport. 5. Multistep reactions. 6. Adsorption of intermediates. 7. Transient methods for kinetically controlled reactions. 8. Transient methods for reactions controlled by diffusion as well as electrode reaction. 9. The electrical double layer. 10. Effect of double layer on electrode kinetics. 11. Kinetics of hydrogen evolution. 12. Metal deposition. 13. Metal dissolution. 14. Corrosion at a homogeneous electrode. 15. Corrosion at an inhomogeneous surface. 16. Prevention of corrosion.

1. INTRODUCTION: GENERAL CHARACTERISTICS OF ELECTRODE REACTIONS

Reactions of electrodes are similar to catalytic reactions in that they occur at interfaces and are surface reactions. One group of electrode reactions shares many characteristics with catalytic reactions and consequently is often grouped under the title 'electrocatalysis'. The major difference from catalytic reactions is that all electrode reactions are accompanied by a net transfer of charge between the phases adjoining the interface.

The simplest type of electrode reaction involves merely the transfer of an electron between a metal electrode and a redox couple in an electrolytic solution as typified by the reaction:

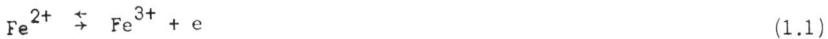

$$Fe^{2+} \rightleftharpoons Fe^{3+} + e \qquad (1.1)$$

Such a reaction has the fewest analogies to a catalytic reaction because the reactants do not interact significantly with the metal electrode more than is required for the transfer of the electron. Gas evolution reactions such as

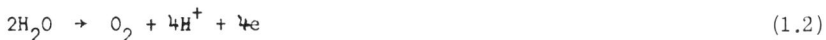

$$2H_2O \rightarrow O_2 + 4H^+ + 4e \qquad (1.2)$$

are more closely related to catalytic reactions because they involve adsorbed intermediate species.

Metal deposition and dissolution:

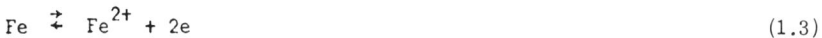

$$Fe \rightleftharpoons Fe^{2+} + 2e \qquad (1.3)$$

97

are of great practical importance both in electroplating and in its converse, electromachining, or in the less desirable reaction of corrosion. Such reactions involve the material of the electrode itself and are more closely related to crystallization and the dissolution of crystals. Reactions which involve solid transformations also occur at electrodes and an important example is the reaction:

$$PbSO_4 + 2H_2O \rightarrow PbO_2 + 4H^+ + SO_4^= + 2e$$

which occurs at the positive plate when a lead accumulator is charged.

Examples could be multiplied to show the widespread practical importance of electrode reactions as well as their intrinsic interest. The aim here is to outline the way in which electrode reactions may be described and understood in order that this understanding may be used to control these reactions. The basic problem is the understanding of the kinetics since in batteries and fuel cells the faster the electrode reactions the more efficient is the system. In corrosion the aim is to retard the electrode reaction as much as possible. In synthesis of chemical compounds it is important that the reaction is highly selective; this means that one reaction should be fast while its competitors are slow. Great strides have been made in understanding these problems in the last two or three decades but it must be admitted that our knowledge remains incomplete.

2.   HOW DO WE MEASURE THE KINETICS OF ELECTRODE REACTIONS?

We have seen that electrode reactions always involve charge transfer to or from the electrode. Thus, if a reaction like

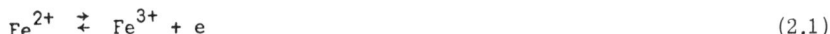

$$Fe^{2+} \rightleftarrows Fe^{3+} + e \qquad (2.1)$$

proceeds at a platinum electrode we can follow how much reaction has occurred by measuring the amount of charge transfer. (Faraday expressed this quantitative relation in his Laws of Electrolysis.) It follows that the rate of the reaction is given by the rate of charge transfer which is the current flowing through the electrode. This is the most usual and convenient way of measuring the rate of an electrode reaction. However, it must be noted that it is not a specific measure; if two electrode reactions are occurring simultaneously at one electrode we shall measure the algebraic sum of both rates.

Suppose a cell is made up consisting of two platinum electrodes dipping into a solution containing ferric and ferrous ions in dilute sulphuric acid as shown schematically in Fig.1. In such a system the current I measured by the ammeter is proportional to the rate of $Fe^{2+} \rightarrow Fe^{3+} + e$ at the anode and to the rate of $Fe^{3+} + e \rightarrow Fe^{2+}$ at the cathode and the total rates of these processes are constrained to be equal although, because the areas of the electrodes may not be equal,the rate per unit area of each electrode may not be the same. If each point on an electrode is equivalent, the rate which is of interest is the rate on unit area which is given by the current density:

$$j = I/A \qquad (2.2)$$

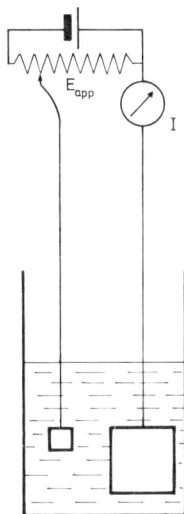

FIG.1.   Schematic diagram of a two-electrode electrolysis experiment.

For a given system the first observation of great importance is that the current density j depends markedly upon the potential difference E_{app} applied between the electrodes. This turns out to be the most important feature distinguishing electrode reactions from ordinary chemical reactions or heterogeneous catalytic reactions. The rate of the reaction is strongly dependent on the electrode potential.

With the apparatus of Fig.1 it is not possible to define the effect of potential on the rate properly because the applied potential is distributed in an unknown way between the two interfaces and the solution between them (the potential drop in the connecting wires can usually be neglected). This distribution depends on the whole geometry of the cell. A better defined system can be obtained by introducing a third electrode (Fig.2) which does not carry any current and whose potential therefore remains constant. This may be any stable reference electrode such as a calomel electrode.

The potential difference between the electrode under study — the test electrode — and the reference electrode is measured by a high input impedance device so that negligible current flows through the reference electrode. This measured potential difference E thus indicates changes occurring at the test electrode only. Changes in the potential drop in the solution are largely eliminated by the device known as a Luggin capillary, shown in Fig.2. (This arrangement is closely analogous to that used for measuring a p.d. across a given resistor in a network except that one connection is made by an electrolyte rather than a wire.)

The dependence of the current density passing through the test electrode on the potential difference E is now a well-defined function that can be analysed further to characterize the behaviour of the electrode reaction.

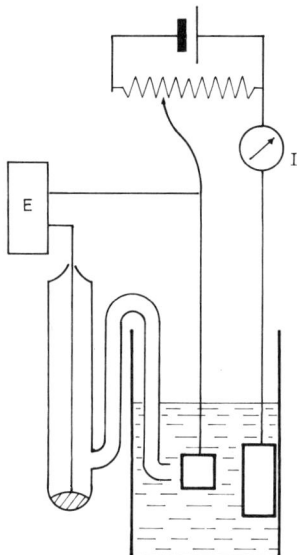

FIG.2.   Schematic diagram of a three-electrode electrolysis experiment.

Besides the study of this relationship we can also study the usual relationships of chemical kinetics: the dependence on reactant concentration,
temperature, etc. This type of analysis for the Fe^{2+}/Fe^{3+} reaction shows that
the anodic process is first order in the Fe^{2+} concentration while the cathodic
process is first order in the Fe^{3+} concentration. Using this information, we
shall now discuss the fundamental relation between the rate and the potential E
in the simplest type of system.

3.   CONSEQUENCES OF CHARGE TRANSFER IN KINETICS

A reaction like (2.1) can occur only when the reacting species is close
to the electrode surface because the electron must make a radiationless
transition from ion to electrode or vice versa. This can occur over a
distance of probably not more than 1 nm (10 Å). On the other hand, there
need not be any strong bonding between the ion and the metal and for the
moment we shall assume that none exists. The rate of the electrode
reaction may then be written in the way which is usual for a single-step
chemical reaction. The rate of the anodic reaction is

$$v_+ = k_+ \left[Fe^{2+} \right] \tag{3.1}$$

and that of the cathodic reaction

$$v_- = k_- \left[Fe^{3+} \right] \tag{3.2}$$

where k_+ and k_- are the rate constants and $[Fe^{2+}]$ and $[Fe^{3+}]$ are the concentrations of Fe^{2+} and Fe^{3+} at the reaction site, which may be called the pre-electrode layer. If the concentrations are expressed in moles per unit volume, the rate will be in terms of moles of reactant oxidized (or reduced) per unit area of the electrode per unit time, and the current density due to the anodic reaction is

$$j_+ = F \, v_+ \tag{3.3}$$

where F is Faraday's constant. Similarly

$$j_- = F \, v_- \tag{3.4}$$

Provided that no other reaction occurs at the electrode, the observed current density is

$$j = j_+ - j_- \tag{3.5}$$

or, from Eqs (3.1) - (3.4),

$$j = F \{ k_+ \left[Fe^{2+} \right] - k_- \left[Fe^{3+} \right] \} \tag{3.6}$$

Since we know that j is a function of the potential difference E in circumstances where the reactant concentrations in the pre-electrode layer are known to be constant, it follows that the rate constants must be potential dependent.

We can set limits on this potential dependence by invoking the equilibrium behaviour of the electrode. At equilibrium there is no net current:

$$j = 0 \tag{3.7}$$

The equilibrium is dynamic and the common value of the partial currents is called the exchange current:

$$j_o = k_{+,e} \left[Fe^{2+} \right]_b = k_{-,e} \left[Fe^{3+} \right]_b \tag{3.8}$$

In Eq.(3.8) the subscript e has been given to the rate constants to indicate that these are their values at equilibrium. Similarly the subscript b is given to the concentrations because these must now be equal to the bulk concentrations.

It is known that electrodes at equilibrium obey the Nernst equation, which for this electrode takes the form:

$$E_e = E^{\ominus} - (RT/nF) \ln \{ \left[Fe^{2+} \right]_b / \left[Fe^{3+} \right]_b \} \tag{3.9}$$

if we assume that the activity coefficients are unity. Here n is the number of electrons transferred in the electrode reaction which for the present example is one and $E^{\ominus}$ is the standard potential. From (3.8) and (3.9) it follows that

$$k_{+,e}/k_{-,e} = \exp \{ n(E_e - E^{\ominus})F/RT \} \tag{3.10}$$

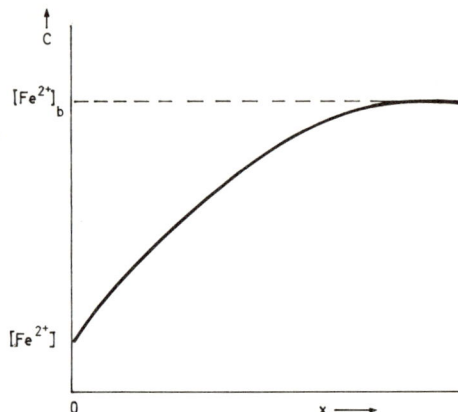

FIG. 3. Profile of concentration of Fe^{2+} near an electrode at which Fe^{2+} is oxidized. The electrode is at
$x = 0$ and the x axis is perpendicular to the electrode surface with the positive direction towards the solution.
$[Fe^{2+}]_b$ is the bulk concentration of Fe^{2+} and $[Fe^{2+}]$ is the concentration at the electrode surface.

which gives us the potential dependence of the ratio of the rate constants
under equilibrium conditions. To obtain the separate potential dependence
of the two rate constants we must construct a model or appeal to experiment.
We defer the model and take the latter course. Experiment shows that
usually anodic reactions are accelerated as the potential is made more
positive, and cathodic reactions are accelerated as the potential is made
more negative, and that there is frequently an exponential relation between
j and E. This behaviour is expressed by

$$k_{+,e} = k_{+,o} \exp \{(1-\alpha) \; n \; E_e \; F/RT\} \tag{3.11}$$

$$k_{-,e} = k_{+,o} \exp \{- \; \alpha \; n \; E_e \; F/RT\} \tag{3.12}$$

if α is a constant between zero and unity. It is known as the transfer
coefficient.

$$k_{+,o} = \exp \{- \; (1-\alpha) \; n \; E^{\ominus} \; F/RT\} \tag{3.13}$$

$$k_{-,o} = \exp \{\alpha \; n \; E^{\ominus} \; F/RT\} \tag{3.14}$$

It is clear that (3.11) and (3.12) satisfy the condition (3.10), although of
course many other types of relation would also satisfy (3.10). They may be
generalized to non-equilibrium conditions:

$$k_+ = k_{+,o} \exp \{(1-\alpha) \; n \; f \; E\} \tag{3.15}$$

$$k_- = k_{-,o} \exp \{- \; \alpha \; n \; f \; E \} \tag{3.16}$$

Here we have used the abbreviation:

$$f = F/RT \qquad (3.17)$$

which will be convenient in the following.

The effect of potential can now be made explicit in the rate equation by using (3.15) and (3.16) in (3.8):

$$j = nF \left[k_{+,o} [Fe^{2+}] \exp\{(1-\alpha)nfE\} - k_{-,o} [Fe^{3+}] \exp\{-\alpha\, nfE\} \right] \qquad (3.18)$$

This relation can be clarified by using the equilibrium condition:

$$j_o = nF\, k_{+,o} \left[Fe^{2+} \right]_b \exp\{(1-\alpha)\, nfE_e\} \qquad (3.19)$$

$$= nF\, k_{-,o} \left[Fe^{3+} \right]_b \exp\{-\alpha\, n\, f\, E_e\} \qquad (3.20)$$

and introducing the overpotential η as the deviation of the potential E from its equilibrium value E_e:

$$\eta = E - E_e \qquad (3.21)$$

Thus if we divide (3.18) by j_0 and use Eqs (3.19) - (3.21) we obtain

$$j/j_o = ([Fe^{2+}]/[Fe^{2+}]_b)\exp\{(1-\alpha)nf\eta\} - ([Fe^{3+}]/[Fe^{3+}]_b)\exp(-\alpha nf\eta) \qquad (3.22)$$

This is the general form of the current-voltage curve although it cannot be used to interpret experimental curves until we know the relation of the concentration of the reactants in the pre-electrode layer to that in the bulk. This depends on the transport of matter to the electrode.

4. MASS TRANSPORT

4.1. Steady state

It is frequently possible to ignore the transport of reactants to the interface in a gas-phase catalytic reaction because this tends to be rapid compared with the rate of the surface reaction itself. Since diffusion coefficients in liquids are usually several orders of magnitude smaller, this cannot always be done with electrode reactions. If the rate of transport is comparable with the rate of the electrode reaction, the arrival of reactant at the electrode cannot keep pace with the rate at which it is consumed by the electrode reaction, and the local concentration near the electrode drops.

We shall consider this problem first in the simplest conditions. First, that the reactants (Fe^{2+}, Fe^{3+}) are present at a much lower concentration than that of another electrolyte (H_2SO_4), the base electrolyte. The ions of the base electrolyte screen the electric field in the solution from the reactant ions so that their movement is almost entirely by diffusion and can be described by Fick's laws. Second, we shall begin by assuming that a steady state can be set up. This is most simply achieved by using a rotating disc electrode.

The electrode is the lower surface of a horizontal disc rotating about its centre on a vertical axis. Provided the rotation speed is not too high, the flow of fluid past the surface of such a disc is laminar. The fluid is drawn upwards towards the surface of the disc and flung outwards radially across the surface. The unique feature of this system is that the state of the fluid can be described in terms of a hydrodynamic boundary layer of uniform thickness over the surface of the disc. To a first approximation, the reactant concentration is constant outside the boundary layer while within it the fluid may be considered stationary. The concentration of the reactant will vary as a function of distance in a way similar to that shown in Fig.3. In the steady state the current passing through the electrode must be related to the flux at the electrode surface by the continuity equation:

$$j/nF = D (d c/d x)_{x=o} \qquad (4.1)$$

where the co-ordinate x is perpendicular to the disc surface with the latter placed at x = 0 and D is the diffusion coefficient. The slope of the curve in Fig.3 at x = 0 may be expressed in terms of a characteristic length δ known as the diffusion layer thickness (not equal to the thickness of the hydrodynamic boundary layer mentioned above). From Fig.3 it follows that

$$(d c/d x)_{x=o} = ([Fe^{2+}]^b - [Fe^{2+}])/\delta \qquad (4.2)$$

whence

$$j/nF = D([Fe^{2+}]^b - [Fe^{2+}])/\delta \qquad (4.3)$$

The value of δ is a function of the hydrodynamics, and for a rotating disc electrode Levich obtained

$$\delta = 1.61 \, D^{1/3} \, \nu^{1/6} \, \omega^{-\frac{1}{2}} \qquad (4.4)$$

if δ is in cm and D is in $cm^2 \cdot s^{-1}$. ν, the kinematic viscosity, is in $cm^2 \cdot s^{-1}$, and ω, the angular velocity, is in s^{-1}.

It follows from (4.3) that the current must reach a limiting value $j_{L,+}$ as the concentration of Fe^{2+} in the pre-electrode layer drops to a value small compared with that in the bulk:

$$j_{L,+} / nF = D[Fe^{2+}]^b/\delta \qquad (4.5)$$

From (4.3) and (4.5),

$$j/j_{L,+} = ([Fe^{2+}]^b - [Fe^{2+}]) / [Fe^{2+}]^b \qquad (4.6)$$

This enables us to obtain the concentration ratio required for (3.22) in terms of the current and the limiting current:

$$[Fe^{2+}] / [Fe^{2+}]^b = 1 - j/j_{L,+} \qquad (4.7)$$

A similar discussion for the flux of Fe^{3+} leads to the result that

$$[Fe^{3+}] / [Fe^{3+}]^b = 1 - j/j_{L,-} \qquad (4.8)$$

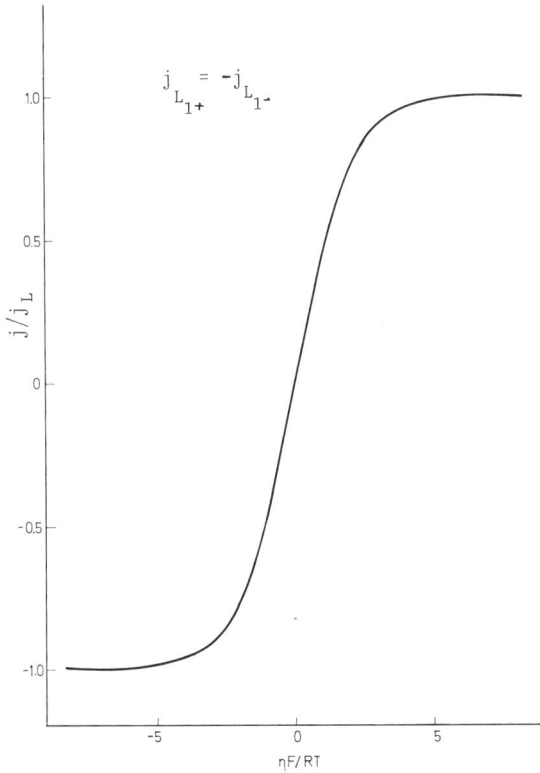

FIG.4.   Current-voltage curve for a one-electron (n = 1) reaction whose rate is controlled only by the diffusion of reactants to the electrode surface. The limiting currents of anodic and cathodic reactions are assumed to be equal.

where the limiting cathodic current is

$$j_{L,-} / nF = - D' \left[Fe^{3+} \right]^b / \delta'$$ (4.9)

the primes referring to the Fe^{3+} ion.

Thus the current-voltage current in the steady state can be expressed from (3.22), (4.7) and (4.9) as

$$j/j_o = (1-j/j_{L,+}) \exp\{(1-\alpha)nfn\} - (1-j/j_{L,-}) \exp\{-\alpha nfn\}$$ (4.10)

In this equation there are four parameters: the transfer coefficient α; the exchange current j_0, expressing the rate of the electrode reaction itself; $j_{L,+}$ and $j_{L,-}$, the limiting currents expressing the rate of transport of the reactants to and from the electrode surface. We may note in passing the similarity of this equation to that for the current at a p-n junction where $\alpha = 1$ and the limiting currents are related to the drift velocities of minority carriers.

It is useful to consider the limiting forms of Eq.(4.10).

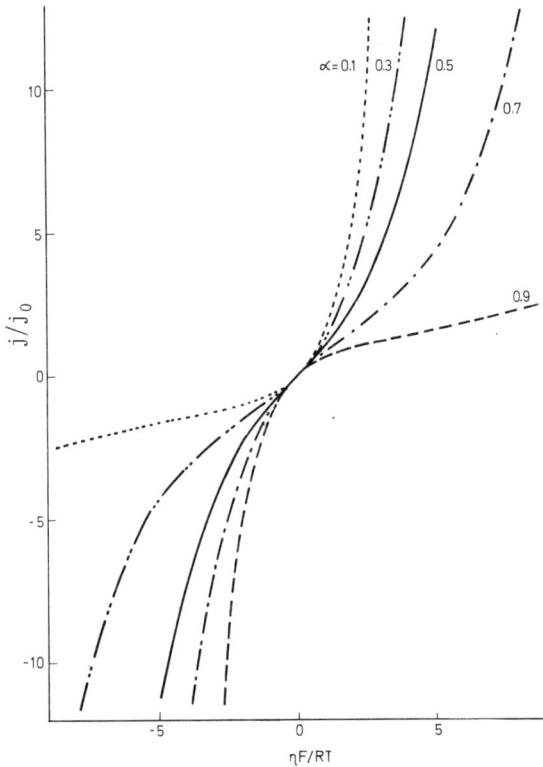

FIG. 5.   Current-voltage curve for a one-electron ($n = 1$) electrode reaction whose rate is controlled by the kinetics of the electrode reaction itself.   Diffusion of reactants is assumed to be very much faster than the rates depicted here.   The value of the transfer coefficient α is shown by each curve.

4.2. Fast electrode reaction, slow mass transport

$$j_o \gg j_{L,+}, \; j_{L,-}$$

Consequently $j \ll j_0$ and the two terms on the right-hand side of (4.10) are approximately equal, whence

$$\exp nf\eta \; = \; (1 - j/j_{L,-}) \, / \, (1 - j/j_{L,+}) \tag{4.11}$$

This current-voltage curve has the form shown in Fig.4. Both parameters (j_0 and α) characteristic of the kinetics of the electrode reaction itself have disappeared from (4.11) and consequently no information about the kinetics can be obtained under these conditions.

4.3. Fast mass transport, slow electrode reaction

$$j_o \ll j_{L,+}, \; j_{L,-}$$

4.3.1. $j \ll j_{L,+}, j_{L,-}$

Equation (4.10) becomes

$$j/j_o = \exp\{(1-\alpha)nf\eta\} - \exp\{-\alpha nf\eta\} \qquad (4.12)$$

The double exponential form has a slope dependent only on α, as shown in Fig.5. Experimentally it is useful to consider the two further limiting cases:

(a) $j \gg j_o$

$$j/j_o = \exp\{(1-\alpha)nf\eta\} \qquad (4.13)$$

so that a plot of log j against η readily yields α from the slope and j_0 from the intercept. This is known as a Tafel plot. The corresponding plot for the cathodic reaction is exactly analogous.

(b) $j \ll j_o$

This corresponds to small values of η so that the exponentials may be expanded and the terms higher than first order in η in the series neglected:

$$j/j_o = nf\eta \qquad (4.14)$$

Thus at small overpotentials the electrode behaves in an Ohmic way. Note that α has disappeared from (4.13).

4.3.2. $j \sim j_{L,+}$

The second term of (4.10) may be neglected and the result re-arranged to

$$(1-\alpha)nf\eta = \ln (j/j_o) - \ln (1-j/j_{L,+}) \qquad (4.15)$$

Thus η may be considered as the sum of two terms, one characteristic of the electrode reaction and the second of the transport process. Here we may draw the analogy to resistances in series, although we must note that both the resistance associated with the electrode reaction and that associated with the transport process are non-Ohmic and are different functions of potential. When $j \ll j_{L,+}$ the reaction resistance dominates, but as j approaches $j_{L,+}$ the transport resistance increases and eventually becomes infinite. (In practice the limiting current $j_{L,+}$ is exceeded because as the potential increases, some other reaction begins to carry current.) Similar behaviour is observed for the cathodic reaction.

4.4. More general steady-state treatment of an electrode reaction controlled by transport and reaction rate

If Eq.(4.10) is divided by $j \exp\{(1-\alpha)nf\eta\}$ it becomes

$$\frac{1}{j_o \exp\{(1-\alpha)nf\eta\}} = \frac{1}{j} - \frac{1}{j_{L,+}} - (\frac{1}{j} - \frac{1}{j_{L,-}}) \exp(-nf\eta) \qquad (4.16)$$

The right-hand side of (4.16) is directly obtainable from experiment, and by plotting the log of this against η the parameters α and j_0 can be obtained. At low overpotentials where $nf\eta \ll 1$, the exponentials can be expanded and if only the leading term of the series is retained (4.16) reduces to

$$nf/j = 1/j_0 + 1/j_{L,+} - 1/j_{L,-} \tag{4.17}$$

The electrode behaves Ohmically close to equilibrium but as the sum of three resistances, one related to the electrode reaction and one to the transport of each reactant.

4.5. Exchange current and rate constant

This relation may be obtained by solving (3.19) and (3.20) for E_e:

$$\exp(nf\,E_e) = k_{-,o}[Fe^{3+}]_b\,/\,k_{+,o}\,[Fe^{2+}]_b \tag{4.18}$$

and substituting this into (3.19) or (3.20):

$$j_o = nF\,k_{+,o}^{\alpha}\,k_{-,o}^{1-\alpha}\,[Fe^{2+}]_b^{\alpha}\,[Fe^{3+}]_b^{1-\alpha} \tag{4.19}$$

Thus from the concentration dependence of j_0 the value of α can be determined and also the product of rate constants:

$$k_{+,o}^{\alpha}\,k_{-,o}^{1-\alpha} = k^{o} \tag{4.20}$$

which is known as the standard rate constant.

k^0 is a more useful quantity than $k_{+,0}$ or $k_{-,0}$ because it depends only on the electrode being studied while the latter depend on the nature of the reference electrode. k^0 may be regarded as the common value of the rate constants of the anodic and cathodic reactions at the standard potential of the electrode reaction.

5.   MULTISTEP REACTIONS

So far, the electrode reaction has been assumed to occur in a single step. Many reactions actually occur in a series of such elementary reactions. A simple example is a linear sequence which may be represented as

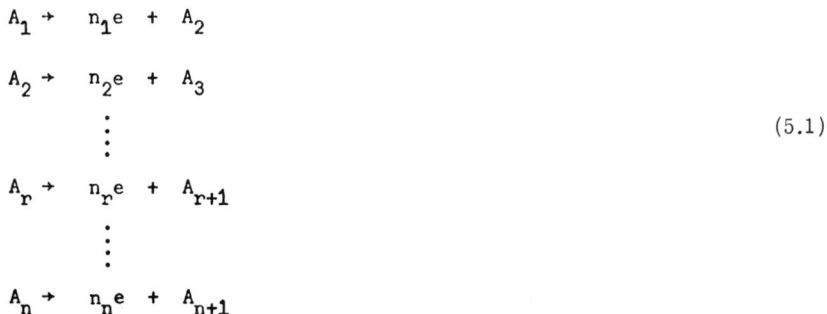

$$A_1 \rightarrow n_1 e + A_2$$

$$A_2 \rightarrow n_2 e + A_3$$

$$\vdots$$

$$A_r \rightarrow n_r e + A_{r+1} \tag{5.1}$$

$$\vdots$$

$$A_n \rightarrow n_n e + A_{n+1}$$

If it is assumed that one reaction (the r-th) is slower than any of the others, the simplest approach is to assume that equilibrium is set up in all the reactions preceding this rate-determining step. If this is done, the ratio of concentrations of intermediates in each preceding reaction is given by the appropriate form of the Nernst equation (3.9) in terms of the prevailing electrode potential E, e.g.

$$[A_r] / [A_{r-1}] = K_{r-1} \exp(n_{r-1}fE) \tag{5.2}$$

where

$$K_{r-1} = \exp(-n_{r-1}fE_r^o) \tag{5.3}$$

By considering such equations for each preceding reaction we may obtain the concentration of the reactant in the rate-determining step in terms of that in the initial step:

$$[A_r] / [A_1] = K_+ \exp(n_+fE) \tag{5.4}$$

where

$$K_+ = K_1 K_2 \ldots\ldots K_{r-1} \tag{5.5}$$

$$n_+ = n_1 = n_2 + \ldots\ldots n_{r-1} \tag{5.6}$$

The rate of the r-th step in the anodic direction is written in the same form as that of the elementary step previously considered:

$$v_{r,+} = k_{r,+} [A_r] \exp\{n_r(1-\beta_r) fE\} \tag{5.7}$$

where β_r is the transfer coefficient of the r-th step. From (5.4) and (5.7),

$$v_{r,+} = v_+ = k_{r,+} K_+ [A_1] \exp\{n_+ + n_r(1-\beta_r)fE\} \tag{5.8}$$

This takes the same form as the rate equation of an elementary step if we define the transfer coefficient for the reaction as a whole as

$$n(1-\alpha) = n_+ + n_r(1-\beta_r) \tag{5.9}$$

We shall in general use α for the overall observable transfer coefficient and β for the transfer coefficient of an elementary step. Provided that the cathodic reaction has the same rate-determining step, the current-voltage curve again has the form of (3.22). The effect of mass transport can be introduced as for a simple reaction if it is assumed that no transport of the intermediate species occurs.

It will be noted that all reactions so far have been assumed to be first order. More complex schemes may be accommodated using the equilibrium hypothesis. The reactions may be of higher order and the intermediate reactions may involve stable species as well as unstable intermediates. The rate-determining step may occur more than once for the transfer of n electrons in the overall reactions. If it occurs ν times, the n in Eq.(3.22) must be replaced by n/ν and ν is known as Horiuti's stoichiometric number. This type of scheme readily includes chemical as distinct from electrochemical reactions at the interface since the number of electrons transferred in a chemical reaction is zero and so the appropriate n_i is put equal to zero. Chemical reactions in solution must be taken into account in the solution of Fick's law.

6.   ADSORPTION OF INTERMEDIATES

In the above treatment of multistep reactions it is assumed that the intermediates do not diffuse. The only reasonable way in which this condition can be satisfied is by adsorption of the intermediates. However, if this occurs there is a limited region in which the intermediate may be accommodated, such as a monolayer, because only in a monolayer can the intermediate form a chemical bond with the electrode.

We consider a two-step scheme with a single adsorbed intermediate:

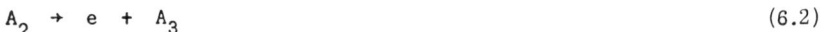

$$A_1 \rightarrow e + A_2 \tag{6.1}$$

$$A_2 \rightarrow e + A_3 \tag{6.2}$$

This is the simplest model for an electrocatalytic reaction. If the maximum amount of A_2 which can be accommodated on unit area of the electrode is $[A_2]_{sat}$ the fractional occupation θ can be defined as

$$\theta = [A_2] / [A_2]_{sat} \tag{6.3}$$

If the surface is uniform and formation of A_2 can occur only on parts of the surface not already occupied by A_2, the rate equations become:

$$j_{1,+} = F k_{1,+} [A_1] (1-\theta) \exp \{(1-\beta_1)fE\} \tag{6.4}$$

$$j_{1,-} = F k_{1,-} \theta \exp(-\beta_1 fE) \tag{6.5}$$

$$j_{2,+} = F k_{2,+} \theta \exp \{(1-\beta_2)fE\} \tag{6.6}$$

$$j_{2,-} = F k_{2,-} [A_3] (1-\theta) \exp (-\beta_2 fE) \tag{6.7}$$

If the electrode is assumed to be at equilibrium, these equations give the partial exchange currents $j_{1,0}$ and $j_{2,0}$ (θ and E become θ_e and E_e). The resulting equations lead to expressions for θ_e:

$$\theta_e (1-\theta_e)^{-1} = (k_{1,+}/k_{1,-}) [A_1] \exp fE_e \tag{6.8}$$

and

$$\theta_e (1-\theta_e)^{-1} = (k_{2,-}/k_{2,+}) [A_3] \exp (- fE_e) \tag{6.9}$$

These are the same form as the Langmuir isotherm because the assumptions
made about a uniform surface and no interactions between the adsorbed
species are the same as those which lead to a Langmuir adsorption isotherm.

If each equation (6.4) to (6.7) is divided by its equilibrium value, these
equations are obtained in a convenient dimensionless form:

$$j_{1,+}/j_{1,o} = \{(1-\theta) / (1-\theta_e)\} \exp\{(1-\beta_1)f\eta\} \tag{6.10}$$

$$j_{1,-}/j_{1,o} = (\theta/\theta_e) \exp(-\beta_1 f\eta) \tag{6.11}$$

$$j_{2,+}/j_{2,o} = (\theta/\theta_e) \exp\{(1-\beta_2)f\eta\} \tag{6.12}$$

$$j_{2,-}/j_{2,o} = \{(1-\theta) / (1-\theta_e)\} \exp(-\beta_2 f\eta) \tag{6.13}$$

In the steady state the observed current is given by

$$j/2 = j_{1,+} - j_{1,-} = j_{2,+} - j_{2,-} \tag{6.14}$$

From Eqs (6.10) - (6.14) the general expression for the surface occupation
can be found:

$$\frac{\theta(1-\theta)^{-1}}{\theta_e(1-\theta_e)^{-1}} = \frac{j_{1,o} \exp\{(1-\beta_1)f\eta\} + j_{2,o} \exp(-\beta_2 f\eta)}{j_{1,o} \exp(-\beta_1 f\eta) + j_{2,o} \exp\{(1-\beta_2)f\eta\}} \tag{6.15}$$

and the value of θ/θ_e or $(1-\theta)/(1-\theta_e)$ obtained from this may be substituted
into Eqs (6.10) - (6.14) to obtain the current in terms of the observable
quantities η and the exchange currents. If the current-potential curves can
be studied over a wide enough region, the shape can be used to deduce the
mechanism as well as values of the transfer coefficients and exchange
currents of the component reactions.

The exchange currents themselves can be expressed as functions of
the standard rate constants, concentrations of stable reactants and the
equilibrium concentration of the adsorbed species. This result obtained by
eliminating E_e between (6.8) or (6.9) and the equilibrium form of (6.4) - (6.7) is

$$j_{o,1} = F k_1^o [A_1]^{\beta_1} (1-\theta_e)^{\beta_1} \theta_e^{1-\beta_1} \tag{6.16}$$

$$j_{o,2} = F k_2^o [A_3]^{1-\beta_2} \theta_e^{\beta_2} (1-\theta_e)^{1-\beta_2} \tag{6.17}$$

The equilibrium coverage θ_e is a measure of the intrinsic strength of the
adsorption bond. Equations (6.16) and (6.17) therefore show that as the
strength of adsorption increases, the effectiveness of electrocatalysis first
increases (at low θ_e), goes through a maximum and then decreases (at
high θ_e). The maximum in $j_{0,1}$ occurs at $\theta_e = \beta_1$ while that in $j_{0,2}$ occurs at
$\theta_e = 1 - \beta_2$.

This simple model indicates a way in which the properties of the electrode material affect the kinetics of the electrode reaction. The model can be fairly easily extended to account for heterogeneous surfaces or the interaction between adsorbed species. This extension makes no qualitative change in the predictions but only relatively minor quantitative changes. On the other hand, many real systems occur with more than two partial reactions and more than one adsorbed intermediate. Apart from the increased algebraic complexity, which will obviously cause difficulties, there are two difficulties in principle. One is the problem of dealing with mixed adsorption for which only rather approximate methods are so far available; the other is the introduction of new parameters for each extra step in the overall process. With the limited number of parameters available from steady-state measurements, it becomes impossible to determine a unique reaction scheme. This is a general problem of reaction kinetics. The solution to it is to find ways of isolating individual steps in the reaction scheme so that their kinetics can be studied. This can be done either by using direct observation of the intermediate species, especially by spectroscopy, or by using non-steady states or transient methods to investigate steps with different time constants. Here we shall discuss mainly the latter method because it is concerned with methods that are essentially electrochemical.

7.   TRANSIENT METHODS FOR KINETICALLY CONTROLLED REACTIONS

7.1. Experimental method

Transient methods in electrochemistry involve the observation of one electrical variable (current or voltage) as a function of time while the other electrical variable (voltage or current) is subjected to a prescribed time function. A wide variety of prescribed time functions has been used and the one used depends on the particular problem to be solved and experimental convenience. Control of the current implies control of the flux of reaction at the electrode whereas control of the potential implies control of the rate constants of the electrode reaction. The latter is often simpler from the theoretical point of view and it will be discussed here.

The experimental arrangement for controlled potential experiments has been developed particularly in the last two decades. The three-electrode cell of Fig.2 is combined with a high gain amplifier such as an operational amplifier, known as a potentiostat. A schematic circuit is shown in Fig.6. The potential between the reference electrode and the test electrode is connected in opposition to a voltage source E and the sum of these two is fed into the input of the amplifier. The output is fed through a resistance into the auxiliary and test electrodes of the cell, which are therefore in the feedback loop of the amplifier. In this circuit the output current supplied to the cell by the amplifier is just that required to make the potential of the test electrode with respect to the reference electrode equal and opposite to the potential difference supplied by E. The amplifier must have a high input impedance so that the current flowing through the reference electrode is extremely small. Hence the value of E determines the potential of the test electrode. E may be constant or follow a prescribed time variation. When

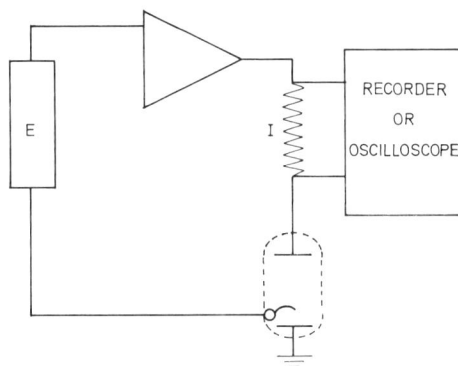

FIG. 6.   Schematic diagram of an electrolysis experiment with a three-electrode cell under potentiostatic control.

rapid time variation is to be studied (e.g. times of the order of 1 μs) very careful design of the circuit and the cell is necessary.

Among the many possible time functions we select three as important examples of the method: (a) potential step, (b) linear potential sweep, and (c) sinusoidal potential variation.

(a)   Potential step

We consider the effect of an abrupt imposition of an overpotential η on an electrode at which the reaction described by (6.1) and (6.2) is occurring. The electrode is considered to be at equilibrium until time t = 0, when the potential jumps to η. The observation then consists of the variation of the current density j with time. Under these transient conditions the total current density is the sum of the currents of the two partial reactions:

$$j = j_1 + j_2 \tag{7.1}$$

where

$$j_1 = j_{1,+} - j_{1,-} \tag{7.2}$$

$$j_2 = j_{2,+} - j_{2,-} \tag{7.3}$$

and the current components are given by Eqs (6.10) - (6.13) in terms of η.

At times t > 0, η is constant and the time variation of j arises because θ changes with time. This change is expressed by

$$F [A_2]_{sat} \; (d\theta/dt) = j_1/n_1 - j_2/n_2 \tag{7.4}$$

We note that $F[A_2]_{sat}$ is the charge passed through the circuit to form or remove a monolayer of the intermediate A_2. Thus this quantity divided by the partial current of a particular reaction is the time required for that reaction to form or remove a monolayer. It is therefore convenient to express the rate equations (6.10) etc. in terms of this type of time constant. Thus (6.10) becomes

$$j_{1,+} / n_1 F [A_2]_{sat} = (1-\theta) / \tau_1 \tag{7.5}$$

where

$$\tau_1^{-1} = j_{1,0} (1-\theta_e)^{-1} \exp \{(1-\beta_1)f\eta\} / F[A_2]_{sat} \tag{7.6}$$

and τ_1 is the time required to fill the empty sites present on the electrode at equilibrium by reaction 1 occurring in the anodic direction at an overpotential η. Similarly (6.12) becomes

$$j_{2,+} / n_1 F [A_2]_{sat} = \theta / \tau_2 \tag{7.7}$$

where

$$\tau_2^{-1} = (j_{2,0}/\theta_e) \exp\{(1-\beta_2)f\eta\}/F[A_2]_{sat} \tag{7.8}$$

and τ_2 is the time required to remove the amount of A_2 present on the electrode at equilibrium by reaction 2 occurring in the anodic direction at an overpotential η.

If we assume that η is sufficiently large that $j_{1,-}$ and $j_{2,-}$ are negligible at all $t > 0$, then (7.4) simplifies to

$$d\theta/dt = (1-\theta)/\tau_1 - \theta/\tau_2 \tag{7.9}$$

The solution of this is obtained by integrating and noting that at $t = 0$, $\theta = \theta_e$:

$$\frac{\tau_2 - (\tau_1 + \tau_2)\theta}{\tau_2 - (\tau_1 + \tau_2)\theta_e} = \exp - (\frac{\tau_1 + \tau_2}{\tau_1 \ \tau_2}) t \tag{7.10}$$

We note here that the time constant governing the behaviour of the intermediate species in this reaction scheme is $\tau_1 \tau_2 /(\tau_1 + \tau_2)$, i.e. it will be determined by the smaller of the two time constants or the faster of the two reactions. This contrasts with the steady-state behaviour, which is determined by the slower of the two reactions.

We may also note that the way in which the intermediate concentration changes with time depends on the relative values of the time constants. If $\tau_2 > \tau_1$, (7.10) becomes

$$(1-\theta) / (1-\theta_e) = \exp (-t/\tau_1) \tag{7.11}$$

and θ increases with time, whereas if $\tau_1 > \tau_2$, (7.10) becomes

$$\theta/\theta_e = \exp (-t/\tau_2) \tag{7.12}$$

and θ decreases with time.

This difference of behaviour is reflected in the current. If $\tau_2 > \tau_1$, the observed current is approximately $j_{1,+}$ and from (7.5) and (7.11)

$$j \approx j_{1,+} = n_1 F [A_2]_{sat} (1-\theta_e) \tau_1^{-1} \exp (- t/\tau_1) \qquad (7.13)$$

whereas if $\tau_1 > \tau_2$, the observed current is approximately $j_{2,+}$ and from (7.7) and (7.12)

$$j \approx j_{2,+} = n_2 F [A_2] \theta_e \tau_2^{-1} \exp (- t/\tau_2) \qquad (7.14)$$

Although the current falls with time in each case the time constant of this fall is characteristic of the faster of the two reactions, which can therefore be obtained directly from the potential step experiment. A study of the potential dependence of the time constant will yield the transfer coefficient of the faster reaction, as we can see from (7.6) or (7.8). Once the time constant is determined (7.13) or (7.14) can be used to find the amount of intermediate adsorbed at equilibrium and hence the exchange current of the faster reaction. Thus a combination of a steady-state experiment and a potential step experiment would yield a complete characterization of a simple two-step reaction of this type.

(b) Linear potential sweep

In this experiment the voltage E follows a ramp function so that the potential of the test electrode is constrained to vary linearly with time at a rate v:

$$v = dE/dt \qquad (7.15)$$

The initial point and final point of the ramp can be selected so that the range studied provides the most useful information. For most reactions this would span the equilibrium potential. However, if this range is chosen for the simple two-step reaction (6.1), (6.2), this means that the simplifications used in (a) above cannot be used. The use of the full set of equations (6.10) to (6.13), together with the greater complexity of the imposed condition, means that an analytical solution to the general problem cannot be obtained. The problem can be solved using a computer and a simulation technique but the result is not very illuminating.

An analytical solution can be obtained if it is assumed that the overpotential η_i at which the sweep starts is already large enough for the back reactions to be neglected. If it is also assumed that $\beta_1 = \beta_2$, a solution can be obtained by replacing τ_1^{-1} and τ_2^{-1} by

$$\tau_1^{-1} = \tau_{1,i}^{-1} \exp \{(1-\beta) fvt\} \qquad (7.16)$$

$$\tau_2^{-1} = \tau_{2,i}^{-1} \exp \{(1-\beta) fvt\} \qquad (7.17)$$

PARSONS

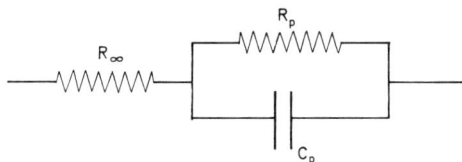

FIG. 7.   Equivalent circuit of an electrode at which a two-step reaction with an adsorbed intermediate is occurring. R_p, R_∞ and C_p are defined in the text. The double layer capacity is omitted.

where $\tau_{1,i}$ and $\tau_{2,i}$ are the time constants given by (7.6) and (7.8) respectively when the overpotential has its initial value η_i. When (7.16) and (7.17) are substituted in (7.9) and the latter is integrated, we obtain

$$\ln \left\{ \frac{\tau_{2,i} - (\tau_{1,i} + \tau_{2,i})\theta}{\tau_{2,i} - (\tau_{1,i} + \tau_{2,i})\theta_i} \right\} = \frac{\tau_{2,i} + \tau_{1,i}}{\tau_{2,i} \tau_{1,i}} \frac{\exp\{(1-\beta)fvt\}}{(1-\beta) fv} \tag{7.18}$$

(where θ_i is the value of θ at $t = 0$) which has the following limiting values:

when $\tau_{2,i} \gg \tau_{1,i}$:

$$\ln \frac{1-\theta}{1-\theta_i} = \frac{1}{\tau_{1,i}} \frac{\exp\{(1-\beta_1) fvt\}}{(1-\beta_1) fv} \tag{7.19}$$

and when $\tau_{1,i} \gg \tau_{2,i}$:

$$\ln \frac{\theta}{\theta_i} = \frac{1}{\tau_{2,i}} \frac{\exp\{(1-\beta_2) fvt\}}{(1-\beta_2) fv} \tag{7.20}$$

If these results are combined with (7.5) and (7.7) (taking note of (7.16) and (7.17)), we obtain the current as a function of time:

when $\tau_{2,i} \gg \tau_{1,i}$:

$$\ln (j/j_i) = (1-\beta_1)fvt + \{\tau_{1,i}(1-\beta_1)fv\}^{-1} \exp\{(1-\beta_1)fvt\} \tag{7.21}$$

when $\tau_{1,i} \gg \tau_{2,i}$:

$$\ln (j/j_i) = (1-\beta_2)fvt + \{\tau_{2,i}(1-\beta_2)fv\}^{-1} \exp\{(1-\beta_2)fvt\} \tag{7.22}$$

where j_i is the current at $t = 0$.

   While the characteristics of the reaction could be extracted from this type of experiment, it is clear that this is not as straightforward as from the potential step method in spite of the fact that the sweep v provides an additional variable. We shall see later that the linear sweep method does provide valuable information in other circumstances.

(c) Sinusoidal perturbation

A small sinusoidal signal is superimposed on the steady-state potential of the electrode. This results in a small sinusoidal current superimposed on the steady-state current. It may be represented by a Taylor expansion of the form:

$$j_1 = \bar{j}_1 + (\frac{\partial j_1}{\partial \theta})_\eta \, \hat{\theta} \, e^{i\omega t} + (\frac{\partial j_1}{\partial \eta})_\theta \, \hat{\eta} \, e^{i\omega t} \tag{7.23}$$

$$j_2 = \bar{j}_2 + (\frac{\partial j_2}{\partial \theta})_\eta \, \hat{\theta} \, e^{i\omega t} + (\frac{\partial j_2}{\partial \eta})_\theta \, \hat{\eta} \, e^{i\omega t} \tag{7.24}$$

Here $\bar{j}_1 = \bar{j}_2$ is the steady-state current at the mean overpotential η; $\hat{\theta}$ and $\hat{\eta}$ are the amplitudes of the oscillations of θ and η; i is $\sqrt{-1}$ and ω is the angular frequency.

Substituting these two equations into (7.4), we obtain an equation for θ:

$$F[A_2]_{sat} \frac{d\theta}{dt} = \left[(\frac{\partial j_1/n_1}{\partial \theta})_\eta - (\frac{\partial j_2/n_2}{\partial \theta})_\eta\right]\hat{\theta} \, e^{i\omega t}$$

$$+ \left[(\frac{\partial j_1/n_1}{\partial \eta})_\theta - (\frac{\partial j_2/n_2}{\partial \eta})_\theta\right]\hat{\eta} e^{i\omega t} \tag{7.25}$$

which yields

$$F[A_2]_{sat} \, \hat{\theta} = \frac{\left[(\frac{\partial(j_1/n_1)}{\partial \eta})_\theta - (\frac{\partial(j_2/n_2)}{\partial \eta})_\theta\right]\hat{\eta}}{j\omega - \left[(\frac{\partial(j_1/n_1)}{\partial \theta})_\eta - (\frac{\partial(j_2/n_2)}{\partial \theta})_\eta\right]\frac{1}{F[A_2]_{sat}}} \tag{7.26}$$

after integration.

It is convenient to define the relaxation time by

$$\tau^{-1} = \left[(\frac{\partial(j_2/n_2)}{\partial \theta})_\eta - (\frac{\partial(j_1/n_1)}{\partial \theta})_\eta\right] \frac{1}{F[A_2]_{sat}} \tag{7.27}$$

The observed sinusoidal current is given by

$$j = j_1 + j_2 - \bar{j}_1 - \bar{j}_2 \tag{7.28}$$

Substituting (7.23) and (7.24) into this and using (7.26) to eliminate $\hat{\theta}$, we obtain

$$\frac{j}{\hat{\eta} \, e^{i\omega t}} = (\frac{\partial j_1}{\partial \eta})_\theta + (\frac{\partial j_2}{\partial \eta})_\theta$$

$$+ \frac{\tau}{j\omega\tau+1} \left[(\frac{\partial j_1}{\partial \theta})_\eta + (\frac{\partial j_2}{\partial \theta})_\eta\right]\left[(\frac{\partial(j_1/n_1)}{\partial \eta})_\theta - (\frac{\partial j_2/n_2}{\partial \eta})_\theta\right] \frac{1}{F[A_2]_{sat}} \tag{7.29}$$

This is the expression for a complex admittance. At very large frequencies the second term vanishes and the admittance becomes resistive with a resistance R_∞ defined by

$$\frac{1}{R_\infty} = \left(\frac{\partial j_1}{\partial \eta}\right)_\theta + \left(\frac{\partial j_2}{\partial \eta}\right)_\theta \tag{7.30}$$

At very low frequencies the admittance also becomes resistive and there is a term additional to $1/R_\infty$ which arises from a resistance R_0 defined by

$$\frac{1}{R_0} = \frac{\tau}{F[A_2]_{sat}} \left[\left(\frac{\partial j_1}{\partial \theta}\right)_\eta + \left(\frac{\partial j_2}{\partial \theta}\right)_\eta\right] \left[\left(\frac{\partial(j_1/n_1)}{\partial \eta}\right)_\theta - \left(\frac{\partial(j_2/n_2)}{\partial \eta}\right)\right] \tag{7.31}$$

This behaviour may be described by an equivalent circuit of the form shown in Fig.7, with components independent of frequency and having the value

$$R_p = -R_\infty^2 / (R_0 + R_\infty) \tag{7.32}$$

$$C_p = -R_0 \tau / R_\infty^2 \tag{7.33}$$

This theory is valid in general for a two-step reaction and this is one advantage of small-amplitude methods: that the specific details of an assumed model can be introduced at a late stage of the derivation. We now introduce the model defined by (6.10)-(6.13) to discuss what information can be obtained from experiment. To avoid undue complication we consider two limiting conditions:

First, measurement at the equilibrium potential $\bar{\eta}$ and $\bar{j} = 0$:

$$\tau^{-1} = \frac{1}{F[A_2]\theta_e(1-\theta_e)} \left\{\frac{j_{1,o}}{n_1} + \frac{j_{2,o}}{n_2}\right\} \tag{7.34}$$

$$R_\infty^{-1} = n_1 f j_{1,o} + n_2 f j_{2,o} \tag{7.35}$$

$$R_0^{-1} = \frac{f(j_{2,o} - j_{1,o})(j_{1,o} - j_{2,o})}{j_{1,o}/n_1 + j_{2,o}/n_2} \tag{7.36}$$

from which $j_{1,0}$, $j_{2,0}$ and θ_e can be obtained if n_1, n_2 and $[A_2]_{sat}$ are known.

Second, measurement at an overpotential η large enough for the rate of the back reactions to be neglected:

$$\tau^{-1} = \frac{1}{F[A_2]_{sat}} \left\{\frac{j_{1,+}}{n_1(1-\theta)} + \frac{j_{2,+}}{n_2\theta}\right\} \tag{7.37}$$

$$R_\infty^{-1} = (1-\beta_1) n_1 f j_{1,+} + (1-\beta_2) n_2 f j_{2,+} \tag{7.38}$$

$$R_o^{-1} = \frac{f(j_{2,+}/\theta - j_{1,+}/(1-\theta))}{(j_{1,+}/n_1(1-\theta) + j_{2,+}/n_2\theta)} \left[(1-\beta_1)j_{1,+} - (1-\beta_2)j_{2,+} \right] \quad (7.39)$$

Again, similar information can be obtained if n_1, n_2 and $[A_2]_{sat}$ are known, i.e. $j_{1,+}$, $j_{2,+}$ and θ, although a complete analysis would probably require the study of the potential dependence of the admittance.

Here, as in each of the methods described in this section, we have ignored the presence of the electrical double layer. The nature of the double layer at the electrode will be discussed later but its presence means that the admittance described here is in parallel with an admittance that is normally purely capacitative. The analysis of the observed properties is thus somewhat more complicated.

8.   TRANSIENT METHODS FOR REACTIONS CONTROLLED BY DIFFUSION AS WELL AS ELECTRODE REACTION

8.1. General

Although transient methods are introduced here by considering purely kinetically controlled reactions, in the greater part of work with transients they have been used to distinguish kinetic control from control by diffusion of reactants to and from the electrode. We have seen in Section 4 that when the exchange current of the electrode reaction j_0 becomes comparable to or larger than the diffusion limiting currents no kinetic information is obtainable from steady-state measurements. However, fast transients enable information to be obtained about faster reactions.

We shall continue to assume that there is an excess of a base electrolyte present so that transport of the reactant ions is by diffusion. Also, we shall consider short times so that the bulk motion of the electrolyte can be neglected and there is no transport of the reactants by convection. If a plane electrode is assumed and edge effects are ignored, the transport of a reactant is described by Fick's second law in one dimension:

$$\frac{\partial c}{\partial t} = D\frac{\partial^2 c}{\partial x^2} \quad (8.1)$$

where x is the distance perpendicular to the electrode surface. The solution of Fick's second law is a boundary value problem with the boundary conditions usually

$$c_R = \left[Fe^{2+} \right]_b \qquad \text{at all } x \text{ when } t = 0 \quad (8.2)$$

$$c_R = \left[Fe^{2+} \right]_b \qquad \text{at large } x, \text{ at all } t \quad (8.3)$$

$$(\frac{\partial c_R}{\partial x})_{x=o} = \frac{j}{nFD_R} \quad (8.4)$$

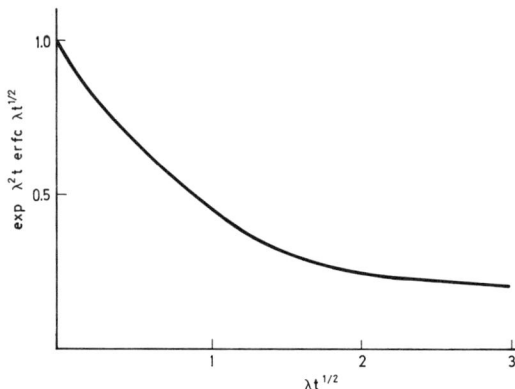

FIG. 8.   The function $\exp \lambda^2 t \ \mathrm{erfc} \ \lambda r$ as a function of $\lambda t^{\frac{1}{2}}$.

where D_R is the diffusion coefficient of the reduced species (Fe^{2+} in the Fe^{2+}/Fe^{3+} system) whose concentration at any value of x and t is c_R. A similar diffusion equation with similar boundary conditions must be solved for c_O (the oxidized species).

Condition (8.4) relates the problem to the imposed potential through Eq.(3.22). The two diffusion equations are then solved for the values of c_R and c_O at the electrode surface x = 0 and this value is inserted back into (3.22) to obtain the current as a function of time. The solution of the diffusion equation is analogous to the problem of the conduction of heat in matter and may be solved in just the same way.

8.2. Potential step

The potential is stepped at t = 0 from its equilibrium value $\eta = 0$ to an overpotential η. Thus at all t > 0 the value of η in (3.22) is constant. The diffusion equation, which is of course a partial differential equation, can be converted to an ordinary differential equation by the technique of Laplace transformation. After the solution is obtained it is inverted; the concentration at the electrode is obtained as

$$\left[Fe^{2+}\right] = \left[Fe^{2+}\right]_b - \frac{A}{\lambda}(1-e^{\lambda^2 t} \ \mathrm{erfc} \ \lambda t^{\frac{1}{2}}) \tag{8.5}$$

$$\left[Fe^{3+}\right] = \left[Fe^{3+}\right]_b + \sqrt{\kappa} \frac{A}{\lambda}(1-e^{\lambda^2 t} \ \mathrm{erfc} \ \lambda t^{\frac{1}{2}}) \tag{8.6}$$

where

$$\lambda = \frac{j_o}{nF} \left\{ \frac{e^{(1-\alpha)nf\eta}}{D_R^{\frac{1}{2}} \left[Fe^{2+}\right]_b} + \frac{e^{-\alpha nf\eta}}{D_O^{\frac{1}{2}} \left[Fe^{3+}\right]_b} \right\} \tag{8.7}$$

$$A = \frac{j_0}{nF} \left\{ \frac{\exp(1-\alpha)nf\eta}{D_R^{\frac{1}{2}}} - \frac{\exp(-\alpha nf\eta)}{D_O^{\frac{1}{2}}} \right\} \tag{8.8}$$

$$\kappa = D_R/D_O \tag{8.9}$$

When (8.5) and (8.6) are substituted in (3.22) the current time curve can be expressed in the compact form:

$$j = j_{CT} \exp(\lambda^2 t) \, erfc(\lambda t^{\frac{1}{2}}) \tag{8.10}$$

where

$$j_{CT} = j_0 \{\exp[(1-\alpha)nf\overline{\eta}] - \exp(-\alpha nf\eta)\} \tag{8.11}$$

is the current that would flow at this value of η if the concentrations at the electrode surface remained equal to the bulk concentration, i.e. the current which would flow if the charge transfer process were slow and the mass transport very much faster.

Equation (8.10) is the equation of a falling transient; the form of $\exp \lambda^2 t \, erfc \, \lambda t^{\frac{1}{2}}$ is shown in Fig.8. At values of $\lambda t^{\frac{1}{2}} > 1$ this function may be expanded in the series:

$$\exp \lambda^2 t \, erfc \, \lambda t^{\frac{1}{2}} = \frac{1}{\pi^{\frac{1}{2}}\lambda t^{\frac{1}{2}}} \left\{ 1 - \frac{1}{2\lambda^2 t} + \frac{1 \times 3}{(2\lambda^2 t)^2} - \frac{1 \times 3 \times 5}{(2\lambda^2 t)^3} \cdots \right\} \tag{8.12}$$

Thus at long times,

$$j = j_{CT}/\lambda\pi^{\frac{1}{2}}t^{\frac{1}{2}} \tag{8.13}$$

or, if we make use of (8.7) and (8.11),

$$j = \left(\frac{1}{nF\pi^{\frac{1}{2}}t^{\frac{1}{2}}}\right) \frac{1 - \exp(-nf\eta)}{\{D_R^{-\frac{1}{2}}[Fe^{2+}]_b^{-1} + D_O^{-\frac{1}{2}}[Fe^{3+}]_b^{-1}\exp(-nf\eta)\}} \tag{8.14}$$

in which we note that j_0 and α have vanished. This means that at long times the process has become completely controlled by the mass transport and we can obtain no information about the kinetics of the electrode reaction itself in this region. Kinetic information can be obtained at shorter times; this is best done by nomograms but it is also possible to use an alternative series solution. At small values of $\lambda t^{\frac{1}{2}}$ we may use the series expansion:

$$\exp \lambda^2 t \, erfc \, \lambda t^{\frac{1}{2}} = 1 - 2\lambda(t/\pi)^{\frac{1}{2}} + \lambda^2 t - 2\lambda^3 t^{3/2}/\pi^{\frac{1}{2}} \tag{8.15}$$

Thus at short times a plot of the observed current j against $t^{\frac{1}{2}}$ may be used to extrapolate to obtain the value of j_{CT} as the intercept at $t = 0$. However, there are severe practical difficulties in making measurements at sufficiently short times when j_0 is large. Some of these are concerned with the properties

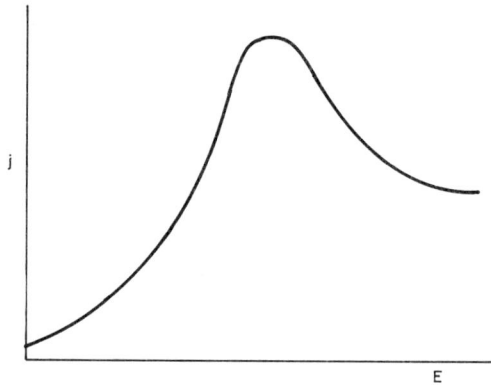

FIG. 9.   Sketch of the current transient produced by a linear voltage sweep at an electrode where a simple
reaction occurs.

of the potentiostat and the physical arrangement of the cell but an intrinsic
difficulty is due to the existence of the electrical double layer at the electrode
interface. This will be discussed later; here we merely point out that it
behaves as an electrical capacity in parallel with the electrode impedance.
It is therefore impossible to maintain the electrode potential constant in a
potentiostatic experiment until times larger than the time constant of the
double layer capacity and the charging circuit. Consequently, although the
potentiostatic method has the advantage of relative simplicity, it cannot be
used to study very fast reactions.

8.3. Linear potential sweep

If we now consider the potential to be controlled in the form of a ramp
function or a linear variation with time, the boundary value problem is
considerably more complicated and we shall outline the solution for two
cases. The boundary conditions (8.2) - (8.4) still hold, but it is convenient to
simplify the situation by assuming that the concentration of the oxidized
species Fe^{3+} is close to zero in the bulk solution so that the analogues of
(8.2) and (8.3) are

$$c_o = 0 \qquad \text{at all x when t = 0} \tag{8.16}$$

$$c_o = 0 \qquad \text{at large x, at all t} \tag{8.17}$$

8.3.1. Fast electrode reaction: $j \ll j_0$ under all circumstances

The potential of the electrode is then related to the local concentration
of Fe^{2+} and Fe^{3+} at the electrode surface by an equation like (3.9) (the
Nernst equation):

$$E = E^{\ominus} - (RT/nF) \ln [Fe^{2+}]/[Fe^{3+}] \tag{8.18}$$

This provides part of the boundary condition at the electrode although (8.4) is used with its analogue for Fe^{3+} in the form:

$$D_R \left(\frac{\partial c_R}{\partial x}\right)_{x=o} = -D_O \left(\frac{\partial c_O}{\partial x}\right)_{x=o} = \frac{j}{nF} \qquad (8.19)$$

The surface concentrations are obtained in the form of an integral equation by Laplace transformation and the final result may be expressed in the form:

$$j = nF \left[Fe^{2+}\right]^b (\pi D_R a)^{\frac{1}{2}} \chi(at) \qquad (8.20)$$

where

$$a = nfv = (nf/t)(E - E_i) \qquad (8.21)$$

E_i being the electrode potential at $t = 0$ and

$$\int_o^{at} \frac{\chi(z)dz}{(at-z)^{\frac{1}{2}}} = \left[1 + \kappa^{\frac{1}{2}} \exp \{(nf)(E^O - E_i) \exp (at)\right]^{-1} \qquad (8.22)$$

where $\kappa = D_O/D_R$.

The form of the current-voltage curve is shown in Fig.9. It is usual to characterize this by the peak potential and current:

$$E_p = E^O + (RT/nF) \left[1.109 + \ln \kappa^{\frac{1}{2}}\right] \qquad (8.23)$$

$$j_p = 0.4463 \; nF \; (D_R a)^{\frac{1}{2}} \left[Fe^{2+}\right]^b \qquad (8.24)$$

from which it is evident that E^0 and the bulk concentration of the reduced species can be obtained. It follows from the assumption made in deriving these relations that they tell us nothing about the kinetics of the reaction at the electrode surface.

8.3.2. Slow electrode reaction: $j \gg j_o$

Equation (8.18) may then be replaced by (3.18) with the second term neglected:

$$j = nF \; k_{+,o} \left[Fe^{2+}\right] \exp \{(1-\alpha) \; nf\epsilon\} \qquad (8.25)$$

This problem can be solved in a similar way to the previous problem and the result may be expressed:

$$j = nF \left[Fe^{2+}\right]^b (\pi D_R b)^{\frac{1}{2}} \; \chi(bt) \qquad (8.26)$$

where

$$b = (1-\alpha) \; nfv = \left[(1-\alpha)nf/t\right] (E-E_i) \qquad (8.27)$$

and

$$1 - \int_0^{bt} \frac{\chi(z)dz}{(bt-z)^{\frac{1}{2}}} = \exp(u-bt)\chi(bt) \tag{8.28}$$

where

$$\ln u = \ln\left[(\pi D_R b)^{\frac{1}{2}}/k_{+,o}\right] + (1-\alpha)nfE \tag{8.29}$$

Again the current-voltage curve can be characterized by the values at the peak:

$$E_p = \left[(1-\alpha)nf\right]^{-1}\left[0.780 + \ln D_R b - \ln k_{+,o}\right] \tag{8.30}$$

$$j_p = 0.227\ nF\left[Fe^{2+}\right]_b k_{+,p} \tag{8.31}$$

where $k_{+,p}$ is the rate constant at the potential E_p given by

$$k_{+,p} = k_{+,o}\exp\left[(1-\alpha)nfE_p\right] \tag{8.32}$$

Since b is a linear function of sweep speed v, it is evident that $(1-\alpha)$ may be obtained by plotting E_p against $\ln v$ and the exchange current can be found from (8.31) if the equilibrium potential of the electrode is known.

Although the mathematical difficulties of the linear sweep method are great, many possible reaction sequences have been worked out numerically and the experimental simplicity of the method makes it very useful, especially for qualitative studies of complex systems. Here a repetitive triangular signal is used and the method is called cyclic voltametry.

8.4. Sinusoidal perturbation

The application of a sinusoidal potential variation about the equilibrium value results in the flow of a sinusoidal current through the electrode. The boundary condition (8.4) may then be written:

$$\left(\frac{\partial c_R}{\partial x}\right)_{x=o} = \frac{j}{nF\ D_R} = \hat{j}e^{i\omega t} \tag{8.33}$$

where $\hat{j}$ is the amplitude of the current. The solution of the diffusion equation is then

$$\left[Fe^{2+}\right] = \left[Fe^{2+}\right]_b - \frac{1-i}{nF(2\omega D)^{\frac{1}{2}}}\hat{j}e^{i\omega t} \tag{8.34}$$

Thus the concentration at the electrode surface also varies in a sinusoidal manner, lagging the current by $\pi/4$.

If we expand (3.22) about the equilibrium potential we can express it in the form:

$$j/j_o = \frac{\left(\left[Fe^{2+}\right]_b - \left[Fe^{2+}\right]\right)}{\left[Fe^{2+}\right]_b} + \frac{\left[Fe^{3+}\right]_b - \left[Fe^{3+}\right]}{\left[Fe^{3+}\right]_b} + nf\eta \tag{8.35}$$

From (8.3 5), (8.3 4) and the analogous equation for Fe^{3+}, we can obtain the overpotential:

$$nf\eta = \frac{\hat{j}e^{i\omega t}}{j_o} + \frac{1-j}{nF(2\omega)^{\frac{1}{2}}}\left\{\frac{1}{D_R^{\frac{1}{2}}[Fe^{2+}]_b} + \frac{1}{D_O^{\frac{1}{2}}[Fe^{3+}]_b}\right\}\hat{j}e^{i\omega t} \qquad (8.36)$$

Thus the electrode behaves as an impedance:

$$Z = \frac{\eta}{\hat{j}e^{i\omega t}} = \frac{1}{nfj_o} + \frac{1-j}{n^2Ff(2\omega)^{\frac{1}{2}}}\left\{\frac{1}{D_R^{\frac{1}{2}}[Fe^{2+}]_b} + \frac{1}{D_O^{\frac{1}{2}}[Fe^{3+}]_b}\right\} \qquad (8.37)$$

The first term in (8.37) is a resistance (R_{CT}) known as the charge transfer resistance, which represents the rate of the electrode reaction itself. The second term is an impedance, which may be regarded as having equal resistive and capacitative components (phase angle $\pi/4$) and which depends on the inverse square root of the frequency. This impedance is given the name of Warburg and the symbol –W– and it represents the effect of the mass transport of reactants to and from the electrode. The equivalent circuit of this electrode is then –ᴍᴍ–W–.

Kinetic information can be obtained from the measurement of the impedance of the electrode as a function of frequency. If the measurement is made in the form of an equivalent resistance R and capacitance C in series, both R and $1/\omega C$ will vary linearly with $\omega^{-\frac{1}{2}}$ but the difference between them will be constant and equal to $(nfj_0)^{-1}$ so that j_0 can be obtained. The transfer coefficient α may then be determined by varying the reactant concentration and using (4.19). Experimental limitations to this method arise because the charge transfer resistance is small for a fast reaction so that higher fre-quencies are required to reduce the Warburg impedance sufficiently to detect R_{CT}. However, the experiment also includes some contribution to the impedance from the electrolytic solution which cannot be indefinitely reduced. There is also a parallel branch in the circuit with the double layer capacity and the impedance of this becomes small at high frequencies. Nevertheless, because the a.c. bridge is a null instrument of high precision, this method is able to measure the fastest reaction rate among the simple relaxation techniques.

9.   THE ELECTRICAL DOUBLE LAYER

Until now we have described the formal kinetics of electrode reactions and the methods of studying them with the minimum of description of the location in which these reactions occur. We now outline the model of the structure of the interface between a metal and an electrolyte solution. This is a region usually about 10^{-9} m thick where the properties differ from those of either bulk phase. Much of the information about this region has been obtained by studying electrodes under conditions in which no electrode reaction can occur. Thus no continuous current can flow through the electrode. How-ever, if the potential of the electrode is changed, a transient current flows and the electrode behaves as a condenser. The plates of this 'molecular condenser' consist of the metal surface and the adjoining few molecular layers of the electrolyte. Owing to the high density of charge carriers in the metal, the charge on the metal is very close to the surface even on a molecular scale and may with good approximation be regarded as a surface charge in the classical electrostatic sense. On the other hand, the density of charge carriers in the electrolyte is lower and they are ions of radius about 10^{-10} m or larger. Consequently the charge in the electrolyte is distributed in the form of a space charge with the highest charge density at the distance of an ionic diameter from the metal surface. It is the structure of this part of the metal-electrolyte double layer which has attracted most attention.

The behaviour of the interface as a condenser suggests that its properties could be studied simply by measuring the capacity of this condenser using an a.c. bridge technique. Much of this work has been done using mercury as the metal for several reasons. There is a wide potential region on mercury in which the occurrence of electrode reactions may be neglected. The liquid metal surface is smooth and structureless. The surface may be readily renewed and therefore can be studied in a clean state; this is exploited particularly in the dropping mercury electrode. The liquid interface also has the possibility of easy measurement of the interfacial tension.

It may be shown by a thermodynamic analysis that the interfacial tension γ is related to the differential double layer capacity of unit area of the interface C by

$$- \partial^2 \gamma / \partial E^2 = C \qquad\qquad (9.1)$$

if E is the potential of the electrode with respect to a suitable reference electrode. The differential capacity may also be related to the charge σ on unit area of the metal side of the double layer by

$$C = \partial \sigma / \partial E \qquad\qquad (9.2)$$

Consequently, by measuring C as a function of potential and integrating, the charge and interfacial tension may be found. Integration constants may be found by direct measurement of γ as a function of E. This relation is known as an electrocapillary curve and is approximately parabolic. Since it follows from (9.1) and (9.2) that

$$\sigma = - \partial \gamma / \partial E \qquad\qquad (9.3)$$

the maximum of the electrocapillary curve corresponds to the potential at which $\sigma = 0$ (potential of zero charge (p.z.c.)). The importance of the interfacial tension is that its dependence on the concentration of the constituents of the electrolytic solution (both ionic and non-ionic) may be analysed thermodynamically to obtain the composition of the surface layer. Thus it is possible to find the contribution of each of the ions in the solution to the charge on the solution side of the double layer, as well as the way in which these contributions depend on potential.

In the simplest examples, typified by mercury in contact with aqueous NaF solution, the interaction between the ions and the metal is purely an electrostatic one between their charges and the charge on the metal. The part of the double layer in solution can then be divided into two regions: the one closest to the metal is called the inner layer and is composed entirely of solvent; the one further from the metal is a diffuse layer or space charge of ions. The boundary between the inner and diffuse layers is called the outer Helmholtz plane (OHP). The potential drop between the metal and the solution ϕ^m may be divided into a part across the inner layer $\phi^m - \phi_2$ and a part across the diffuse layer ϕ_2. Here ϕ is used because these are local average potentials. Superscripts indicate potentials of phases and subscripts potentials in the interfacial region; the potential of the solution is arbitrarily set at zero. Then we may write

$$\frac{1}{C} = \frac{\partial E}{\partial \sigma} = \frac{\partial \phi^m}{\partial \sigma} = \frac{\partial (\phi^m - \phi_2)}{\partial \sigma} + \frac{\partial \phi_2}{\partial \sigma} \tag{9.4}$$

$$\frac{1}{C} = \frac{1}{C^i} + \frac{1}{C^d} \tag{9.5}$$

where C^i and C^d are the differential capacities of the inner and diffuse layers respectively. Thus we see that under these simplified conditions the electrode behaves as if it consisted of two capacitances in series.

The theory of the diffuse layer capacitance is the same as that for a space charge in a semiconductor or for the ionic atmosphere round an ion (Debye-Hückel theory). We consider a uni-uni-valent electrolyte (equivalent to an intrinsic semiconductor) in which the cation and anion concentration is n in unit volume of the bulk. Near a plane metal electrode the potential is ϕ at a distance x from the metal surface. Hence the local charge density is

$$\rho = en \{\exp(-\phi f) - \exp(\phi f)\} \tag{9.6}$$

where e is the protonic charge and $f = F/RT = e/kT$. The potential can then be obtained by combining (9.6) with the one-dimensional form of Poisson's equation:

$$d^2\phi/dx^2 = -\rho/\epsilon \tag{9.7}$$

where ϵ is the permittivity:

$$d^2\phi/dx^2 = -(en/\epsilon) \{\exp(-\phi f) - \exp(\phi f)\} \tag{9.8}$$

To integrate this we use the identity:

$$d^2\phi/dx^2 = \tfrac{1}{2} d(d\phi/dx)^2/d\phi \tag{9.9}$$

and the limits $\phi = 0$, $d\phi/dx = 0$ as $x \to \infty$. Hence

$$d\phi/dx = -(2kT\, n/\epsilon)^{\frac{1}{2}} 2 \sinh(\phi f/2) \tag{9.10}$$

It is convenient to introduce the characteristic Debye length defined as

$$L_D = (\epsilon kT/2\, e^2 n)^{\frac{1}{2}} = (\epsilon/2kT\, n)^{\frac{1}{2}}/f \tag{9.11}$$

so that (9.10) may be written:

$$d\phi/dx = 2 \sinh(\phi f/2)/L_D\, f \tag{9.12}$$

This expression for the field applies to the outer Helmholtz plane, which is
at x_2 where the potential is ϕ_2. The field $(d\phi/dx)_{x_2}$ here can also be expressed
in terms of the charge density σ on the metal surface using Gauss's law; thus

$$(d\phi/dx)_{x_2} \;=\; -\,\sigma/\epsilon \;=\; 2\sinh(\phi_2 f/2)/L_D\ f \tag{9.13}$$

This gives us a relation between the experimentally measurable σ and ϕ_2
from which the latter may be calculated. If (9.13) is differentiated we
obtain the diffuse layer capacity:

$$C^d \;=\; d\sigma/d\phi_2 \;=\; (\epsilon/L_D)\cosh(\phi_2 f/2) \tag{9.14}$$

We note that C^d behaves as a parallel plate capacitance with plate
separation $L_D/\cosh(\phi_2 f/2)$. In water at 25°C, ϵ/L_D is 228 $c^{\frac{1}{2}}\mu F\cdot cm^{-2}$ where
c is the concentration of the uni-uni-valent electrolyte in $mol\cdot dm^{-3}$. This
capacity is therefore quite large under usual working conditions. Values of
C^d obtained at given values of σ using (9.13) and (9.14) may be combined with
experimental values of the total capacity C to obtain C^i from (9.5). The
result varies from about 15 $\mu F\cdot cm^{-2}$ when σ is negative to about 30 $\mu F\cdot cm^{-2}$
when σ is positive. If C^i is considered as a parallel plate condenser, this
gives the value of ϵ/x_2, or, expressed in terms of relative permittivity, x_2/ϵ_r
varies from 6×10^{-11} to 3×10^{-11} m. Since x_2 must be at least equal to an
ionic diameter, the relative permittivity must be greater than unity. On the
other hand, it is unlikely to be as large as the value in bulk water since this
would lead to values of x_2 of 47×10^{-10} to 24×10^{-10} m. The detailed description
of this region is still in dispute but it seems most likely that x_2 is about
$3\text{-}4\times10^{-10}$ m and that the minimum value of the relative permittivity is about
5.5, which is the high-frequency value for bulk water corresponding to intermolecular distortions only. There will also be a contribution due to the
rotation of the water dipoles which may increase the relative permittivity to 15.
It must be emphasized that this simple model is valid in only a very
limited number of systems. In most systems one or more of the ions present
in the solution can approach the metal close enough for it to be no longer
possible to assume that the interaction between ion and metal is of the
simple electrostatic type described in (9.6). In addition to this interaction,
there is a short-range interaction that may be of the type involved in the
formation of chemical bonds and may include components due to the replacement of solvent molecules near the metal. In general, this interaction is of a
nature specific to the ion being studied and so the phenomenon is described
as specific adsorption. Because of the short range of this type of interaction
the specifically adsorbed ions are often assumed to form a partially filled
monolayer next to the metal with their centres on the inner Helmholtz plane
at $x = x_1$, and this layer is treated by methods similar to those used in gas
adsorption (though allowing for solvent displacement). The diffuse layer
still exists at greater distances from the metal surface and the ions in it are
now considered to interact with the sum of the charge on the metal and that
on the inner Helmholtz plane σ^1. Thus in (9.13) σ must be replaced by $\sigma + \sigma^1$.
This means that it may be possible to modify or even reverse the charge on
the diffuse layer (and hence ϕ_2) at a given charge on the electrode by introducing appropriate specifically adsorbed ions.

TABLE I. POTENTIAL OF ZERO CHARGE
OF SOLID METALS WHICH DO NOT ADSORB
HYDROGEN (recommended values)

Metal	Solution	P.z.c./V
Bismuth	0.002 M KF	-0.39 ± 0.02
Lead	0.001 M NaF	-0.56 ± 0.02
Cadmium	0.001 M NaF	-0.75 ± 0.02
Antimony	0.002 M KClO$_4$	-0.15 ± 0.02
Tin	0.002 M KClO$_4$	-0.38 ± 0.02
Thallium	0.001 M NaF	-0.71 ± 0.04
Indium	0.003 M NaF	-0.65 ± 0.02
Copper	0.001 M NaF	0.09 ± 0.02
Gold (polycryst.)	0.002 M NaF	0.18 ± 0.01
Gold (110)	0.005 M NaF	0.19 ± 0.01
Silver (polycryst.)	0.001 N Na$_2$SO$_4$	-0.7 ± 0.05
Silver (single cryst.)	0.01 N Na$_2$SO$_4$	-0.66 ± 0.03[a]
Silver (111)	0.001 M KF	-0.46 ± 0.02
Silver (100)	0.005 M KF	-0.51 ± 0.02
Silver (110)	0.005 M NaF	-0.77 ± 0.01

[a] Corrected for the electrolyte asymmetry.

The amount of ions specifically adsorbed can be calculated from experiment if diffuse layer theory is assumed. The total surface excess of an ion of a given species (Γ_i) is obtainable from the measurement of interfacial tension at constant potential with respect to an electrode at equilibrium with the solution using Gibbs equation. Diffuse layer theory can be used to calculate the contributions to the total diffuse layer charge σ^d due to the various ionic species in solution σ_i^d. It must be assumed that at least one of the ionic constituents is not specifically adsorbed (such an ion can actually be introduced for this purpose). The total surface excess of this ion is then put equal to $\sigma_i^d/z_i e$ and from this the values of σ_i^d for the other constituents are obtained from diffuse layer theory. The calculated value of $\Gamma_i - \sigma_i^d/z_i e$ is then the amount of ion of species i which is specifically adsorbed.

The adsorption of non-ionic species from solution is also important in the structure of the double layer at electrodes. Since this adsorption occurs only as a result of short-range forces, it is usually assumed that all adsorbed molecules are present in a monolayer next to the metal surface (although in a few examples there is evidence of polymolecular layer formation). The inner layer of the original double layer may be distorted because most uncharged species are larger than water molecules. The replacement of water molecules by larger, usually less polar, molecules means that the larger molecules are usually less stable in the interface at high fields. Hence the typical behaviour for non-ionic adsorption is that it occurs strongly at low charges σ and decreases with increase of charge in either direction, the molecule often being completely desorbed at high positive or high negative charges.

It can be seen from this description of the double layer that a primary role in the control of adsorption is played by the charge on the metal surface. Consequently it is of great importance to know the potential at which the charge is zero (p.z.c.). This quantity is a characteristic of the metal and

varies quite widely in a way which has little relation to the standard potential
of the metal/metal ion electrode. It is most closely related to the electronic
work function.

Some values of the p.z.c. are shown in Table I, which was collected by
Frumkin[1]. As far as possible these are values for the metal in contact with
a simple electrolyte whose ions are not specifically adsorbed in the region of
the p.z.c. Specific adsorption shifts the point of zero charge by an amount
depending on the ion and the amount adsorbed. Specific adsorption of anions
shifts the p.z.c. to more negative potentials (as a result of the dipolar effect
of the specifically adsorbed anions and the cationic excess in the diffuse
layer) and cations shift it in a positive direction. The adsorption of uncharged
molecules may also shift the p.z.c. if they are adsorbed with a dipolar component perpendicular to the interface.

To a first approximation, the study of the structure of the double layer
on one metal may be used to provide results for use on another metal if the
origin of the potential scale is taken as the p.z.c. on each metal. In this way
the large amount of data accumulated for mercury electrodes may be used
to interpret effects on other metals. However, effects like specific adsorption
are specific to the metal as well as to the ion, so that this approximation must
be used very cautiously. The understanding of specific adsorption has not
yet reached the stage that enables us to describe the way in which it depends
on the nature of the metal theoretically.

10. EFFECT OF DOUBLE LAYER ON ELECTRODE KINETICS

In considering the kinetics of electrode reactions we have so far ignored
the difference in the environment of the reacting species at the electrode
surface and in the bulk of the electrolyte. In fact, the reacting species must
come within molecular dimensions of the metal surface before charge
transfer can take place, i.e. it must be within the electrical double layer
discussed in the previous section. The simplest situation occurs when the
reacting ion is located in the outer Helmholtz plane at the potential ϕ_2 and
charge transfer occurs from this site. At equilibrium the local concentration
of a reacting ion of charge ze is then

$$[A^z]_{OHP} = [A^z]_b \exp(-zf\phi_2) \tag{10.1}$$

and it is this concentration which should appear in the rate equation. This
equilibrium can be maintained up to quite high rates of electrode reaction
and it is rarely necessary to consider the kinetics of transfer across the
outer part of the double layer. It should, however, be noted that under
steady-state conditions the dimensions of the double layer (~ 1 nm) are very
much less than the dimensions of the diffusion layer ($\delta \sim 10\ \mu m$). It is therefore possible to treat these two problems quite independently. Under fast
transient conditions this independence may break down when the effective
thickness of the diffusion layer shrinks to a value comparable to the Debye

---

[1] Table from LEIKIS, D.I., RYBALKA, K.V., SEVASTYANOV, E.S., FRUMKIN, A.N., J. Electroanal.
Chem. 46 (1975) 161.

length of the diffuse part of the double layer. It is then necessary to solve
the diffusion equation taking into account the change in potential across the
diffuse layer. We shall ignore this possibility here.

It is also necessary to modify the rate equation for the electrode reaction. From (3.15) onwards we have assumed that the whole of the potential
drop between the metal and the solution affects the rate of the electrode
reaction. In fact if the reaction is at the outer Helmholtz plane, the potential
difference ϕ_2 across the diffuse layer should be subtracted from E in the
exponential of (3.15) etc. Hence this equation should read:

$$k_+ = k_{+,o} \exp\{(1-\alpha)\ nf\ (E-\phi_2)\} \tag{10.2}$$

and similarly (3.16) should read:

$$k_- = k_{-,o} \exp\{-\alpha nf\ (E-\phi_2)\} \tag{10.3}$$

The concentration of the reactant in the anodic reaction at the OHP is given
by (10.1), but it must be remembered that under non-equilibrium conditions
the concentration appearing on the right-hand side of (10.1) is that just
outside the diffuse layer and at the inner edge of the diffusion layer. We
shall retain the unsubscripted symbol for this:

$$\left[A^z\right]_{OHP} = \left[A^z\right] \exp\left(-z_i f\phi_2\right) \tag{10.4}$$

Similarly, the concentration of the reactant in the cathodic reaction at the
OHP is given by

$$\left[A^{z+n}\right]_{OHP} = \left[A^{z+n}\right] \exp\left[-(z_i + n)\ f\phi_2\right] \tag{10.5}$$

If these equations are developed in the same way as (3.15) and (3.16), (3.22)
is replaced by

$$j/j_o = \left[\left(\left[A^z\right]/\left[A^z\right]_b\right) \exp\{(1-\alpha)nfn\} - \left(\left[A^{z+n}\right]/\left[A^{z+n}\right]_b\right) \exp(-\alpha nfn)\right]$$

$$\times \exp\{-\left[z + (1-\alpha)n\right]\ f(\phi_2 - \phi_{2,e})\} \tag{10.6}$$

where $\phi_{2,e}$ is the value of $\phi_{2,e}$ at the equilibrium potential. The concentration
factors may be treated now exactly as discussed in Sections 4 and 7 provided
that the condition holds that the Debye length is small compared with the
diffusion layer thickness.

The expression for the exchange current is also modified. Equation (4.19)
is replaced by

$$j_o = nF\ k^o\ \left[A^z\right]_b^\alpha\ \left[A^{z+n}\right]^{1-\alpha} \exp\{-\left[z + (1-\alpha)n\right]f\phi_{2,e}\} \tag{10.7}$$

The importance of the double-layer effect on electrode reactions can be
judged from (10.6) and (10.7). When the reaction involves a single electron
transfer (n = 1) between multivalent ions ($|z| > 1$), it is possible for the

coefficient of ϕ_2 in the exponential in (10.6) to be markedly larger than the coefficient of η; it can also be of opposite sign. In dilute solutions much of the metal-solution potential drop occurs across the diffuse layer so that $d\phi_2/d\eta$ can be close to unity. In these circumstances the potential dependence of the electrode kinetics depends more on the double-layer factor than on the intrinsic potential dependence of the reaction rate. There are corresponding effects on the concentration, temperature dependence, etc. Further, it is possible, by adding to the solution surface-active material, to modify the ϕ_2 potential in the way discussed in Section 9. Thus the kinetics of the reaction may be profoundly altered by the addition of surface-active materials which play no direct role in the electrode reaction.

We have described here briefly the simplest example of double-layer effects, in which the reactant is located at the OHP. It is also possible for the reactant to be located at other points in the double layer. If it is further from the metal, the analysis is closely similar to that described above. If it is closer to the metal, the electrostatic problem is more complicated and specific interactions must also be taken into account.

Another type of effect which may be classed with double-layer effects is the so-called blocking effect. If an ion or molecule is adsorbed at the interface, it may prevent the reactant approaching close to the metal surface at that point. The local rate of reaction may therefore be reduced considerably and may even be negligible. This situation may be treated by assuming that the effective area of the electrode is reduced to the area not covered by adsorbed species, and this approximation is often adequate. However, there are simultaneous modifications to the double-layer structure which make this model only a first approximation.

11. KINETICS OF HYDROGEN EVOLUTION

Hydrogen evolution is a common cathodic reaction in aqueous solution and it occurs at all types of metal electrodes. The overall process may be represented by

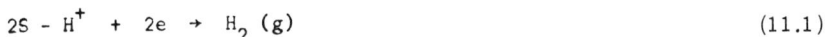

$$2S - H^+ + 2e \rightarrow H_2 \ (g) \tag{11.1}$$

where S represents the solvation shell of the proton which may be a number of water molecules in aqueous acid solution or OH^- together with a number of water molecules in aqueous alkaline solution. On a few inactive metals the reverse process of anodic dissolution of hydrogen can also be studied. On these metals, linear potential sweep (Section 7) experiments, at potentials positive with respect to the equilibrium hydrogen potential, demonstrate clearly the existence of adsorbed hydrogen atoms. On metals like platinum these form an almost complete monolayer. From such experiments, as well as approximate theoretical calculations of the energy of the reaction path, it is almost certain that the hydrogen atom adsorbed on the metal is an intermediate whenever hydrogen evolution occurs. The most probable reaction sequence is either

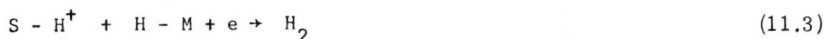

$$S - H^+ + e \rightarrow H - M \tag{11.2}$$

$$S - H^+ + H - M + e \rightarrow H_2 \tag{11.3}$$

or

$$S - H^+ + e \rightarrow H - M \qquad (11.4)$$

$$2 H - M \rightarrow H_2 \qquad (11.5)$$

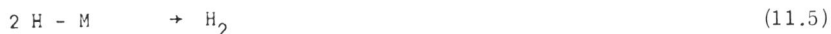

Reaction (11.2) is identical to reaction (11.4) and is known as the 'discharge' (or Volmer) reaction. Reaction (11.3) is known as the 'ion + atom' (or Horiuti-Heyrovsky) reaction. Reaction (11.5) is known as the 'recombination' (or Tafel) reaction. The discharge reaction is often considered in an analogous way to an atom transfer chemical reaction. The 'bond' between the proton and its solvation sheath is stretched so that the electron from the metal can partially transfer to the proton with the formation of a metal-hydrogen bond. A similar process is envisaged for the ion + atom reaction except that the electron participates in the existing metal-hydrogen species. Some authors consider that this process involves formation of an adsorbed H_2^+ ion. The recombination reaction is closely similar to the type of surface reaction considered in gas-phase catalysis.

The rate of hydrogen evolution varies enormously with the nature of the metal electrode. On lead, the exchange current density is about 10^{-13} A·cm^{-2} whereas on platinum it is more than 10^{-3} A·cm^{-2}. This wide difference is due to the difference in the strength of the bond between the atomic hydrogen and the metal. The reaction at metals where the exchange current is low (less than 10^{-5} A·cm^{-2}) can be studied by steady-state methods with the assumption that mass transfer limitations may be ignored. A considerable amount of information about the reaction can be obtained from the current-voltage curves, and in some cases the mechanism can be established. However the non-uniform nature of most solid metal surfaces means that the value of the transfer coefficient and reaction order are not unambiguous guides to the mechanism. The extra evidence provided by transient and other methods is therefore of great importance. When the exchange current is high the characteristics of the electrode reaction itself are obscured by the mass transport phenomena, and transient methods provide the only reliable route. However, much work on hydrogen evolution was carried out before transient methods were fully developed, and their use to study this reaction has not been very systematic. When the reaction is slow the simple theory developed in Section 7 can be used for mechanism (11.2) and (11.3). It requires modification to allow for a non-uniform surface. Further modification is required to allow for the second-order non-electrochemical reaction (11.5), and for fast reactions the treatment of mass transport of Section 8 must be combined with the treatment of two-step reactions of Section 7.

Although arguments still occur about the details of this reaction, the results of its extensive study may be summarized as follows. As the metal-hydrogen bond strength increases, the exchange current first increases to a maximum and then decreases. At this maximum the equilibrium surface coverage of hydrogen atoms is approaching a complete monolayer. In the region of the maximum the reaction probably follows mechanism (11.4) and (11.5), with the recombination reaction as the rate-controlling step near the equilibrium potential. On either side of the maximum, the mechanism is probably (11.2) and (11.3) with the ion + atom reaction as the rate-controlling step. In this region, however, as the potential is made more negative, the discharge and ion + atom reactions may be regarded as equally determining the rate. For metals with very low exchange currents this is the only region of potential accessible to measurement because stray currents due to impurities prevent measurements at very low current densities.

12. METAL DEPOSITION

The reactions of metal deposition and dissolution differ from the reactions considered up to now in two respects: the charge transfer is accomplished by the transfer of the charged ion from the solution to the metal phase, and in most practical examples the interfacial structure changes as the reaction proceeds. The problems associated with the latter may be eliminated by studying metal deposition on mercury with the formation of amalgams (other liquid metals could be used but they are, of course, less accessible).

The formation and dissolution of simple metal amalgams like those of the alkali and alkaline earth metals are rapid processes conforming to the models of Section 8, although the observations are complicated by the parallel evolution of hydrogen. The deposition of Zn^{2+} and Cd^{2+} is slower with the possible participation of an adsorbed intermediate species. The reactions of the iron group metals are still slower and the kinetics may be studied by steady-state methods. There is at present no clear picture of the details of the process by which a cation loses its solvation sheath and is transferred into an amalgam. When polyvalent ions are deposited, the intermediate valence states may exist as transient species in solution or adsorbed on the mercury.

The deposition of metals onto solid metal surfaces has close analogies with the crystallization of solids from vapour or solution. In the early theories, two-dimensional nucleation was considered to occur on a perfect lattice plane, and the probability of nucleus formation was dependent on the supersaturation of the solution with respect to the species being deposited. This dependence arises from a consideration of the free energy of formation of a nucleus of critical size and is of the form $RT \ln(c_{ss}/c)$. In an electrochemical system exactly the same concepts were used but it was noted that this free energy was ηF where η is the overpotential required to maintain the concentration of the reacting ion at the value c_{ss} if the reaction is diffusion-controlled. Once the nucleus is formed it may grow two or three dimensionally. If the former, the nucleation process must repeat for each subsequent layer.

Solid phase depositions following this model have been established in the last two decades, particularly in the growth of anodic films on mercury or on amalgams where the surface is initially perfectly uniform. Examples are the growth of TlCl or $Cd(OH)_2$ on Tl or Cd amalgams respectively. In electrochemical experiments it is possible to separate the processes of nucleation and growth by potential step experiments. A short step at a high potential forms a number of nuclei which may then be grown at a lower potential. Under these conditions, growth may be expressed by

$$j = nF \, kS \qquad\qquad (12.1)$$

when k is the rate constant $(\mathrm{mol \cdot cm^{-2} \cdot s^{-1}})$ for the deposition process and S is
the area on which deposition occurs. k is potential-dependent like all electrochemical rate constants (cf. (3.11),(3.12)) but is maintained constant during
a potentiostatic growth experiment. Since material is being deposited, the
rate of growth can also be expressed in terms of the volume of material
deposited, $d(V/V_m)/dt$ where V_m is the molar volume of the deposit given by
M/ρ where M is the molar mass and ρ the density. Thus

$$I \;=\; nF \, kS \;=\; nF(\rho/M)(dV/dS)(dS/dt) \qquad (12.2)$$

The value of S and of dV/dS depends on the geometry of the growing nucleus,
and various forms of result are obtained for different forms of nucleus. For
example, if the nucleus is assumed to be a cylinder with its axis perpendicular to the metal surface with height h and radius r, on which deposition
occurs only on the cylindrical surface, then $S = 2\pi rh$ and $V = \pi r^2 h$ so that

$$I \;=\; 2\pi n Fh \; k^2 \,(M/\rho)\, t \qquad (12.3)$$

If this is multiplied by the number of nuclei per unit area, the total current
density is obtained. However, this assumes that the growth of all nuclei is
independent; in fact, as they grow they overlap, eventually covering the whole
surface of the electrode. For random overlap of cylindrical nuclei of radius r,
the surface of the electrode actually covered by deposit S_1 can be expressed as

$$S_1 \;=\; 1 - \exp\,(-\,\pi r^2) \qquad (12.4)$$

This clearly leads to a decrease in the surface available for deposition S as
time goes on. Consequently the current-time characteristic at constant
potential has a maximum.

This simple approximate model may be extended to allow for progressive
nucleation, i.e. formation of new nuclei during the process of growth, for
different geometries and for growth of further layers overlapping the growth
of the sublayers. In simple cases such as growth on liquid metal surfaces,
it may be possible to identify with some certainty the type of growth occurring,
provided that the deposition process is sufficiently slow that control of the
rate by diffusion may be ignored. When the overall process is controlled
both by the deposition process and by diffusion of reactants to the electrode
surface, the problem becomes difficult, and only approximate solutions are
available at present. When the deposition process is much faster than
diffusion, the problem again simplifies and reduces to the situation discussed
in Section 4.2 (steady state) or Section 8 (transient) for diffusion-controlled
reactions.

The models described above are less suitable for the description of
deposition on solid metal surfaces because it is only with the utmost difficulty
that a perfect surface can be obtained on a solid metal. More often a real
metal surface contains numerous imperfections of various types which can
form nuclei for growth. In particular, the emergence of a screw dislocation
at the surface can form a self-repeating edge on which continuous growth can
occur without the necessity for further nucleation. This results in the now
well-known spiral growth pattern. Under these conditions the deposition may
be idealized into the consideration of a set of parallel step lines like a
staircase whose tread is one atom high. Two models have been proposed for

deposition on such a surface. In the 'adatom model' the charge transfer takes place uniformly over the metal surface with the formation of adsorbed atoms which then diffuse across the surface to the edge where they are incorporated into the lattice structure. The alternative is a 'direct deposition' model in which charge transfer occurs only at the growing edges and the discharged atom is incorporated simultaneously into the lattice. It is difficult to distinguish between these models because the simple metal deposition processes are sufficiently fast for it to be essential to consider diffusion to the electrode at the same time. In the adatom model this can be simple linear diffusion of the type described in Section 8, whereas in the direct deposition model it must be an approximately cylindrical diffusion with the edge as axis. Detailed analysis shows that the effect of this perturbation in the diffusion field cannot be distinguished experimentally from the effect of surface diffusion of adatoms in the alternative model. The mechanism of this process is therefore still a subject of discussion.

In practical metal deposition dissolution we are usually interested in steady-state behaviour, and this can be approximately attained although the detailed configuration of the surface is continually changing. Under such conditions the current-voltage curve usually approximates to that described by (4.10) in that there may be an exponential section succeeded at large deviations from equilibrium by a diffusion limited current. In metal plating the aim is usually to deposit a uniform layer of metal over the whole object. Thus the current density must be uniform over the whole surface. This may be difficult to achieve if the object has a convoluted shape, when the total current through the plating bath is controlled (or the total voltage across it). The resistive path to the more accessible parts of the surface is less, so the potential drop across the interface at these points is greater, thus leading to higher current densities if these correspond to the exponential part of the current-voltage characteristic. This unequal distribution of metal may be avoided by working on the limiting current region of the characteristic. Under these conditions the potential across the interface can vary over quite wide limits while the current density remains constant at all points of the surface.

Diffusion limiting currents may be inconveniently high and other types of limiting current are frequently used. If the metal ion in solution is complexed (e.g. $Ag(CN)_2^-$) then the complex ion may react sufficiently slowly at the electrode that the most favourable mechanism is the dissociation of the complex followed by deposition of a lower complex or the aquo-ion. The kinetics of such a process may then be determined by the rate of dissociation of the complex. Since this is a homogeneous process its rate is independent of the electrode potential, and the overall deposition rate is independent of potential. Another method of achieving a limiting current density which is not too high without reducing the reactant concentration is to add a surface-active agent which forms a coherent film on the electrode surface. Under some conditions the deposition rate may then be controlled by the rate of ionic penetration through the film, which is a modified diffusion process and again independent of potential.

However, the effect of surface-active agents on the deposition of metals is complicated by many factors. The type of crystal growth may be greatly modified, for example, by preferential adsorption on different crystal faces. In the extreme a single face may grow with the formation of a whisker. The surface-active compound may be 'buried' in the metal as it deposits and this may lead to a modification of the properties of the deposited metal. At the same time, this removal of the adsorbed compound means that it must be replaced from the solution. If the bulk concentration of the surface-active agent is low, this replacement may be diffusion controlled. The stationary concentration of adsorbed compound may then be higher on the more accessible parts of the surface than on the inaccessible parts, which can lead to the 'levelling' of a rough surface.

Finally, it must be noted that metal deposition is often accompanied by simultaneous evolution of hydrogen, some of which may enter the lattice.

13. METAL DISSOLUTION

In the simplest situations metal dissolution is the reverse of metal deposition. On real metal surfaces, dissolution occurs at the surface imperfection and can lead simply to the aquated ion. The different stabilities of the crystal faces may result in preferential etching and again this can be modified by the addition of surface-active agents. As in the case of deposition, the continual modification of the configuration of the surface in the course of the process means that it is difficult to carry out macroscopic experiments with known current density, but the current-voltage curve observed in the approximate steady state follows the usual exponential relation.

In many examples of metal dissolution the stable product is not the aquo-ion in solution but an insoluble compound which may be deposited on the electrode. The final stable product may be judged from the results of equilibrium measurements summarized in thermodynamic tables or more graphically in the potential-pH diagrams prepared by Pourbaix. However, kinetic conditions may result in the formation of a metastable intermediate which reacts very slowly to form the thermodynamically stable product. Hence the thermodynamic predictions must be treated with caution. Further-more, the behaviour of an insoluble product at the electrode surface may differ widely. If the solid formed coheres poorly to itself and to the electrode surface, it will tend to separate from the electrode with the result that formation of this product will continue. On the other hand, if the solid product coheres well and adheres to the electrode surface, a compact film is formed on the electrode surface. If this film is ionically conducting it will continue to grow, but as the field within it diminishes (at constant potential) the rate of growth decreases and eventually becomes so slow that the film may be considered to have reached a constant thickness. Alternatively, some adherent films may undergo a transition to an electronically conducting form. This prevents further growth since ions are no longer transported across the film and another electrode reaction may occur at the film/electrolyte interface which is now the electrode interface. These adherent films protect the metal from dissolution and the metal is said to be in a 'passive' state. Strictly the term passivity is used when the film is invisible to the naked eye and the metal appears to be unaltered. The film causing passivity may even be as thin as a monolayer and the distinction between retardation of the dissolution by an adsorbed layer (as in a double-layer effect) and by a film which can be considered as a thin solid phase becomes difficult. There has been much argument between proponents of the two extreme points of view, but it is probable that systems exist which tend to either extremes as well as those occupying intermediate positions.

The formation of a solid phase in the anodization of a metal has many features in common with the cathodic deposition of metals. A new phase is being formed and therefore similar problems of nucleation and growth are involved.

14. CORROSION AT A HOMOGENEOUS ELECTRODE

In the previous section we have briefly described the phenomena that occur when metals dissolve and this is the important reaction in electro-chemical corrosion. However, this reaction was described as if it occurred under the conditions of a laboratory experiment where it may be studied without being affected by other reactions and the energy required to perform the reaction comes from some convenient external source. Under practical conditions, corrosion occurs with the coupling of two electrochemical reactions which themselves provide the energy for the overall process. In other words, they form an electrochemical cell that operates spontaneously. It follows from this that a corrosion reaction must be spontaneous. For this reason a thermodynamic analysis has considerable use. This type of information is summarized conveniently in the Pourbaix diagrams to which we have already referred. From such diagrams it is possible to deduce regions of potential and pH under which the metal cannot corrode because it cannot undergo any spontaneous reaction in the system described. In the remaining regions this type of analysis can only tell us that a reaction may occur but it can tell us nothing about the rate of any possible reaction. Here we are interested primarily in the rate, and the discussion will therefore be from a kinetic point of view. Another limitation of Pourbaix's diagrams should be mentioned: they represent a selection of the most important reactions which may occur in aqueous solutions (most often the formation of the various oxides) but they cannot represent all possible reactions so that care must be exercised in their use for this reason.

We consider first the simple example of zinc dissolving in a dilute acid in the absence of oxygen. Under these still rather idealized conditions there are two possible reactions:

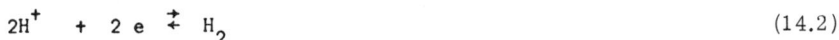

$$Zn^{2+} + 2e \rightleftharpoons Zn \tag{14.1}$$

$$2H^+ + 2e \rightleftharpoons H_2 \tag{14.2}$$

It is convenient to plot the four partial currents for these two reactions on a diagram of E (the electrode potential against some suitable reference electrode) against log j. We use E rather than η because we wish to represent both reactions on the same potential scale. Such a plot is shown in Fig.10. From such a diagram it is evident that the rate of zinc dissolution is controlled by the rate of hydrogen because the curve representing the reverse direction of reaction (14.1) cuts the curve of (14.2) nearly at the zinc potential. The co-ordinates of the intersection of these curves give the corrosion current j_c and the corrosion potential E_c; the latter is a special case of a 'mixed' potential, i.e. a potential controlled by more than one type of chemical reaction. Since j_c is large compared with the exchange current density for the hydrogen reaction, the overpotential for hydrogen evolution is high whereas that for zinc dissolution is very low because j_c is less than the exchange

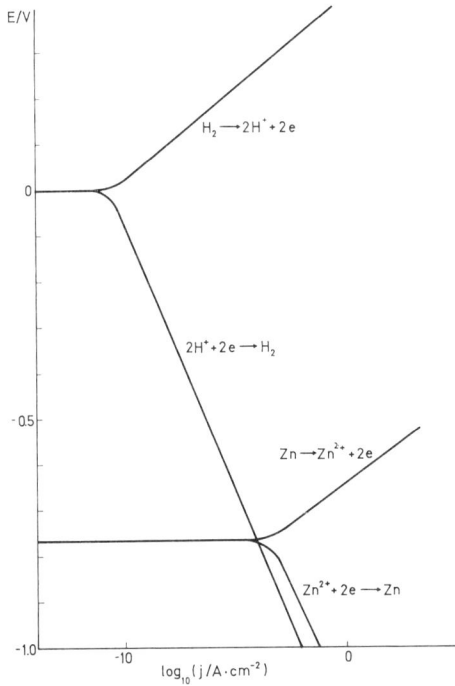

FIG.10.   Current-voltage curves for a mixed electrode. Zinc electrode at which hydrogen evolution can occur.

current density for zinc dissolution. This is again an expression of the fact
that the corrosion rate (j_c) is controlled by the rate of hydrogen evolution.

When the exchange currents of the two reactions are more similar in
magnitude, as e.g. for the reactions:

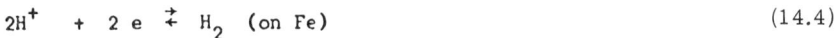

$$Fe^{2+} + 2 e \rightleftarrows Fe \qquad\qquad\qquad (14.3)$$

$$2H^{+} + 2 e \rightleftarrows H_2 \text{ (on Fe)} \qquad\qquad (14.4)$$

the corrosion potential lies between the two equilibrium potentials, and the
rate of corrosion depends on the rate of both processes (Fig.11). In
Figs 10 and 11 we have assumed that the individual reaction rates are
controlled entirely by the charge transfer reaction. In fact, control by
diffusion is also important and limiting currents may be evident.

The total current density at a uniform electrode at which reactions
1...i...N are occurring is the sum of currents given by Eq.(4.10):

$$j = \sum_{i=1}^{i=N} j_{i,o} \left[(1-j_i/j_{L,i,+}) \exp \{(1-\alpha_i)n_i\, f\,(E-E_{i,e})\} \right.$$

$$\left. - (1-j_i/j_{L,i,-}) \exp \{ -\alpha_i n_i\, f\,(E-E_{i,e})\} \right] \qquad (14.5)$$

where j_i is the partial current due to the i-th reaction.

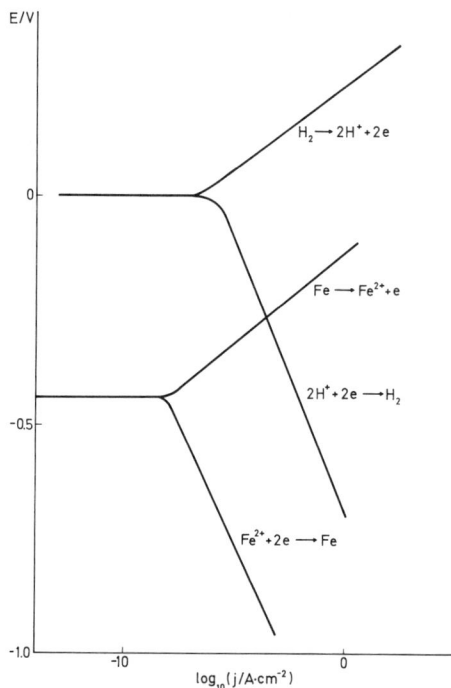

FIG. 11.   Current-voltage curves for a mixed electrode.   Iron electrode at which hydrogen evolution can occur.

The corrosion potential E_c is given by the value of E obtained by setting the total current equal to zero. The corrosion current is then the value of the partial current corresponding to the net rate of metal dissolution. When N = 2 this is given by either of the two terms on the right-hand side of (14.5). The two types of behaviour represented in Figs 10 and 11 are obtained by making the two limiting assumptions about E_c. If $|E_c - E_{1,e}| \gg$ nf and $|E_c - E_{2,e}| \gg$ nf then only one partial current need be retained for each reaction; this corresponds to comparable exchange currents for the two reactions:

$$j_c = j_{1,o} (1-j_c/j_{L,1,+}) \exp \{(1-\alpha_1)n_i f(E_c-E_{1,e})\}$$

$$= j_{2,o} (1-j_c/j_{L,2,-}) \exp \{-\alpha_2 n_2 f (E_c -E_{2,e})\} \qquad (14.6)$$

Hence, if $j_c \ll |j_{L,1,+}|, |j_{L,2,-}|$

$$E_c = \frac{(1-\alpha_1)n_1 E_{1,e} + \alpha_2 n_2 E_{2,c}}{(1-\alpha_1)n_1 + \alpha_2 n_2} + \frac{1}{f\{(1-\alpha_1)n_1 + \alpha_2 n_2\}} \ln\left\{\frac{j_{1,o}}{j_{2,o}}\right\} \qquad (14.7)$$

It is clear from (14.7) that the corrosion potential, like all mixed potentials, has some characteristics of an equilibrium potential in that it arises

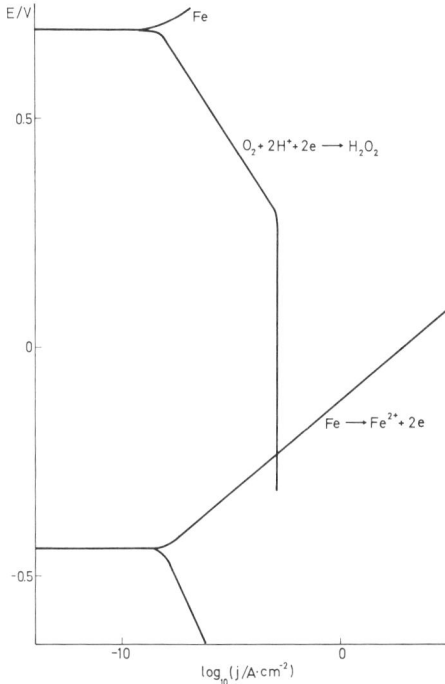

FIG. 12.   Current-voltage curves for a mixed electrode. Iron electrode at which the reduction of oxygen can occur.

from the balance of two reactions. However, it must be clearly distinguished from an equilibrium potential since the reactions which are balanced are not identical as they are in the equilibrium case. Hence the properties of the intermediate states do not cancel and E_c depends strongly on the kinetic parameters j_0, α and j_L. Consequently E_c may differ widely for two apparently identical experiments as a result, for example, of small amounts of surface contamination. The deviation of potential from the corrosion potential may be treated in much the same way as deviations from the reversible potential. Thus under conditions again that $j_c \ll \left| j_{L,1,+} \right|$, $\left| j_{L,2,-} \right|$ we find

$$j = j_c \{(1-j/j_{L,1,+}) \exp \{(1-\alpha_1)n_1 \; f \; \Delta E \}$$

$$- (1-j/j_{L,2,-}) \exp (-\alpha_2 n_2 \; f \; \Delta E)\} \tag{14.8}$$

where $\Delta E = E - E_c$. Thus the current-voltage curve of a corroding electrode has marked similarities to that of a simple electrode but differs in that the slopes of the cathodic and anodic branches are unrelated; $1-\alpha_1$ and α_2 are transfer coefficients for different processes.

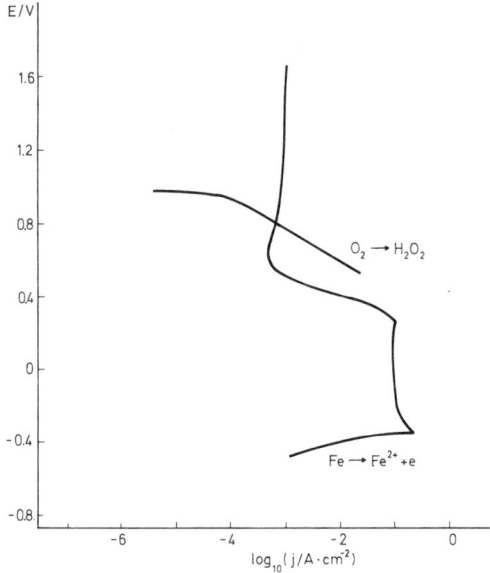

FIG. 13. Current-voltage curves for a mixed electrode. Iron electrode in the active, passive and transpassive regions at which the reduction of oxygen can occur.

If $j_c \simeq j_{L,2,-}$,(14.6) simplifies to

$$E_c = E_{1,e} + \frac{1}{(1-\alpha_1)n_1 f} \ln (\frac{j_{L,2,-}}{j_{1,o}})$$ (14.9)

and the corrosion potential does not depend on the characteristics of the interfacial reaction 2 but only on its mass transfer characteristics. A similar result would be obtained for the complementary case when $j_c \simeq j_{L,1,+}$ but this is a less probable situation because the anodic limiting current of metal dissolution is very high.

This type of behaviour is illustrated in Fig.12 where the reaction of iron dissolution is coupled with the reduction of oxygen to hydrogen peroxide. Since the solubility of oxygen in aqueous solution is relatively low, the limiting current for this reaction is reached well before the potential region corresponding to iron dissolution. It is clear from Fig.12 and Eq.(14.9) that the corrosion potential may vary strongly with the kinetics of metal dissolution but that the rate of corrosion depends only on the rate of oxygen diffusion to the corroding sample.

The third type of behaviour is that represented for Zn in Fig.10 where the deviation of one electrode from equilibrium is very small. If this reaction is that labelled 1, then E_c can be taken as equal to $E_{1,e}$ and

$$j_c = (1-j_c/j_{L,2,-}) \exp\{ - \alpha_2 n_2 f (E_{1,e} - E_{2,e})\}$$ (14.10)

Here again the kinetics of the faster reaction 1 play no part in determining the process, which is controlled entirely by the slower process. In contrast to the situation in Fig.12, however, here the corrosion potential is independent of these kinetics provided that the corrosion current remains small compared with the exchange current of the faster process. Essentially similar behaviour would be observed if the cathodic process were diffusion controlled, as for example if oxygen reduction were coupled with zinc dissolution.

The discussion of metal dissolution in corrosion reactions so far has assumed that metal dissolution occurs with the formation of species such as the simple metal ions which are soluble in the solution. As we have seen in Section 13, other types of anodic process are possible and the existence of passivity is particularly important in the discussion of corrosion   The formation of a passive film has two major consequences in terms of a corrosion diagram: the current-voltage curve for the metal dissolution is profoundly modified, as shown in Fig.13, and the kinetics of the coupled reaction are also considerably affected because this reaction is now occurring on a different surface.

15. CORROSION AT AN INHOMOGENEOUS SURFACE

In the previous section we have discussed corrosion in circumstances when the two coupled reactions occurred uniformly over the surface of the corroding specimen. It is rare for this to occur in practice for various reasons: the metal may itself be inhomogeneous either chemically or physically or the geometric arrangement may be such that parts of the surface are readily accessible whereas others are much less so. The specimen then consists of areas on which the reaction is predominantly anodic and areas on which it is predominantly cathodic and there is a flow of electrons through the specimen from the former to the latter. Since the resistance of the specimen may be finite this may lead to a potential difference between the anodic and cathodic regions. Also, the areas of the different regions may not be be the same, so that the anodic and cathodic current densities may differ. Both of these factors complicate the condition for solving the rate equations, and exact solutions can be obtained only for rather simplified models. On the other hand, useful qualitative conclusions can be obtained using corrosion diagrams of the type already discussed.

When two metals are in close contact, e.g. if a copper pipe is joined directly to an iron pipe, the one with the more positive standard potential (the more 'noble' metal) will become the cathodic area and if the rate of the cathodic reaction (H^+ or O_2 reduction) is faster on this metal than on the other, the rate of corrosion of the latter may be accelerated. In any case, the corrosion may be concentrated at the region close to the junction between the metals so that failure will occur more rapidly than if the corrosion were more uniformly distributed.

Physical differences such as differences in the amount of cold work done on different parts of the same metal can also result in different electro-chemical activity. The surface region with a greater number of dislocations will be the anodic area and so corrosion will be concentrated here. This phenomenon is particularly important when the model is subjected to stress while in a corrosive environment. In such conditions stress-corrosion cracking can occur and result in failure of metal parts well below the normal yield point of the metal.

An important example of the effect of the geometric arrangement is that known as differential aeration. When the rate of corrosion is controlled by a limiting current due to oxygen diffusion as in Fig.12, the accessibility of the surface determines the rate of oxygen reduction at each element of surface. The high rate of reduction of oxygen at the accessible areas leads to a local increase in pH because the reaction also involves the removal of H^+. Under these conditions iron is passivated and dissolution is almost stopped in these areas. On the other hand, no passivation occurs in the inaccessible areas and these become predominantly anodic. Such corrosion is particularly insidious because it affects concealed regions.

16. PREVENTION OF CORROSION

The most straightforward electrochemical method of preventing corrosion is to adjust the potential of the corroding metal to a more negative value so that the rate of corrosion is reduced to a negligible value. This is known as cathodic protection. It can be done with an external power source and a non-corrodible anode. It then requires a constant power input and preferably some more or less elaborate control system, and the anode may also be expensive. Further, the distribution of current over a large object or one of complex form may make complete protection difficult. If the protection is not complete, corrosion may be confined to particularly vulnerable areas. An alternative method of cathodic protection is the use of a sacrificial anode. The object to be protected is electrically connected to a piece of active metal such as zinc or magnesium. This forms the anodic area and imposes a more negative potential on the object to be protected, which therefore becomes the cathodic area. The cost here is essentially that of the anode, which is of course eventually lost by corrosion. The problems of current distribution are similar in both methods of cathodic protection.

Inhibition is a widely used method of corrosion protection. An inhibitor is a species which is adsorbed on the metal surface and consequently retards the electrode reaction occurring there. The inhibitor should be a very strongly adsorbed material so that a large concentration in the environment is not required. It is also important that it should inhibit the right reaction. For example, an inhibitor which retarded the oxygen reduction reaction in Fig.12 would be useless because this electrode reaction rate has no effect on the corrosion rate, which is controlled by diffusion of oxygen. Inhibitors are described as anodic or cathodic depending on which reaction they inhibit. Their effect on the electrode reactions may be understood in the way discussed in Section 10 and may be a primarily electrostatic effect or a blocking effect. The extent of adsorption and hence of inhibition depends on the potential, as discussed in Section 9, and consequently the nature of the double layer is important in understanding inhibition. In particular, the relation between the point of zero charge and the corrosion potential can suggest what type of inhibitor will be most successful. Thus the charge on the inhibitor should be opposite to the charge on the metal surface at the corrosion potential, but if the charge is close to zero an uncharged inhibitor may be more successful. When both anodic and cathodic reactions control the corrosion rate it is usually preferable to use a cathodic inhibitor. This is because an inadequate amount of such an inhibitor merely gives a rather smaller reduction in corrosion rate. On the other hand, an inadequate amount of an anodic inhibitor may be distributed non-uniformly over the corroding surface and corrosion may then be concentrated on the uninhibited regions which may then undergo more serious corrosion if the cathodic reaction is able to proceed at its full rate.

The adsorption of inhibitors that block the reaction rate may be considered to form a continuous range of behaviour to the formation of much thicker films which protect the surface mechanically. Paints, bitumens, etc., act primarily in this way although the constituents may also inhibit the electrode reactions when the thick film is broken. Passive films may also be formed by adjustment of potential (the converse of cathodic protection) although this obviously requires careful control. It is usually preferable to promote passivity by the addition of oxidizing anions like chromate to the environment or to alloy the material with the object of forming a good passive layer as in stainless steels. Thin metal films are widely used for corrosion protection but again caution is required in their use. A thin layer of copper on iron gives excellent protection so long as the layer is intact, but corrosion is intense at any crack which forms a concentrated anode area of exposed iron while the copper provides a large cathodic area. Conversely, a thin layer of zinc (galvanizing) also protects when the layer is ruptured because the zinc then acts as a sacrificial anode and maintains the exposed iron as a cathodic area until all the zinc has dissolved.

Although there is a large empirical content to corrosion protection, an understanding of the electrochemical basis of the process has led to greater efficiency in the development of good protection.

BIBLIOGRAPHY

GENERAL TEXTBOOKS

(a) Introductory

[1] BAUER, H.H., Electrodics, Thieme (1972).
[2] FRIED, I., Chemistry of Electrode Processes, Academic Press (1974).
[3] BOCKRIS, J.O'M, DRAZIC, D., Electrochemical Science, Taylor and Francis (1972).
[4] WEST, J.M., Electrodeposition and Corrosion Processes, 2nd edn, van Nostrand (1971).
[5] FONTANA, M.G., GREEN, N.D., Corrosion Engineering, McGraw-Hill (1967).

(b) Standard texts

[6] VETTER, K.J., Electrochemical Kinetics, Academic Press (1967).
[7] ERDEY-GRUZ, T., Kinetics of Electrode Processes, Hilger (1972).
[8] NEWMAN, J., Electrochemical Systems, Prentice-Hall (1973).
[9] SPARNAAY, M.J., The Electrical Double Layer, Pergamon (1972).
[10] DELAHAY, P., Double Layer and Electrode Kinetics, Interscience/Wiley (1965).
[11] DAMASKIN, B.B., PETRII, O.A., BATRAKOV, V.V., Adsorption of Organic Compounds on Electrodes, Plenum (1971).
[12] MYAMLIN, V.A., PLESKOV, Yu.V., Electrochemistry of Semiconductors, Plenum (1967).
[13] THIRSK, H.R., HARRISON, J.A., A Guide to the Study of Electrode Kinetics, Academic Press (1972).
[14] DELAHAY, P., New Instrumental Methods in Electrochemistry, Interscience (1954).
[15] CONWAY, B.E., Electrode Processes, Ronald (1965).

[16] HUSH, N.S., Reactions of Molecules at Electrodes, Wiley/Interscience (1971).

[17] EYRING, H., HENDERSON, D., JOST, W., Physical Chemistry IXA, IXB, Academic Press (1970).

[18] KABANOV, B., Electrochemistry of Metals and Adsorption, Freund (1969).

[19] LEVICH, V.G., Physicochemical Hydrodynamics, Prentice-Hall (1962).

SPECIFIC REFERENCES

Sections 1, 2 and 3

Covered in most of the textbooks referred to.

Section 4

See particularly [6], [8], [14] and [19].

Section 5

RIDDIFORD, A.C., J.Chem.Soc. (1960) 1175.

PARSONS, R., Croat.Chim.Acta 42 (1970) 281.

OLDHAM, K.B., J.Am.Chem.Soc. 77 (1955) 4697.

LOSEV, V.V., in Modern Aspects of Electrochemistry (CONWAY, B.E., BOCKRIS, J.O'M., Eds) 7 (1972) 314.

Section 6

See [15].

GILEADI, E., CONWAY, B.E., in Modern Aspects of Electrochemistry (BOCKRIS, J.O'M., CONWAY, B.E., Eds) 3 (1964) 347.

PARSONS, R., Surf.Sci. 4 (1964) 418; 18 (1969) 28; Discuss. Faraday Soc. 45 (1968) 40.

Section 7

See particularly [6], [8] and [14].

GERISCHER, H., MEHL, W., Z.Elektrochemie 59 (1955) 1049.

ARMSTRONG, R.D., J.Electroanal.Chem. 39 (1972) 81.

Section 8

See particularly [6], [8], [13] and [14].

GERISCHER, H., VIELSTICH, W., Z.Phys.Chem.N.F. 3 (1955) 16.

NICHOLSON, R.S., SHAIN, I., Anal.Chem. 36 (1964) 706.

ERSHER, B.V., Zh.Fiz.Khim. 22 (1948) 683.

RANDLES, J.E.B., Discuss. Faraday Soc. 1 (1947) 11.

Section 9

See [9], [10], [11] and [12].

PARSONS, R., in Modern Aspects of Electrochemistry (BOCKRIS, J.O'M., CONWAY, B.E., Eds) 1 (1954) 103.

Section 10

See [9].

PARSONS, R., in Advances in Electrochemistry and Electrochemical Engineering (DELAHAY, P., TOBIAS, C.W., Eds) 1 (1961) 1.

PARSONS, R., J.Electroanal.Chem. 21 (1969) 35.

Section 11

See [6], [7] and [15].

PARSONS, R., Trans.Faraday Soc. 54 (1958) 1053.

FRUMKIN, A.N., in Advances in Electrochemistry and Electrochemical Engineering (DELAHAY, P., TOBIAS, C.W., Eds) 1 (1961) 65; 3 (1963) 287.

KRISHTALIK, L.I., ibid 7 (1970) 283.

Section 12

See [6], [7] and [13].

HARRISON, J.A., THIRSK, H.R., in Electroanalytical Chemistry (BARD, A.J., Ed.) 5 (1971) 67.

DESPIC, A.R., POPOV, K.I., in Modern Aspects of Electrochemistry. (BOCKRIS, J.O'M., CONWAY, B.E., Eds) 7 (1972) 199.

Section 13

As for Section 12.

HOAR, T.P., in Modern Aspects of Electrochemistry (BOCKRIS, J.O'M., CONWAY, B.E., Eds). 2 (1954) 262.

Sections 14, 15 and 16

See [4] and [5].

POURBAIX, M., Lectures on Electrochemical Corrosion, Plenum (1973).

EVANS, U.R., The Corrosion and Oxidation of Metals, Arnold (1960).

UHLIG, H.H., The Corrosion Handbook, Wiley (1948).

FRICTION

D. TABOR
Department of Physics,
Cavendish Laboratory,
University of Cambridge,
Cambridge,
United Kingdom

Abstract

FRICTION.
1. Nature and topography of surfaces. 2. Methods for studying surfaces. 3. Nature of real surfaces.
4. Contact between surfaces: single asperity. 5. Contact between surfaces: multiple asperities. 6. Contact
between multiple asperity surfaces: elastic-plastic transition. 7. Experimental measurement of real area of
contact. 8. Meaning of real area of contact. 9. Mechanism of friction. 10. Junction growth. 11. Effect
of surface films. 12. Summary of metallic friction. 13. Friction of ionic solids. 14. Friction of polymers.
15. Friction of rubber. 16. Intermittent motion. 17. Frictional heating. 18. Adhesion. 19. Rolling
friction. 20. Lubricated sliding.

REFERENCE BOOKS

[1] BOWDEN, F.P., TABOR, D., The Friction and Lubrication of Solids, Oxford University Press,
     Part I (1954); Part II (1964).
[2] BOWDEN, F.P., TABOR, D., Friction, an Introduction to Tribology, Doubleday, New York (1973),
     Heinemann, London (1974) - a simple text.
[3] (The Methuen Monograph by BOWDEN, F.P., TABOR, D. (1964) is now out of print.)

For a recent critical review at a fairly specialist level see the chapter "Friction, lubrication and wear"
by D. Tabor in Surface and Colloid Science 5 (E. Matijević, Ed.), Wiley, New York (1972). This gives over
150 references including titles of books by Kraghelsky, Voyutskii, Akhmatov, E. Rabinowicz, Dowson and
Higginson, and Ragnar Holm.

INTRODUCTION

The general problems of friction, lubrication and wear are of great
antiquity and also of very real current interest. The part they play in modern
industry and technology is very large and in recent years a new word has been
coined to describe the field: tribology, from the Greek word tribos = rubbing.
Tribology is defined as the "science and technology of interacting surfaces
in relative motion and of the practices relating thereto".

This paper does not attempt to deal with the technological aspects nor
indeed with the whole field. It attempts to provide some insight into the
main physical processes involved in friction, adhesion and wear, but not
with lubrication. The emphasis is on basic ideas.

HISTORICAL

(See Ref.[1], Part II, Chapter XXIV)

The first qualitative study was made by Leonardo da Vinci (1452-1519) who wrote: "Friction produces double the amount of effort if the weight is doubled", i.e. frictional force F proportional to load W. (N.B. this was 200 years before Newton had given a clear definition of force.) Leonardo also observed that the area of apparent contact had little influence on friction. He wrote: "Friction made by the same weight will be of equal resistance at the beginning of movement though the contact may be of different breadths and lengths." This was not based on idle speculation but on experiment. Leonardo's work was largely forgotten for about 200 years; the two laws were rediscovered by Amontons (1663-1705) who gave his first paper on friction in 1699. The two laws are usually known by his name:

(i)   Frictional force F is proportional to load (with the coefficient of friction $\mu = F/W$ about $1/3$ for most solids. The $1/3$ is not to be taken too seriously; it probably represents the coefficient of friction between dirty surfaces).

(ii) Frictional force is independent of the size of the bodies. This was received by the French Academy with some scepticism and is still regarded with some suspicion. It is, however, roughly true.

Amontons' laws were rediscovered by Coulomb nearly 100 years later. Like his predecessors he realised that his surfaces were rough but he thought that the roughnesses fitted together rather like the parts of a jig-saw puzzle. Thus doubling the area of the solids would, in his view, double the area of true contact. Since the friction was scarcely affected by the size of the bodies he deduced that adhesion played a minor part in the frictional process. The main part was due to climbing up the asperities or bending them over. No clear idea of how such a process can dissipate energy was given. Nevertheless the idea of roughness as a major cause of friction remains embedded in the current archaic terminology of applied mechanics (rough = frictional surface; smooth = frictionless surface).

Only in the last thirty years or so has it been realized that the true area of contact between solids depends not only on the geometry of the surface asperities but more particularly on the detailed way in which the asperities themselves are deformed: this may involve both elastic and plastic deformation.

In what follows we shall describe the nature and topography of solid surfaces and the way in which asperities deform. This leads both theoretically and experimentally to the conclusion that the area of real contact over a very wide range of conditions is approximately proportional to the load and depends little on the gross size of the bodies. We shall then show that friction is mainly due to adhesion at the regions of real contact, the frictional force being that required to shear the junctions so formed. Another factor arises if the asperities on a hard surface produce grooving, ploughing or cutting of the other surface. This part of the friction, often called the deformation component, is usually small compared with the adhesion component of friction. The adhesion component, with thoroughly clean metals, is greatly augmented by the mechanism of junction growth: this can lead to enormous coefficients of friction or even to gross seizure. The friction can be greatly reduced in two ways: either by using materials of limited ductility (this reduces junction growth) or by introducing surface films (oxides, sulphides or films of soft metals) which reduce the shear strength of the interface. These concepts lead to a general account of the friction of some typical non-metals such as ionic solids, polymers (crystalline and amorphous), textile fibres and lamellar solids.

A brief account will then be given of the mechanism of intermittent motion, the role of frictional heating in polishing, skiing and skating. A few remarks will be included on the general mechanism of wear.

Most of the earlier sections which deal with the friction of unlubricated surfaces are explained in terms of the adhesion component of friction. This major part of the paper will therefore be concluded with an account of the mechanism of adhesion itself. This is, of course, related to surface forces and bond formation. Several theoretical papers have recently appeared attempting to calculate the adhesive strength of metals in terms of electronic structure. We shall, as a corrective to this approach, emphasise the importance of released elastic stresses and ductility. With non-metals, adhesion often involves van der Waals forces, and reference will be made to recent direct studies of these forces.

Finally, we shall turn to the problem of rolling friction, why it is so small and why it is reduced so little by the application of lubricants. We shall show that this is because, although some interfacial effects are involved, the major part of rolling friction is due to the work of deformation. With rubber-like materials the energy expended in rolling is directly related to the hysteretic properties of the rubber. Similar processes are involved in the sliding of hard surfaces over lubricated rubber. Consequently it is possible to increase the lubricated sliding friction of rubber by using a rubber of high hysteresis loss. This has a direct application to the problem of improving the skid resistance of tyres on wet or greasy road surfaces.

1. NATURE AND TOPOGRAPHY OF SURFACES
(See Ref.[1], Part II, Chapter I)

It is difficult to prepare surfaces which are really flat. Asperities are nearly always present which are large compared with molecular dimensions. When two surfaces are in contact they touch only at the asperities. Since the range of molecular attractions is of the order of a few Å the area of intimate contact (real rather than apparent area) is quite small.

2. METHODS FOR STUDYING SURFACES

2.1. Stylus methods, e.g. the Talysurf

This involves sliding a stylus (tip diameter $\sim 2 \times 10^{-4}$ cm) over a surface and obtaining a recorded trace greatly magnified in the vertical direction (up to about 40 000 $\times$). If used carefully and with small-diameter stylus the method will detect pits or scratches down to as little as 100 Å. Eventually it is limited by the difficulty of fixing a datum line and the danger of damaging the surface.

2.2. Oblique (taper) sectioning

This involves cutting a section at an angle to a surface. It is mainly useful when irregularities are elongated in one direction. A cut at an angle of about 6° will give a magnification of 10 × (see Ref.[1], Part I, Fig.2). The method will resolve down to about 1000 Å. The top surface is usually protected by electroplating with a metal of equal hardness; for examples see Ref.[1], Part I, Plates I and II.

2.3. Electron microscopy

Electron microscopes are used extensively and can be used in reflection (Ref.[1], Part II, Fig.1, Plate I) or by taking replicas and using shadowing techniques. Resolution down to a few Å is possible. The scanning microscope has poorer resolution (a few 100 Å) but is ideally suited for surface studies. First, the surface can be studied directly without the use of replicas. Second, it gives a very good depth of focus. Third, it causes little specimen heating: this is very important for materials which decompose or explode easily.

2.4. Low-energy electron diffraction (LEED)

In this method a low-energy beam (approximately 100 eV) of electrons is diffracted from a surface. Since the electrons diffract from only the outer layers of the solid, the diffraction pattern gives information about these surface layers. The arrangement of the atoms at the surface can be studied and so can the effect of adsorbed monolayers.

2.5. Interferometry

Newton ring type fringes are relatively broad; in fact, the width of the line at half-maximum intensity is about 50% of the spacing. However, if multiple-beam interferometry is used, the line width can be reduced to about 2-3% of the width between orders. It involves placing an optical flat on the surfaces to be studied and silvering both the contacting surfaces to about 85-95% reflectivity. The light which emerges has undergone multiple reflections, and this increases the resolution in a similar way to that of having more lines on a diffraction grating. The method can detect surface irregularities to less than 100 Å. A modification of this method is to use a spectrograph and white light. The resolution can now be down to a few Å. (This method will be mentioned later in connection with the measurement of van der Waals forces between mica surfaces.) Many of these optical methods have been developed by Tolansky.[1]

---

[1] TOLANSKY, S., "Multiple Beam Interferometry of Surfaces and Films", Oxford University Press (1948).

3. NATURE OF REAL SURFACES

Such studies reveal two main general features about real surfaces:

(a) Geometric

It appears that however carefully surfaces are prepared (polishing, natural growth, cleavage) they are almost invariably covered with roughnesses large on an atomic scale. Recent studies by Greenwood and Williamson show that the asperities are distributed in Gaussian fashion both as regards heights and peak heights. They also find that, to a close approximation, the tips of the asperities are spherical in shape. Finally, their measurements show that for any given type of surface preparation, although the heights of the asperities may vary widely, the radius of curvature of the asperity tips is roughly constant for all the asperities (see below).

(b) Structural

With metal specimens the surfaces will be covered with monolayers of adsorbed gases or vapours. If the metals are reactive they will also be covered with oxide layers. If the surfaces are mechanically polished the subsurface layers will be deformed, the severity increasing as the surface is approached. The outermost layer will consist of a fudge of oxide and metal-polishing powder. On top of this will be oxide films and adsorbed gases.

4. CONTACT BETWEEN SURFACES: SINGLE ASPERITY

Because asperities on anything grosser than an atomic scale are roughly spherical in shape, it is useful to consider first the deformation of a spherical surface of radius R when pressed against a smooth plane. If E_1, E_2 are Young's modulus, ν_1, ν_2 Poisson's ratio, of the asperity and plane respectively, and if the joining load is W, the contact will be a circle of radius a, where

$$a = \left(\frac{3WR}{4E^1}\right)^{1/3} \tag{1}$$

where

$$\frac{1}{E^1} = \frac{1 - \nu_1^2}{E_1} + \frac{1 - \nu_2^2}{E_2}$$

Another quantity which we shall find useful is the depth ϵ by which the asperity tip is pressed in: it is given by

$$\epsilon = \frac{a^2}{R} \quad \text{so that } W = f(\epsilon) \tag{2}$$

We see from Eq.(1) that the area of true contact A may be written:

$$A \text{ proportional to } W^{2/3} \tag{3}$$

Further, contact pressure $p_m = W/A$ increases as $W^{1/3}$. The maximum shear stress occurs below the surface at a depth of about 0.5 a below the centre of the contact region. As W is increased a stage is reached where the elastic limit at this point is exceeded and the onset of plastic deformation takes place. This occurs when

$$p_m = 1.1 \ Y \tag{4}$$

where Y is the yield stress of the material. (We assume the material to be sufficiently ductile to deform plastically under these conditions.) As W is further increased, p_m gradually increases and the plastic zone grows until it extends right up to the free surface. This is called the condition of full plasticity and at this stage

$$p_m \simeq 3 \ Y \tag{5}$$

For an asperity of radius of curvature 10^{-4} cm, the onset of plastic deformation even for tool steel occurs at a load W of only 1 mg. If the asperity reaches full plasticity the contact pressure p_m is constant and the area of contact is simply given by

$$A = \frac{W}{p_m} \tag{6}$$

5. CONTACT BETWEEN SURFACES: MULTIPLE ASPERITIES

Consider the contact between a hard 'ideally smooth' flat surface and a real flat surface. The latter is covered with asperities, the radius of curvature of the tips being roughly constant. The distribution of asperity heights is roughly Gaussian. How do the asperities deform as the load W is increased? For elastic deformation there are two extreme cases:

(a) The number of asperity contacts remains constant so that an increase in load increases the deformation of each contact. Since the area of contact for each asperity is proportional to (load on each asperity)$^{2/3}$, the area of real contact for all the asperities in contact will simply be proportional to $W^{2/3}$.

(b) The average area of each deformed asperity remains constant and increasing the load increases the number of contacts proportionately. Here the total area of real contact is directly proportional to W. Evidently the area of contact for a real situation will be proportional to W^m where m lies between 2/3 and 1.

For plastic deformation of the asperities, to a first approximation the contact pressure supported by each asperity will be constant (average pressure p_m for full plasticity $\simeq 3Y$). Consequently the total area of contact A will be directly proportional to the load W and independent of the size of the bodies $(A = W/p_m)$.

A more analytical treatment for a rough flat in contact with a smooth flat is as follows. For simplicity, instead of assuming a Gaussian distribution

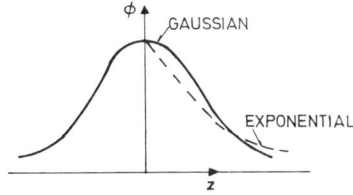

FIG. 1.   Sketch showing probability ϕ of finding an asperity at a height of z above the mean level.
Full curve: Gaussian distribution which corresponds approximately to real surfaces.
Broken curve: exponential distribution as given by Eq. (7). This approximates to the tail of the Gaussian curve.

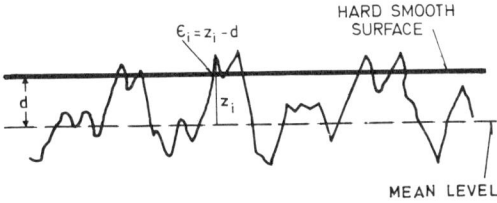

FIG. 2.   Schematic diagram showing typical surface topography of a real surface. The upper horizontal
line represents a hard smooth flat surface which makes contact with all those asperities which are at a height
greater than d above the mean level.

of heights, we assume an exponential law, the probability ϕ that there is
a peak of height z above the mean level being written (see Fig.1):

$$\phi(z) = \frac{1}{\sigma} e^{-z/\sigma} \tag{7}$$

This is a good approximation since only about the top 25% of the asperities
are usually involved in making contact.

We consider the behaviour of asperities within unit area of the hard
smooth surface. If N is total number of peaks/unit area, the number of peaks
in contact,

$$n = \int_{d}^{\infty} N\phi(z)\,dz \tag{8}$$

The compression of an individual asperity $\epsilon_i = z_i - d$ so that

$$z_i = \epsilon_i + d \tag{9}$$

Compression force W_i on an individual asperity $= f(\epsilon) = f(z-d)$ (from Fig.2):

Total load: $$W = \int_{d}^{\infty} f(z-d)\,N\phi(z)\,dz \tag{10}$$

Real area: $$A = \int_{d}^{\infty} g(z-d)\,N\phi(z)\,dz \tag{11}$$

Assuming for simplicity the exponential distribution for ϕ as in Eq. (7), and noting from Eq. (9) that $e^{-z/\sigma} = e^{-(\epsilon+d)/\sigma} = e^{-d/\sigma} \, e^{-\epsilon/\sigma}$,

$$W = N \int_d^\infty \frac{1}{\sigma} \, e^{-z/\sigma} \, f(z-d) \, dz = N \, e^{-d/\sigma} \int_0^\infty \frac{1}{\sigma} \, e^{-\epsilon/\sigma} f(\epsilon) \, d\epsilon \qquad (12)$$

$$A = N \int_d^\infty \frac{1}{\sigma} \, e^{-z/\sigma} \, g(z-d) \, dz = N \, e^{-d/\sigma} \int_0^\infty \frac{1}{\sigma} \, e^{-\epsilon/\sigma} \, g(\epsilon) \, d\epsilon \qquad (13)$$

$$n = N \int_d^\infty \frac{1}{\sigma} \, e^{-z/\sigma} \, dz = N \, e^{-d/\sigma} \qquad (14)$$

The integrals in Eqs (12) and (13) are independent of separation d, hence for each unit area of nominal contact:

The total load $W \propto n$

$$(15)$$

The total real area of contact $A \propto n \propto W$

whence $p_m = W/A = $ const, independent of load and whether the compliance relation $f(\epsilon)$ is elastic or plastic.

The physical meaning of the analytic conclusion is that for an exponential distribution the average size of the asperity contacts remains constant whatever the load, so that doubling the load doubles the number of asperity contacts. Even if, say, 20% of the contacts are plastic and the remainder elastic, the same conclusion holds. There will always be 20% plastic. As the load increases the existing contacts will increase in size but new smaller ones will be formed so that the load-area proportionality will still remain.

6.   CONTACT BETWEEN MULTIPLE ASPERITY SURFACES:
      ELASTIC-PLASTIC TRANSITION

Real surfaces do not, however, show an exponential distribution. A study of surface profiles shows that the distribution is much more nearly Gaussian (Fig.2). The deformation behaviour now depends on two main parameters: (a) the surface topography which can be described by the ratio σ/β where σ is the mean deviation of the asperity heights and β the radius of the tips of the asperities (assumed spherical); and (b) the deformation properties of the material as represented by the ratio of the elastic to the plastic properties. This can be written E'/p_0 where E' is the reduced Young's modulus and p_0 is the pressure at which local plastic deformation occurs. Both parameters are dimensionless and the product is termed the plasticity index:

$$\psi = \left(\frac{E}{p_0}\right)\left(\frac{\sigma}{\beta}\right)^{\frac{1}{2}} \qquad (16)$$

Greenwood shows that if $\psi < 0.6$ the deformation will be elastic over an enormous range of loads. We can see that this corresponds to materials for which p_0 is large compared to E, and/or to smooth surfaces for which σ is small compared to β. In this regime the asperity contact pressure increases somewhat as the load increases but the change is not large. The contact pressure is of order $0.3 \, E'(\sigma/\beta)^{1/2}$ over a very wide range of loads so that the true area of contact is again roughly proportional to the load. For very smooth surfaces the true contact pressure turns out to be between 0.1 and 0.3 p_0.

For most engineering surfaces $\psi > 1$; the deformation is now plastic over an enormous range of loads and the true contact pressure is p_0. This is the simple model described above: the area of contact is again proportional to the load.

7. EXPERIMENTAL MEASUREMENT OF REAL AREA OF CONTACT

The measurement of the real area of contact is very difficult. Only with mica, which can be prepared molecularly smooth, can one equate the real and apparent areas. To some extent this is true of very soft smooth rubbers which deform readily. Again, for a hard sphere pressed into a softer ductile metal where plastic flow can occur around the asperities of the sphere, the geometric and real areas are fairly close. It is for this reason that many of the more fundamental studies of friction (and adhesion) are carried out with this geometry. For contact between 'extended' surfaces, e.g. two nominal flats, the determination is never very satisfactory.

7.1. Electrical resistance

This is an early method which clearly shows that the real area is very much smaller than the apparent area. The inherent difficulty is that the resistance is inversely proportional to the radius a of the contact spot rather than the area (πa^2). Thus if a single contact of given area is broken up into four equal smaller areas the contact resistance of each area will be doubled but there will be four of them in parallel so that the total contact resistance will be halved. Thus the resistance cannot give an unequivocal value for the true area unless the number of contacts is known.

7.2. Thermal resistance

Similar to the electrical resistance approach. An advantage is that thermal conductivity is not so sensitive to oxide films as electrical conductivity is, but on the other hand thermal flow is much more difficult to measure.

7.3. Optical interference

Applicable only to transparent solids and does not reveal lateral discontinuities less than λ.

7.4. Internal reflection

Again only applicable to transparent solids. Light is shone on the interface at such an angle that total internal reflection occurs. However, at contact areas the light is transmitted and the amount transmitted can be measured.

7.5. Talysurf traces

This involves taking a large number of traces of the surfaces after they have been pressed together and then separated. If enough are taken, a 'contour' map of a surface before and after can be obtained.

7.6. Ultrasonic method

Ultrasonic waves are transmitted across the junction and the transmission coefficient measured. This depends on the 'stiffness' of the interface and is itself a measure of the contact area. However, stiffness depends like resistance (see Section 7.1) on a and not a^2, and so again there can be some doubt about the true area of contact unless the number of contact points is known.

8.   MEANING OF REAL AREA OF CONTACT

Real contact between solids pressed together occurs where the atoms on one surface are within the repulsive field of atoms on the other. This is the only way atoms can 'support a load'. Consequently, they are also within the range of their attractive forces. Consequently, to separate the surfaces one ultimately has to overcome the attractive forces between the atoms where atomic contact has occurred (ignoring penumbral regions). This is the basis for the idea that in friction experiments the major part of the frictional force is expended in overcoming adhesion at the interface. However, since even the microcontact regions are large on an atomic scale, it is often more realistic to talk of the bulk properties of the junctions rather than their atomic bonding.

9.   MECHANISM OF FRICTION

Generally the frictional force can be considered to arise from two main processes: (a) the force $F_{adhesion}$ to shear the junctions formed at the regions of real contact (this is equal to As where A is the real area and s the shear strength of the junctions); and (b) the force required to plough the asperities on the harder surface through the surface of the softer. Generally this term is much smaller than the adhesion term.

Typically if we slide (i) a semicircular spade at right angles to its plane, (ii) a sphere, and (iii) a cylinder parallel to its axis over a soft deformable metal, we find that, for a given width of groove formed, the friction of the cylinder > friction of the sphere > friction of the spade. The spade, in fact, gives the ploughing term. We may write

$$F = F_{adhesion} + F_{deformation}$$

$$\simeq F_{adhesion} = As$$

Thus, for a given pair of metals for which s is constant, the friction is proportional to A. In terms of our previous models it is therefore proportional to the normal load and not greatly dependent on the size of the bodies.

10. JUNCTION GROWTH

Consider the simplest geometry for metals — a soft hemisphere, say of copper, sliding over a clean hard smooth surface, say of steel, at a load sufficient to produce plastic deformation of the copper slider. The area of contact is given by

$$A = W/p_{m(copper)}$$

Examination of the steel shows that fragments of copper are left attached to the steel surface. Even at room temperature and at low sliding speeds, the adhesion at the interface is stronger than the copper: consequently shearing occurs in the copper itself. Then

$$F = As_{(copper)}$$

and

$$\mu = \left(\frac{As}{Ap_m}\right)_{copper} = \left(\frac{s}{p_m}\right)_{copper}$$

Now plasticity theory shows that $p_m \simeq 5s$ so that the coefficient of friction should have a value of about 0.2: the observed value is usually of order $\mu = 1$. Part of the reason is that the sliding process greatly work-hardens the surface layers of the copper so that the shear strength of the surface is appreciably greater than the bulk shear strength of the hinterland. This is certainly a real factor. Another, particularly apparent with very ductile metals, is the following.

Plastic flow at the interface during sliding is not produced by the tangential (frictional) stress but by the combined normal and tangential stresses. When the normal load is applied there is plastic flow at a contact pressure p_0. As a tangential stress s is applied, further flow occurs, the surfaces sink together, the area of contact increases and the contact pressure falls to a value p given by a yield criterion of the form

$$p^2 + \alpha s^2 = p_0^2 \tag{17}$$

where α is of order 10. If both surfaces are very clean this process continues indefinitely until the surfaces are fully seized together. For extended surfaces, for example, the whole of the geometric area may be pulled into intimate contact ($\mu > 100$). If, however, some contamination is present so that the shear strength of the interface is $ks_{(metal)}$ where $k < 1$, a stage will be reached where sliding will occur in the interface and no further junction growth will occur. It is easy to show that this gives a finite coefficient of friction of value

$$\mu = \frac{F}{W} = \frac{k}{\alpha^{\frac{1}{2}}(1 - k^2)^{\frac{1}{2}}} \tag{18}$$

Thus if $\alpha = 10$ and $k = 0.95$, μ falls from extremely high values to a value of less than 1. Sliding occurs after junction growth has increased the static area of contact three- or four-fold. For $k = 0.8$, $\mu = 0.4$; $k = 0.6$, $\mu = 0.25$; $k = 0.1$, $\mu = 0.03$.

11. EFFECT OF SURFACE FILMS

We see that friction does not depend appreciably on hardness, since soft metals give a large area of contact but a low shear strength. Low friction may, however, be achieved by using hard metal substrates and interposing a thin soft layer of low shear strength. The load is supported by the substrate, while shearing occurs in the soft film (assuming it is not penetrated during sliding). We have $A = W/p_{substrate}$, $F = As_{film}$. Hence

$$\mu = \frac{F}{W} = \frac{s_{film}}{p_{substrate}} \tag{19}$$

It is not difficult to show that this is equivalent to Eq.(18) for small values of k.

Bearing alloys often operate on this basis. Lubricant films, sulphide films, layers of MoS_2 and of graphite act in this way. The role of oxide is particularly important since most metals are covered with oxide films. The oxide can be penetrated by the load: it is especially vulnerable to the sliding process itself. The relative properties of the oxide and substrate metal are important in determining the viability of the oxide film. For example, on a soft substrate, the oxide may be subjected to excessive deformation and be penetrated by the underlying metal. In contrast, on a hard substrate the oxide may survive under very severe conditions. However, if the oxide itself is brittle it may be disrupted during sliding.

12. SUMMARY OF METALLIC FRICTION

If metal surfaces as normally prepared are placed in contact, the asperities will deform plastically and give an area of contact proportional to the load and independent of the size of the bodies. Friction involves the shearing of junctions at these regions of contact. From this follow the two laws of friction. If the surfaces are thoroughly clean, the sliding process itself will produce junction growth on a large scale and gross seizure may occur. If the surfaces are contaminated by oxide films, junction growth is inhibited and the coefficient of friction will fall to 'reasonable' values of order $\mu = 1$. If there is some penetration of the oxide so that some metal-metal contact occurs, the surfaces will continue to be roughened by the sliding process. Surface contact will still involve plastic deformation and with repeated traversals the surfaces will rapidly fatigue.

If contaminant films are so effective that they are not penetrated and if their own shear strength is low, shearing will occur in the surface film itself and the tangential stresses at the interface will be low. Little junction growth will occur. The asperities may gradually be ironed out and the contacts may then deform elastically. The model for multiple asperity contacts in the elastic range will apply (area of contact again approximately proportional to the load) so that the friction will still be roughly proportional to the load. It should, however, be noted that elastic asperity contacts can persist during sliding only if the tangential stresses are rather small. Under these conditions (e.g. lubricated surfaces) the surface can survive repeated traversal without undergoing rapid fatigue failure.

Another means of preventing large-scale junction growth is to use materials of limited ductility. For example, if the friction of metals is studied as a function of temperature, it is found that the coefficient of friction for bcc clean metals falls from $\mu \simeq 2.5$ in the ductile range to $\mu = 1.5$ below the ductile-brittle temperature, and there is a corresponding change in surface damage. Similarly with cobalt, which shows a phase transformation from hexagonal to cubic at 700 K, the friction is high above 700 K ($\mu = 2.2$) and falls to $\mu = 0.8$ at lower temperatures.

13. FRICTION OF IONIC SOLIDS

Many ionic solids are brittle and surface fragmentation occurs during sliding. Nevertheless the friction is fairly reproducible and Amontons' laws are roughly obeyed. One reason is that at the regions of local contact the contact stresses include a large hydrostatic component so that, although some cracking may occur, the primary deformation observed is plastic flow. For this reason indentation hardness measurements can often be made in brittle solids and these relate to the plastic, not the brittle, properties of the solid. There is also direct evidence to show that strong adhesion can occur, for example, between two pieces of rock-salt. The frictional mechanism thus remembles that of metals. The main difference is that, because of limited ductility, junction growth is greatly inhibited so that, even with very clean surfaces, μ never rises to very high values. In addition, although the frictional processes may be determined by ductility of the solids, the surface damage is usually dominated by cracking.

14. FRICTION OF POLYMERS

With polymers, deformation is neither elastic nor plastic; it is visco-elastic. The area of contact depends on geometry, load and, to a lesser extent, loading time. For a single asperity contact the area of contact for a fixed loading time is proportional to W^m where m lies between about 0.7 and 0.8. The frictional force for a polymer sphere sliding on a 'smooth' surface of, say, glass varies as W^n where n is less than unity implying that the coefficient of friction, $\mu = W^{1-n}$, increases at small loads. Polymer is transferred to the glass so that the friction involves shearing very near the interface in the polymer itself. The shear strength of the polymer increases slightly with contact pressure so that n is a little greater than m. This type of geometry and frictional behaviour applies to the behaviour of textile fibres sliding over one another. For multiple asperity contacts the friction is more nearly directly proportional to the load.

For a wide range of thermoplastics (e.g. nylon, Perspex, Dacron) and over a wide range of experimental conditions the coefficient of friction is of order $\mu = 0.4$ to 0.6. However, with PTFE[2] and high-density polythene the

---

[2] Polytetrafluoroethylene, Teflon, Fluon.

behaviour is very different. If a slider of one of these polymers is slowly slid over a clean glass surface, the initial static friction is 'normal', $\mu_s \simeq 0.3$, and a lump of polymer is left attached to the glass. Subsequent sliding is smooth, the kinetic friction is very low, $\mu_k \simeq 0.05$, and there is transfer of a thin tenuous film of polymer 20 Å thick, with molecular chains oriented parallel to the direction of sliding. On a fresh clean glass surface the static friction is as low as the kinetic friction and transfer remains slight. If the slider is rotated about its vertical axis through 90°, sliding on clean glass now gives a high static friction and lumpy transfer, followed by a low kinetic friction and slight transfer again. Initially, the adhesion with clean glass is strong enough to tear out pieces of polymers from the slider. This process apparently orients material in the contact region to give easy drawing of molecules out of the slider on to the glass (low μ_k and slight transfer). Adhesion between this transferred film and the glass is fairly strong. On the other hand, an unoriented PTFE slider gives low friction and negligible additional transfer if slid over such a transferred film, suggesting that adhesion between PTFE and the film is weak.

If side groups are introduced into the PTFE molecule, the behaviour can be radically changed. For example, suppose we replace one F in $CF_2 - CF_2 -$ by CF_3, the repeat unit becomes hexafluoropropylene and the repeat unit is $- CFCF_3 - CF_2 -$. If we prepare a polymer with only one such group per 400 carbons, the behaviour is identical with PTFE. If, however, we have one such group every 20 carbons (this copolymer is known as FEP) the kinetic friction is as high as the static and the transfer is lumpy. The molecular structure of FEP is the same as that of PTFE. The only difference is the bulky side group CF_3. Experiments with several similar copolymers suggest that the low friction and light transfer of PTFE and high-density polythene are due to their smooth molecular profile.

The sliding of oriented chains (and the drawing of chains out of suitably oriented polymer) involves relatively small stresses. However, these stresses are temperature and rate dependent: at low temperatures and/or at high shear rates, the stresses exceed the bulk shear strength of the polymer. It then becomes easier to tear out lumps of polymer than to slide over the chains. It is indeed found that at low temperatures and/or high sliding speeds the friction of PTFE increases to $\mu_k \simeq 0.3$ and the transfer becomes heavy.

15. FRICTION OF RUBBER

The friction of rubber involves adhesion at the interface as with other solids; but with rubber all the deformation processes are markedly temperature and rate dependent. The friction varies with speed and temperature in a way that reflects these deformation properties. A number of molecular models have been proposed involving bond formation between the rubber molecule and the other surface, dwell-time and stress-activated detachment of the molecule. The treatment is essentially an extension of the Eyring molecular theory of viscous flow: the analysis due to Schallamach gives good qualitative agreement with experiment but rather poor quantitative agreement.

16. INTERMITTENT MOTION

Many sliding systems give intermittent, jerky or stick-slip motion.
Basically this will always tend to occur if the friction decreases with
increasing velocity. The detailed behaviour also depends on the stiffness of
the system, its natural frequency and the damping. In some cases intermittent motion may arise if the surfaces are so supported that the sliding
process deforms one of the members in such a way as to increase the
normal force. This leads to 'spragging'.

17. FRICTIONAL HEATING

Practically all the work performed against friction appears as heat.
Most of it is generated at or very near the sliding interface. Consequently,
although the bulk temperature may be small, the temperature flashes at the
interface can be very high indeed.

The exact value depends on the speed of sliding, the coefficient of friction, the conductivities of both materials and their melting points. The
melting point usually limits the temperature which can be reached but there
are exceptions, e.g. if one or both of the sliding bodies can oxidize and if the
oxidation process is exothermic. This can occur with metals such as
aluminium and magnesium where temperatures of almost 2000°C have been
detected.

Because the temperature rise is so localized, it is very difficult to
measure by conventional means. Some success has been achieved by using
the rubbing surfaces as their own thermocouple and by the use of an infra-red
cell if one of the surfaces is transparent.

Frictional heating is important in polishing where it is essential that
the polishing powder should remain harder than the surface being polished at
the elevated temperature generated between them. In some cases local
heating can cause surface melting and the formation of a smooth flowed
layer. Again there is now a wide body of evidence suggesting that the low
friction of ice and snow is due to the presence of a thin 'lubricating' layer
of water: this is not due to pressure melting but to frictional heating. At
very low temperatures, where this mechanism does not produce sufficient
melting, the friction of ski on snow tends to be like that of ski on sand.
Finally, there are various situations, particularly with powdered explosives,
where the individual grains rub together: if hot spots of sufficient intensity
are generated, initiation of explosion may occur.

18. ADHESION

Most of the preceding sections explain the friction of unlubricated
surfaces in terms of the adhesion which occurs at the regions of real contact.
Nevertheless when two metal surfaces are unloaded to zero load it is often
found that there is no adhesion and no force is required to separate the
surfaces. Contaminant films which may be partially broken down during
sliding may be one cause. It cannot be the entire explanation since the
coefficient of friction is appreciable even between metal surfaces covered
with oxide films. The main reason is that, as the normal load is removed,

elastic stresses in the hinterland are released and this changes the shape of
the interface. Since the junctions are usually highly worked and resemble, in
geometry, a notched bar, they are very brittle and break easily. This may be
demonstrated very simply with metals such as lead or indium which are very
ductile and do not work-harden at room temperature. If their surfaces are
clean and a clean indenter of some harder metal (or glass) is pressed into
them, a strong normal force is required to pull the surfaces apart. The
reason adhesion enters into the frictional mechanism for both ductile and
non-ductile solids is that when friction is measured the normal load is
retained.

Adhesion experiments in high vacuum show that gold, which is fc cubic
and has numerous independent slip planes, sticks strongly to gold: cobalt,
which is hexagonal at room temperature and has fewer slip planes, shows
weaker adhesion; whilst titanium carbide, which has an electrical conductivity like that of a metal (e.g. bismuth) but is brittle, shows extremely small
adhesion. Again, clean germanium shows poor adhesion to clean germanium
at room temperature where it is brittle,but strong adhesion at 700°C where
it becomes ductile. However, if the adhering surfaces formed at 700°C are
allowed to cool to room temperature, the adhesion drops to a very small
value. It is clear that in terms of practical adhesion measurements the
ductility of the material is of great importance. Another aspect of this is
shown by the effect of adsorbed gases on the adhesion of metals. Physisorbed
gases have little effect, but a very thin chemically formed film of, say, oxide
can produce a marked reduction in adhesion. This is because metal-metal
contact is reduced and also because the oxide presents an interfacial layer
of extremely limited ductility.

The bonding which occurs across the interface depends, of course, on
the nature of the solids. For metals, the bonding is due to the interpenetration
of the plasma across the interface. For ionic solids the bonding is Coulombic.
For covalent solids the bonding is partly covalent, partly van der Waals.
For polymers most of the bonding is due to van der Waals forces.

The attractive forces between solids will always include a part arising
from the mutual polarization of the constituent atoms (dispersive
van der Waals forces). For small separations these forces between pairs
of atoms fall off as x^{-7} (x = separation). For larger separations the instantaneous dipole field from one atom takes a finite time to reach its neighbour
and by then its own dipole has changed: this gives 'retarded' van der Waals
forces for which the force falls off as x^{-8}. These forces are roughly additive
so that, by integrating over two half spaces it is possible to calculate the
force between a sphere and a flat or between two crossed cylinders. Experiments carried out with molecularly smooth mica surfaces in the form of
crossed cylinders enable the forces to be measured directly for separations
ranging from 14 Å to 1200 Å and show that the transition between normal
and retarded van der Waals forces occurs as predicted by theory at separations around 300 Å. These observations also provide a simple means of
estimating the adhesive forces between van der Waals solids when they are
brought into atomic contact.

19. ROLLING FRICTION

In the previous sections,friction has been discussed in terms of the
adhesion component, and the deformation or ploughing term has been ignored.

There are two situations where the work done in deforming the surface may dominate the frictional behaviour; one is rolling friction, the other certain types of lubricated sliding.

It has long been known that rolling friction is generally very much smaller than sliding friction and it is this which led to the use of ball bearings, rollers, etc. Why is this so? The explanation first given by Osborne Reynolds is the following. Consider a rigid cylinder rolling over a surface which it deforms elastically. We find that segments within the region of contact are not stretched uniformly. Consequently, as rolling proceeds, these segments will attempt to accommodate themselves to the extensions appropriate to their changing positions within the contact region. This will lead to microslip between the roller and the surface (Reynolds' slip). Thus any sliding friction involved during rolling will occur on a very small scale and dissipate far less energy than would be the case if sliding occurred over the whole contact region. Experiments, however, show that in most situations the microslip is too small to account for the observed rolling friction. Another type of slip within the contact zone occurs if a ball rolls within a groove (Heathcote slip). Here again the slip is usually very small unless the radius of curvature of the groove is only a few per cent greater than that of the ball.

Generally, we may conclude that although these processes involving interfacial slip do, in fact, occur they do not usually contribute an appreciable part to the observed rolling friction. It may also be noted that lubricants, which greatly reduce sliding friction, do not appreciably reduce rolling friction.

Consider first the rolling of a hard steel ball over a flat surface of some soft material such as lead. Even if appreciable adhesion occurs, the junctions within the contact zone are peeled apart. This is a far easier process than the shearing of junctions such as occurs in sliding. Here again we find that lubricants produce only a small reduction in the rolling friction. However, we observe that a plastic groove is formed in the surface of the lead, contacting occurring only over the front half of the nominal circle of contact. If we examine the equilibrium of forces between the ball and the front half of the contact circle we see that the vertical component corresponds to the load, the horizontal force, i.e. the rolling friction, to the force required to displace the soft metal ahead of the ball. Thus the rolling friction for a ball on a soft metal arises primarily from the work required to form a plastic groove in the surface and for this reason lubricants have little effect.

In practical affairs, e.g. in ball bearings, it is not desirable to deform the surface plastically since they would fail by fatigue very rapidly. The loading is generally within the elastic limit of the metal.

To study rolling friction within the elastic limit it is more convenient to use rubber than metals. If a steel ball is rolled over a flat surface of rubber no permanent groove is formed. Nevertheless deformation work is expended in the following way. As the ball rolls forward it performs elastic work on the rubber in the front half of the circle of contact. Similarly the rubber on the rear half of the circle of contact urges the ball forward. If as much energy were restored at the rear as is expended in the front no net energy would be lost during rolling and the rolling friction would be virtually zero. But because rubber is a real material it is not ideally elastic and elastic energy is lost by hysteresis in every loading-unloading cycle. If the elastic input energy per unit distance of rolling is ϕ and if, in each deformation cycle, the rubber loses a fraction α of this energy, the rolling friction F_r is simply $F_r = \alpha \phi$.

Thus, if we compare two rubbers of similar moduli but with widely differing α's, we should find that the rolling friction varies as the ratio of the α's. This is, in fact, observed. We may, however, note that α is usually derived from deformation experiments on rubber involving either pure tension, pure compression or pure shear. For the complex stress path to which every element of the rubber is subjected during rolling, the energy loss is greater than $\alpha\phi$. It is more nearly equal to $3\alpha\phi$.

With rubber and other polymeric materials the hysteretic properties are a function of rate of deformation and temperature. The rolling friction depends on speed and temperature in a manner that reflects these properties. With metals the hysteretic properties show far smaller variation. The absolute values of α are also usually considerably smaller. For example, with a 'dead' rubber α may be as large as 0.8: with a 'bouncy' rubber as low as 0.05: but with ball race steel α is often less than 0.003. For this reason the rolling friction in a ball bearing is very small indeed: in fact, it oftens turns out that the major part of the friction of a ball bearing arises from the rubbing of the balls against the cage.

20. LUBRICATED SLIDING

With unlubricated sliding on rubber the adhesion at the interface is usually high and this term completely swamps any deformation component. It is common to obtain coefficients of friction of order $\mu = 3$. If, however, we slide a hard steel ball on a rubber surface which is extremely well lubricated, we may reduce the adhesion component to a very low level. In that case the friction may be dominated by the deformation component. Simple experiments show that under conditions of effective lubrication the sliding friction of a ball over rubber is, in fact, almost the same as the rolling friction. In both cases it is determined by the hysteretic loss properties of the rubber. This has led to the use of automobile tyres of high hysteresis-loss rubber as a means of improving the skid-resistance of automobiles on wet or greasy road surfaces.

ION BOMBARDMENT AND IMPLANTATION

G. DEARNALEY
AERE Harwell,
Didcot, Oxfordshire,
United Kingdom

Abstract

ION BOMBARDMENT AND IMPLANTATION.
The processes of ion implantation and bombardment are discussed as ways of both altering and measuring the composition and structure of solid surfaces. Practical applications of this subject are described, as well as the physical mechanisms involved in atomic collision phenomena in solids.

1. INTRODUCTION

The interactions between ion beams and solids have a number of important consequences in surface science and in surface materials technology. Perhaps the longest established is the very useful technique of surface cleaning by ion bombardment preparatory to examination, e.g. by LEED. Ion backscattering has become a valuable quantitative method for determining composition as a function of depth and, in conjunction with the phenomenon of ion channelling, for studying disorder and the lattice location of impurities in crystals. Ion implantation, or the introduction of foreign atoms into the surface of a solid in the form of an energetic ion beam, is now a most versatile and controllable means of altering the surface or sub-surface composition. It has been exploited most vigorously in semi-conductors, but we shall see that there are many other areas of solid-state physics and materials technology which are beginning to benefit from the developments in this technique.

Throughout this work I shall be introducing both basic physics and practical technology. I make no apology for this, since it is the mutual interplay between these two which, together with economic factors, provides the driving force for both. Technological application has always been a source of challenging new problems in physics.

2. ATOMIC COLLISIONS IN SOLIDS

An understanding of the interaction mechanisms between a bombarding ion and the atoms of the target is fundamental to the problems to be discussed. The subject can be divided into two broad areas:

(a) The influence of the solid on the motion of the ion, including its rate of energy loss, penetration, charge state, optical and X-ray emission, etc.
(b) The influence of the ion on the solid, generally described as radiation damage and including sputtering, surface topography changes, and the interaction of defects.

167

2.1. The interatomic potential

If we could define, under all circumstances, the interaction potential between an ion and an atom we should be in a good position to calculate such things as ion ranges and damage distributions. Unfortunately, however, this is a complex many-body problem and there is no one analytical expression for the interatomic potential which is valid for all interaction radii. A large number of approximation methods have been devised but these have limited areas of validity and the choice of a potential function is governed by the circumstances of the problem, e.g. the relative velocity and energy transfer.

It is well known that the general form of the interatomic potential is attractive at large radii, due to van der Waals forces, and strongly repulsive at short distances, due to both the Coulomb repulsion and the consequences of Pauli's exclusion principle. The minimum of the potential function corresponds to the equilibrium separation distance, and information on the strength of the interaction near this minimum can be obtained from the compressibility of the solid and other elastic constants. Thus was derived the Born-Mayer function $V(r) = A \exp(-r/B)$, which is relatively successful in describing the potential V for large r though this potential is much too weak at small distances.

At very small separations the atoms experience the Coulomb potential between their nuclei, but screened to some extent by the orbiting electrons. Bohr's potential:

$$V(r) = \frac{Z_1 Z_2 e^2}{r} \; \exp\left(-\frac{r}{a}\right)$$

involves the atomic numbers Z_1 and Z_2 of the ion and target atom, e is the electronic charge, and a is the screening radius which is related to the Bohr radius of the hydrogen atom a_0 (~ 0.53 Å) by

$$a = a_0 (Z_1^{2/3} + Z_2^{2/3})^{-\frac{1}{2}}$$

In most cases that we shall consider, interest lies in the intermediate region, for which a better model of the electron distribution of each atom is required. Here the Thomas-Fermi statistical model of the atom has been very useful, and a good introduction to it is to be found in Schiff's 'Quantum Mechanics' [1]. The interatomic potential is then

$$V(r) = \frac{Z_1 Z_2 e^2}{r} \; \phi_{TF}(r/a)$$

where the Thomas-Fermi screening function has been calculated numerically and tabulated by Gombas [2]. Figure 1 shows the relative magnitudes of these potential functions for a pair of copper atoms, and the weakness of the Born-Mayer potential at small r and of the Bohr function at large r can be clearly seen.

Sometimes for analytical convenience an inverse power-law potential of the form:

$$V(r) = U^{S-1} \frac{Z_1 Z_2 e^2}{Sr^s}$$

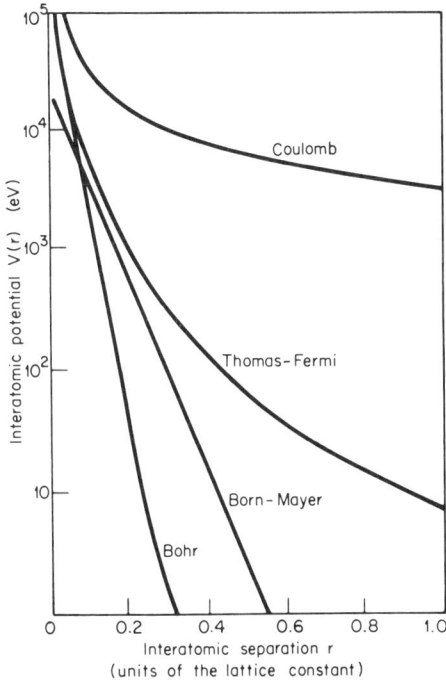

FIG. 1.   Several commonly used approximations to the interatomic potential in copper.

has been used, with U here a parameter of the dimensions of length. It is possible to match the power-law potential, for example, to the Thomas-Fermi potential say at the distance of closest approach, or alternatively to match the slope with that of a more realistic potential and so derive the power S as a function of r. In practice, however, the value of a simple analytical expression is lost unless S = 1 or 2. The excessively long range of these power-law potentials can be artificially corrected by truncating them, say, at the equilibrium separation. In recent years, however, the need for a tractable analytical expression for V(r) has lessened with the development of powerful computers, which are capable of handling the dynamics of atomic collisions with empirical fits to the Thomas-Fermi potential, such as that due to Molière [3]:

$$V(r) = \frac{Z_1 Z_2 e^2}{r} \ (0.1 \ \exp [-6r/a] + 0.55 \ \exp [-1.2r/a]$$

$$+ 0.35 \ \exp [-0.3r/a])$$

At the very lowest relative velocities it can still be useful to consider the collision as one between hard impenetrable spheres. The sphere radius is usually chosen to correspond to the distance of closest approach in a head-on collision, and it is normal to work in the system of co-ordinates which is

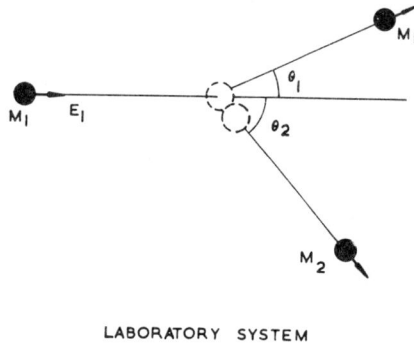

LABORATORY SYSTEM

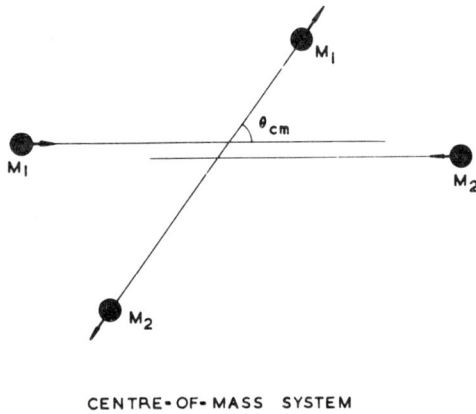

CENTRE-OF-MASS SYSTEM

FIG. 2.   Two-body collision in laboratory and centre-of-mass systems of co-ordinates.

fixed with respect to the centre-of-mass of the two particles (Fig. 2).   If the incident particle energy is E, then for the symmetrical case the centre-of-mass energy is E/2 and hence

$$V(2r_0) = E/2$$

which, if matched to the appropriate Born-Mayer potential, gives the sphere radius r_0 as a function of E:

$$r_0(E) = \frac{B}{2} \ln\left(\frac{2A}{E}\right)$$

so that the radius decreases slowly as E increases.   Ideas based upon hard-sphere collisions are useful in considering the low-velocity impacts which dominate the cascade of recoiling particles around the track of an ion, but they fail to predict the correct angular distribution of scattered particles.

3.   ENERGY LOSS MECHANISMS

Despite much interest throughout the century it must be admitted that our understanding of the processes by which ions lose energy in solids is far from adequate. At best it may be said that we have a rough working knowledge which provides an estimate of ion ranges in amorphous solids.
    Usually, a distinction is made between two major energy loss processes:

    (a)   Elastic interactions between the screened nuclear charges leading
          to energy transfer to the target atom; and
    (b)   Inelastic electron excitation and ionization.

It has been argued that because, on a statistical basis, elastic collisions involving significant energy transfers are rarer than inelastic collisions these two are separable. However, it may not be true to ignore the correlation simply because a close collision is bound to lead to excitation, which will alter the screening of the nuclear charge and hence affect the elastic energy transfer. Also, when the relative velocity is comparable with the orbital electron velocity there is a contribution due to repetitive charge exchange between ion and target atom which can add a further 10% or 20% to the total energy loss. However, since there is as yet no complete treatment of the problem we shall discuss the approximations so far proposed.

3.1. Electronic stopping

At velocities greater than $v = Z_1 e^2/h$, i.e. greater than the K-shell electron velocity, an ion will have a high probability of being completely stripped of its electrons by the impacts it encounters in moving through a solid. The theory of energy loss under these circumstances derives from the work of Bohr [4], Bethe [5] and Bloch [6], who showed that for an ion of velocity v and atomic number Z_1 the linear rate of energy loss:

$$- \frac{dE}{dx} = 4\pi Z_1^2 e^4 \, NB/mv^2$$

where N is the number of target atoms per unit volume and m is the electronic mass. B is a dimensionless 'stopping number' given approximately by

$$B = Z_2 \ln\left(\frac{2mv^2}{I}\right)$$

in which I is the mean ionization potential of the target atoms, given reasonably well by $I = 10.5 \, Z_2(\text{eV})$.
    At lower velocities, the ion begins to accumulate electrons and its effective charge falls from Z_1 to γZ_1 where $0 < \gamma < 1$. However, γ is a function of Z_1, Z_2, v, and even the density of the stopping medium, and so this approach soon founders. There have been some moderately successful attempts to correlate experimentally the effective charge with the energy loss of the ions in this velocity region, but some uncertainty arises from the possibility of auto-ionization of an excited atom in flight between the absorber and the detector.

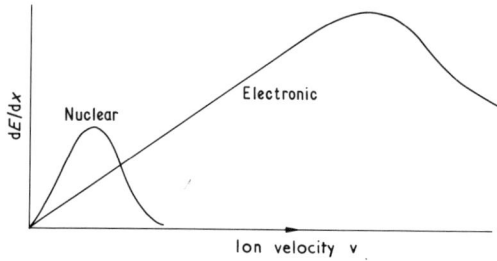

FIG. 3.   Linear rate of energy loss dE/dx due to nuclear and electronic contributions, as a function of ion velocity v.

At the lowest velocities, the energy loss due to electronic excitations is found to be proportional to the ion velocity (Fig. 3). Fermi and Teller [7] were the first to show that this is the consequence of an ion of constant charge moving through a degenerate electron gas. Lindhard and Scharff [8] demonstrated that the problem is equivalent to that of the scattering of an electron gas moving at a drift velocity v in a metal due to the presence of charged centres. The fact that the energy loss is proportional to drift velocity is normally expressed as Ohm's law, and for ions the linear dependence should be valid up to $v \sim Z_1^{2/3} e^2/h$. In 1961 Lindhard and Scharff [9] applied the Thomas-Fermi atomic model to derive the expression:

$$-\left(\frac{dE}{dx}\right)_{elec} = \xi_e \, 8\pi e^2 N a_0 Z_1 Z_2 (Z_1^{2/3} + Z_2^{2/3})^{-3/2} \frac{v}{v_0}$$

in which v_0 is the Bohr velocity $Z_1 e^2/h$ and ξ_e is a dimensionless parameter of magnitude $Z_1^{1/6}$. Lindhard and Scharff found it convenient to introduce the reduced energy and range parameters ϵ and ρ defined by

$$\epsilon = \frac{E a_{TF} M_2}{Z_1 Z_2 e^2 (M_1 + M_2)}$$

where

$$a_{TF} = 0.88 \, a_0 (Z_1^{2/3} + Z_2^{2/3})^{-1/2}$$

and

$$\rho = x \frac{N M_2 \, a_{TF}^2 \, 4\pi M_1}{(M_1 + M_2)^2}$$

In terms of these parameters we may write:

$$-\left(\frac{d}{d\rho}\right)_{elec} = K\epsilon^{1/2}$$

where

$$K = \xi_e \frac{0.0793 \, Z_1^{1/2} \, Z_2^{1/2}(M_1 + M_2)^{3/2}}{(Z_1^{2/3} + Z_2^{2/3})^{3/4} \, M_1^{2/3} A_2^{1/2}}$$

This somewhat cumbersome set of formulae can most conveniently be handled using a small computer.

An alternative approach to the problem has been formulated by Firsov [10] and is surprisingly successful in accounting for the electronic stopping behaviour at low and intermediate ion velocities. Each binary collision of an ion and an atom is viewed as leading to an overlap of their electronic orbitals. A hypothetical interaction plane is conceived which in the symmetrical case lies midway between the two particles and defines the regions of the atomic potentials. When an electron crosses this plane its momentum is transferred abruptly by the relative motion of the ion and atom. If the number density of electrons in the first atom is $n(r)$ and their velocity is $u(r)$, the energy loss on crossing the plane will be

$$\Delta E = m \iint_s \frac{nu}{4} \, \dot{r} \, dr \, ds$$

where the integration is taken over the interaction surface s. Firsov used the Thomas-Fermi statistical model to calculate electronic loss on this basis. Both electron excitation and charge exchange are combined in the Firsov model, which predicts an electronic stopping power proportional to the ion velocity v, as is observed for a variety of ions up to $v \simeq 4 \times 10^8$ cm/s.

3.2. Elastic or nuclear stopping

For the ion energies under consideration we can safely ignore inelastic nuclear excitation, i.e. the absorption of energy resulting in an excited nuclear state, and need only calculate the elastic recoil energy. For this we must know the interatomic potential throughout the scattering process and the most widely useful treatment is that given by Lindhard et al. [11] who considered both inverse power-law and Thomas-Fermi potentials.

Figure 4 illustrates a typical two-body collision process between an ion and an atom, with an impact parameter p. The energy transfer $T(E, p)$ may take any value between zero and a maximum T_m:

$$T_m = \frac{4 M_1 M_2 E_1}{(M_1 + M_2)^2}$$

corresponding to a head-on collision. It is clear that

$$-\left(\frac{dE}{dx}\right)_{elastic} = N \int_0^\infty T(E, p) \, 2\pi p \, dp$$

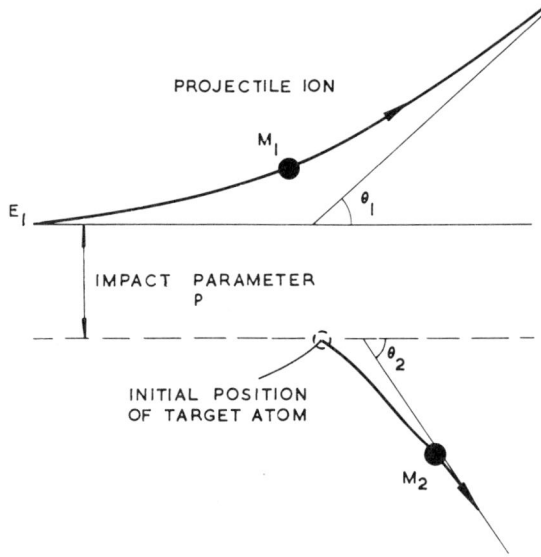

FIG.4.   Typical two-body scattering process with an impact parameter p.

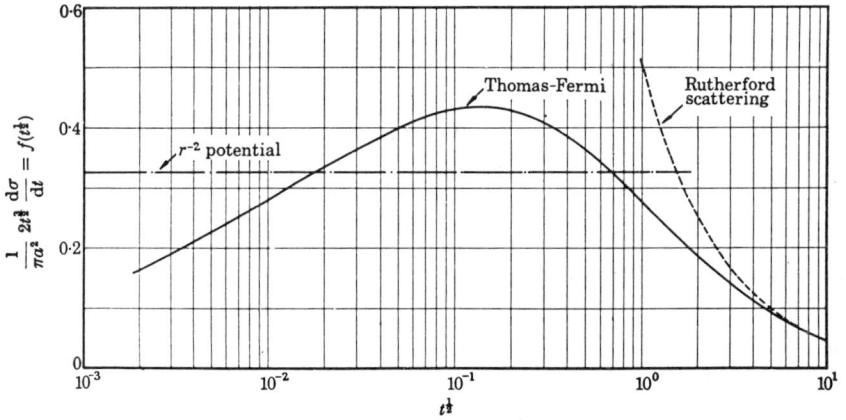

FIG.5.   Universal scattering function $f(t^{\frac{1}{2}})$ defined by Lindhard et al. [12].

An alternative description of the problem [4] is in terms of the probability distribution of energy transfer values, expressed as a differential scattering cross-section:

$$-\left(\frac{dE}{dx}\right)_{elastic} = N \int_{T_{min}}^{T_m} T d\sigma(E, T)$$

where T_{min} is the minimum energy transfer.

To handle the Thomas-Fermi potential in this expression Lindhard et al. [11] adopted an approximation method. A new parameter t was introduced, defined by $t = \epsilon^2 \sin^2(\theta_{cm}/2)$, and t is proportional to the product ET. The approximation formula for the differential cross-section:

$$\frac{d\sigma}{dt} = \frac{\pi a^2 f(t^{1/2})}{2t^{3/2}}$$

is claimed to be valid to an accuracy of about 20%. The function $f(t^{1/2})$ is a 'universal scattering function' which has been evaluated numerically and tabulated by Lindhard et al. [12]. It is shown as a function of $t^{1/2}$ in Fig. 5. The integration necessary to obtain dE/dx can now be carried out (numerically).

It is important to note that the energy transfer values T in the elastic collision process are often far greater than those experienced in electron excitation. Elastic collisions therefore dominate the angular dispersion of a directed ion beam and also account for the bulk of the dispersion in ion range, or 'straggling'. The mean square fluctuation in the elastic energy loss is given by

$$\frac{\overline{E^2}}{E} = N \int_0^{T_m} T^2 d\sigma$$

and this can be calculated in the manner described for the first moment of the energy transfer function.

3.3. Range theory

The range of an ion in a solid is frequently more important than knowledge of its instantaneous rate of energy loss. However, the path of an ion is not straight and so we must distinguish between the total range R_{tot} (measured along the path) and the projected range R_p (measured perpendicular to the target surface). Obviously,

$$R_{tot} = \int_0^E \frac{dE}{-(dE/dx)}$$

and Lindhard and Scharff [9] have provided approximation formulae for the ratio R_p/R_{tot}.

The numerical calculations of Lindhard et al. [11] provide the most useful set of data for estimating ion ranges. This can be done from the 'universal plots' of reduced range ρ versus reduced energy ϵ, for various values of the electronic stopping parameter $K(Z_1, Z_2, M_1, M_2)$. Alternatively, one of the many compilations of range-energy data derived from the so-called LSS theory may be used. The most extensive so far published is that due to Smith and Stephen, given as an appendix in the book 'Ion Implantation' by Dearnaley et al. [13]. This provides the projected range and its standard deviation as a function of energy for all possible ion-target combinations in intervals of 10 units in Z. Other values are obtained by linear interpolation.

3.4. Experimental results

    Most experimental results for range measurement in non-crystalline
materials agree with the LSS predictions to the accuracy which may be
expected in view of the approximations. Thus the data of Davies [14] and
his co-workers using anodic oxidation of metals such as aluminium to achieve
an oxide of controlled thickness, combined with radiotracer techniques,
show reasonably good agreement with theory, especially at the lower energies
where elastic interactions are most important. However, recent results of
Neilson and Thompson [15], who implanted a large variety of ions at 100 keV
into aluminium, indicate that shell effects could be important in elastic
stopping. The smoothed-out Thomas-Fermi model averages out these
effects but in some instances the discrepancy may rise to a factor of two.

4. CHANNELLING

    So far, we have considered only amorphous target materials, but in
1963 came the surprise that correlated small-angle scattering could lead
to a distortion of the normally random distribution of impact parameters,
and hence an anomalously long range. This effect was termed ion 'channelling',
and it arises through the elastic interaction with the nuclear potential, which,
being repulsive, will tend to steer the particles away from an ordered array
or row of scattering centres.
    The systematic increase in mean impact parameters has several
important consequences. Both the elastic and inelastic contributions to the
energy loss are reduced, but the elastic part is more sensitive to impact
parameter and so falls to the greatest extent. Thus for xenon ions channelled
in tungsten crystals, Eriksson et al. [16] showed that electronic stopping
becomes equal to elastic stopping at an energy of 3 keV, whereas in the case
of non-channelled xenon ions this equality occurs at about 2.7 MeV.
    Another important consequence of the steering of ions away from atomic
rows is the fact that cross-sections for nuclear reactions and for wide-
angle scattering, both of which involve a close encounter between the nuclei,
are suppressed. As a crystal is oriented so that either major planes or
axes are in line with a collimated ion beam, so the yield of backscattered
particles and nuclear reaction products passes through a minimum. This
effect can be used to determine the crystal orientation to a very high
accuracy. Lindhard [17] showed that the critical angle for channelling,
i.e. within which the phenomenon could occur, is given by $\theta_c = (2Z_1Z_2e^2/ED)^{1/2}$,
where D is the interatomic distance for the appropriate atomic rows. The
centre of the distribution pattern of particles emerging from a thin crystal
in a transmission experiment can be determined to a much higher accuracy
than θ_c (Fig. 6).
    We shall see then that an important consequence of channelling in the
ion implantation of single-crystal semiconductors stems from the possibility
than an ion injected in a direction $< \theta_c$ to any significant atomic row may
suffer scattering into a channelling direction. Its range is then increased
significantly, and such particles contribute to a 'tail' on the range
distribution [18]. Such scattering into channels cannot occur in a perfect
lattice, but in practice there are sufficient defects, strain, etc., to provide

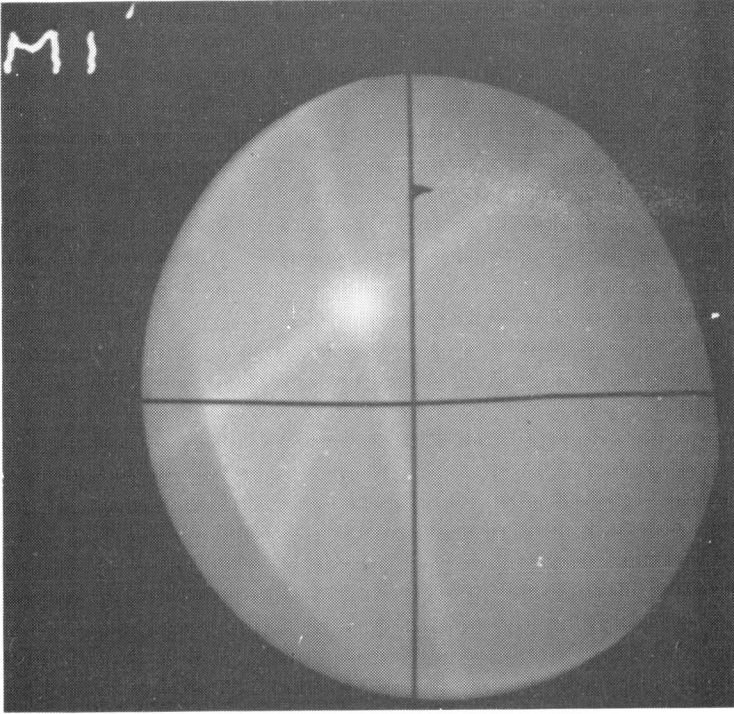

FIG. 6. Transmission of channelled protons through a 300-μm silicon crystal, recorded photographically as a means of determining crystal orientation.

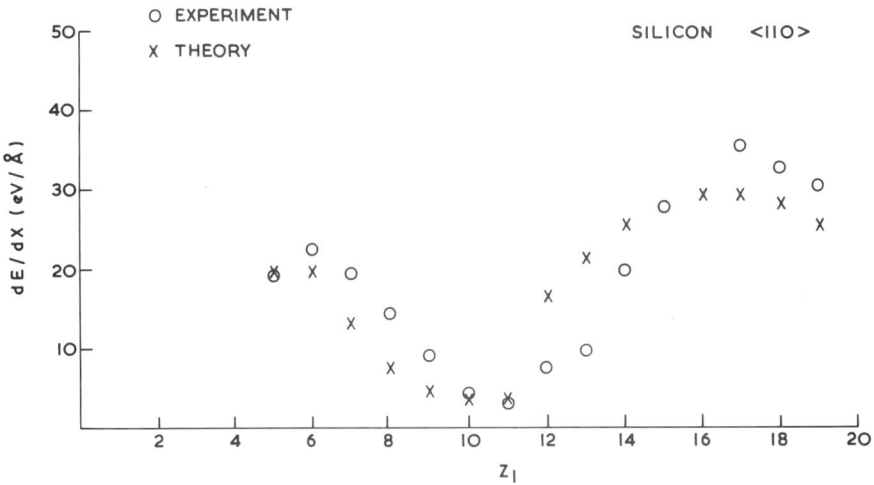

FIG. 7. Results of a calculation of the stopping power of silicon for various ions channelled along the ⟨110⟩ direction, compared with the experimental results of F. H. Eisen (after Briggs and Pathak [20]).

scattering centres which are removed from lattice sites. As is to be
expected, the extent of this effect varies from one crystal to another and
therefore the implanted ion distribution cannot be fully controlled.

There is as yet no complete theory of the range distributions of channelled
ions in crystals. For the case of the 'best-channelled' particles, i.e. those
moving in paraxial trajectories, some success has been achieved [19] by a
modified Firsov theory in which the dominating influence is the extent of the
overlap of electron wave functions for ion and target atoms. A more
sophisticated approach has recently been advanced by Briggs and Pathak [20]
(Fig. 7), who considered the scattering of electrons in the target atoms
caused by the motion of the ion. To arrive at the distribution in range of
channelled ions it is also necessary to calculate the 'de-channelling' or
scattering of ions out of channelled trajectories (the inverse of the effect
mentioned above). This depends upon the precise experimental conditions,
crystal perfection, etc., and so far only a qualitative understanding of the
processes involved has been achieved. Nevertheless, it is possible to
interpret experimental data and to understand the influence of temperature,
ion species, energy, etc.

5.   RADIATION DAMAGE

The large energy transfers from a moving ion to target atoms will
obviously disrupt the order in a bombarded crystalline specimen. Even in
an amorphous material, ion bombardment will induce order-disorder
transformations, strain, gas bubbles, etc., which may be classed as
damage. In relatively insulating media it appears that the intense ionization
along the track of an ion can induce atomic re-arrangement. Such effects
are not necessarily always deleterious, since they can be utilized in particle
track detectors or to control the degree of order within surface layers of a
material.

The simplest damage interaction is the displacement of a single atom
from a lattice site to a nearby interstitial position as a result of an elastic
collision. This combination of two 'point defects' constitutes a 'Frenkel
pair' and, before considering their subsequent behaviour, it is important
to be able to estimate the number and distribution of these primary defects
and the secondaries that result from them.

There is a threshold energy, called the displacement energy E_d, which
must be transferred to the lattice in order to create a Frenkel pair.
Typically $E_d \sim 15$ eV, but it will be anisotropic due to the effects of directed
interatomic bonds, nearest-neighbour distances, etc., in a crystal. The
recoiling target atom will, in general, have enough energy to create a
cascade of secondary displacements, but when the energy of a moving ion
falls below $2E_d$ it becomes possible for the primary atom merely to replace
the struck one (a replacement collision) giving no net displacement. Below
the energy E_d, by definition, no displacements occur. Thus the total number
of Frenkel pairs produced, $\nu(E)$, rises from zero at $E = E_d$ to unity at
$E = 2E_d$. Then, since we are mainly considering interactions at very low
energies (towards the end of the collision cascade) we may assume hard-
sphere collisions to derive the probability $P(E_1, E_2)$ that an ion with energy
E_1 will give rise to a recoil energy between E_2 and $E_2 + dE_2$. Since we are
here dealing with symmetrical collisions between target atoms, the maximum

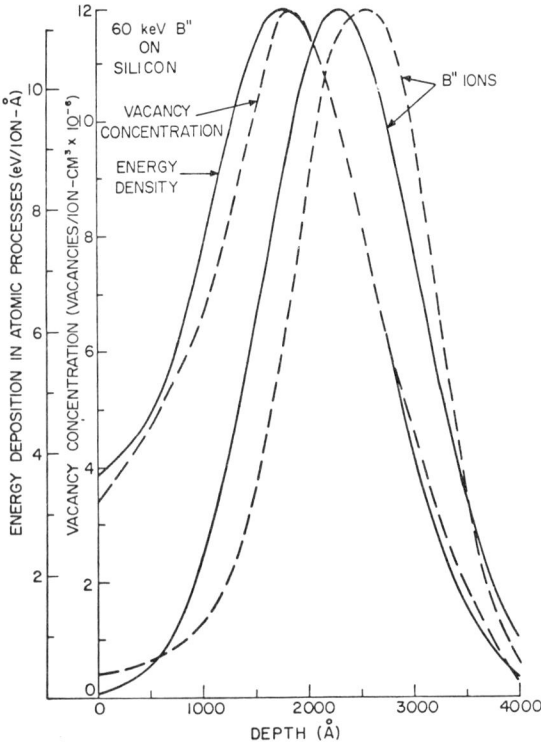

FIG. 8.   Depth distribution of damage in silicon produced by 60-keV boron ions, as calculated by Brice [22].

energy transfer is E_1, the minimum transfer zero. There is a uniform
probability distribution between, for the hard-sphere case, so $P(E_1, E_2)$ =
= dE_2/E_1. One must then (Kinchin and Pease [21]) integrate the number of
displacements produced by a recoil of a given energy multiplied by the
probability of this recoil energy occurring. The average number of further
displacements produced by a recoil of energy $E_1 \gg E_d$ is

$$\int_{E_d}^{E} \frac{\nu(E_1)\, dE_1}{E}$$

Similarly, the other recoiling particle, with energy $E_2 = E - E_1$, contributes

$$\int_{E_d}^{E} \frac{\nu(E_2)\, dE_2}{E}$$

Then the trial solution $\nu(E) = kE$ is found to satisfy the integral equation $\nu(E) = (2/E) \int \nu(x)\,dx$ with $K = 1/2E_d$. Thus Kinchin and Pease arrived at the formula:

$$\nu(E) = \frac{E}{2E_d} \quad \text{for} \quad E \gg E_d$$

Many more sophisticated calculations have since been made, with realistic interatomic potentials, but they lead to essentially the same result.

At higher ion energies it is not permissible to neglect inelastic energy losses. We have seen that Lindhard et al. divide the energy loss into an elastic part and an inelastic part which can be calculated as a function of ion energy. The same thing can be done for the recoil particles, and in this way Brice [22] has derived the depth distribution of displacements for a number of important ion-target combinations (Fig.8). The peak of the damage distribution typically occurs at about 70% of the mean projected ion range.

5.1. Defect migration

After point defects have been created they will, at all but the lowest temperatures, tend to migrate and cluster. Typical activation energies for migration in crystals are 0.5 eV for vacancies and 0.2 eV for interstitials. Within a few atomic layers of the surface, defects interact with an effective 'image' defect via the elastic strain and thereby move to the surface. Within the bulk, material defects may recombine or aggregate into a lower strain-energy configuration generally consisting of small sheets or platelets of defects oriented along a crystal plane.

Covalently bonded semiconductors, such as Si or GaAs, behave quite differently to metals under irradiation. This is probably because the bond lengths and bond angles are rather closely determined and hence it is difficult for a bombarded semiconductor to restore itself. By contrast, metallic bonding is dependent more upon the close packing of positively charged ions in a sea of valence electrons. Hence in the former case the defect clusters surrounding an ion track do not collapse and after heavy irradiation these disordered regions overlap and the semiconductor becomes almost totally amorphous. In contrast, the atoms in a metal can each be displaced many times without the structure becoming amorphous: instead, a large number of dislocations are produced with microcrystalline material between.

Defects may associate with impurity centres in a crystal and, particularly in semiconductors such as silicon, there has been a great deal of investigation of such complexes as the oxygen + silicon vacancy pair (or A-centre) and the phosphorus + silicon vacancy pair (or E-centre). Since these defects, as well as intrinsic ones, are electrically charged and lead to electron energy levels within the band gap, they will have important effects in the electrical properties of semiconductors.

Much of the research on radiation damage has been directed either to achieve radiation-resistant military devices, or to improve the working life of nuclear reactors. Ion implantation now represents a further reason for understanding the phenomena, while ion bombardment is proving a powerful means of simulating in a short time the effects of, say, prolonged neutron irradiation.

5.2. Observation of damage

    In many semiconductors and insulators, radiation-induced defects
modify the optical properties of the material, so that damage is visible to
the eye. Visual observation of the onset of significant damage has frequently
proved a useful guide and has allowed some striking demonstrations [23] of
the reduction of damage in channelled implantations.
    A more sophisticated form of this is to measure the infra-red absorption
and to correlate the frequencies of maximum absorption with local-mode
vibrations of defects or defect complexes. This is particularly useful in the
case of low-mass impurities.
    In silicon and a few other materials, a very important spectroscopic
method has been the technique of electron spin resonance, or electron paramagnetic resonance, in which transitions are induced between Zeeman levels
of a paramagnetic defect in the crystal. Analysis of the EPR spectra under
different conditions of temperature, orientation and uniaxial stress allows
the local configuration of the defect to be deduced [24]. Obviously, this
method is restricted to those defects with unpaired electron spins, and
problems are encountered in polyisotopic materials.
    Transmission electron microscopy offers a very general method of
imaging regions of strain in a lattice by their effect in electron diffraction.
Platelets of interstitials or vacancies give rise to a strain distribution
localized to the periphery and hence appear as 'loops' in electron micro-
graphs. Although progress is being made in image resolution, the technique
is still effectively limited to the study of defect clusters of dimensions 6 Å
or above. It is feasible to create damage by ion bombardment within an
electron microscope, and then to examine the annealing behaviour of visible
defects. One difficulty has arisen from the need in low-voltage microscopes
to use very thin specimens (< 1000 Å thick). Due to image forces, etc.,
there is some migration of defects to the surface over a depth of about 1000 Å,
and hence such thin specimens are not representative of bulk behaviour.
The availability of high-voltage electron microscopes is now removing this
difficulty.
    In metals and semiconductors, defects will modify the electrical
conduction behaviour, and electrical measurements have been used for a
long time as a means of studying bulk damage. It is less easy to apply the
technique to ion-induced damage, which is localized near the surface, due
to the conduction of the substrate. Thin-film specimens must be used and
allowance must be made for the distribution of defects with depth.
    Crystal defects, as already noted, will interfere with the channelled
trajectories of bombarding ions. The energy spectra of backscattered
energetic particles can be used to provide information about the depth and
magnitude of disorder. It is not feasible, as was at first hoped, to derive
quantitatively the number of point defects, since it can be shown that the
strained region which surrounds a vacancy or edge dislocation will contribute
the major part of the dechannelling cross-section. However, since the
technique is sensitive to all defects plus strain it is a useful means of studying
the onset, saturation and annealing of disorder in crystals. Of course, it
is necessary to ensure that the probing bombardment does not itself introduce
significant disorder.
    In some cases, attention is focused upon the strain induced in the
surface of a material rather than on the microscopic defects produced. The

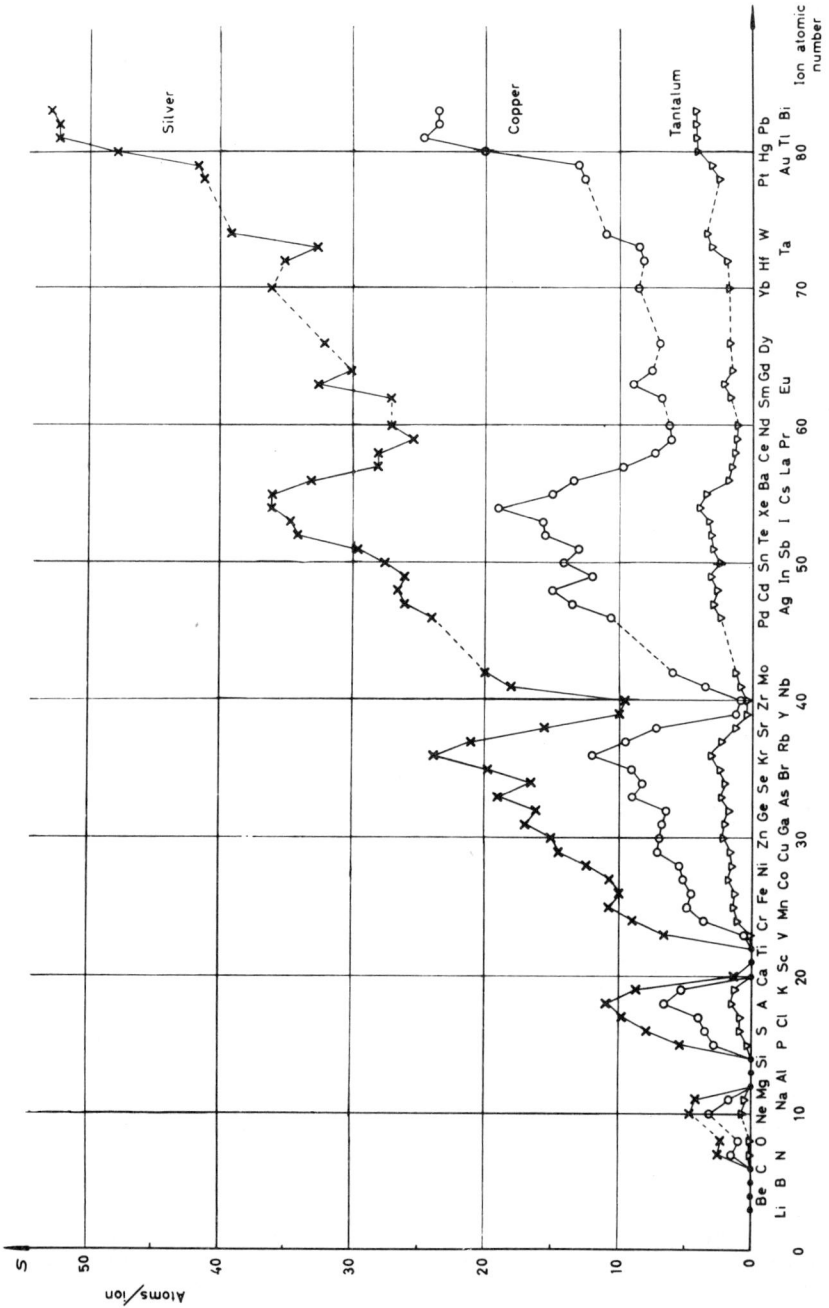

FIG. 9.   The variation of sputtering ratio S as a function of the atomic number of the ion Z_1 for three different target materials (after Almén and Bruce [27]).

volume occupied by irradiated material is usually greater than that of the original, although in silica the order-disorder transformation under irradiation leads eventually to a 20% α-cristobalite component with a volume contraction. These macroscopic volume changes manifest themselves as a lateral compressive (or tensile) stress in the surface of the material. Eernisse [25] has exploited this to observe, by minute deflections of a cantilever reed-like specimen, the stresses induced by ion bombardment. This proves to be an extremely sensitive technique, which is an indication of the large mechanical effects of bombardment, and these have important consequences.

5.3. Bubbles and blisters

When a material is bombarded with an insoluble gaseous species (e.g. argon) and annealed, the gas atoms migrate and aggregate into bubbles, which are easily visible in the electron microscope. Other gas atoms may migrate to the surface, and there have been many studies of the desorption of implanted gases into a high-vacuum test chamber which generally show a variety of desorption peaks corresponding to different original sites and therefore to different activation energies.

In a monoenergetic bombardment there will be a peak concentration of gas bubbles at the depth corresponding to the mean projected range. The consequent swelling imposes a strain which is relieved by the flow of material to form a blister, which may indeed burst and finally flake away from the surface (exfoliation). These effects are observed only under intense bombardment (say 10^{18} ions/cm^2), but these conditions must be expected in fusion reactors and certain other technologically important situations. There is therefore a good deal of study devoted to the mechanism of blister formation and ways of minimizing it.

5.4. Sputtering

This is the name given to the process of ejection of material from a target under ion bombardment. It is widely used for cleaning specimens before examination and as a means of depositing thin foils.

An early suggestion, by von Hippel [26], was that sputtering takes place by thermal evaporation from locally heated spots surrounding each ion impact site. It is now realized that momentum transfer in a series of binary collisions very close to the target surface is the important factor. However, if the material is near to its melting point 'thermally-assisted sputtering' can increase the yield of low-energy emitted atoms.

We define the sputtering yield S as the mean number of atoms ejected per incident ion. S is a function of the ion energy, ion and target masses, angle of incidence and temperature. It rises from a threshold at about 10 eV up to a maximum at an energy which increases with the mass product M_1M_2.

In a classic series of experiments, Almén and Bruce [27] determined S by direct weighing on a microbalance of the material sputtered from a target by a known integrated ion current. Their results (Fig. 9) show marked dips in sputtering yield for certain bombarding species but a general upward trend, in agreement with theory.

The energy distribution of sputtered particles has been measured, by Farmery and Thompson [28], by a time-of-flight experiment using a rapidly rotating collector rotor synchronized with a pulsed ion beam. The ejected

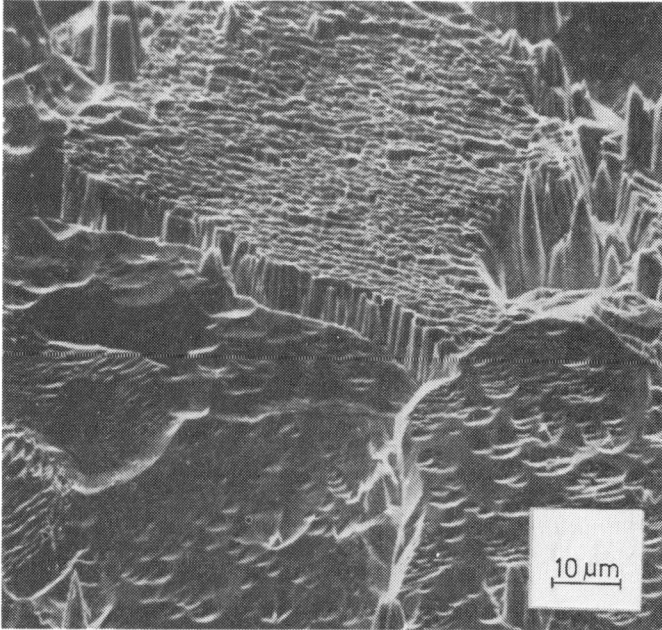

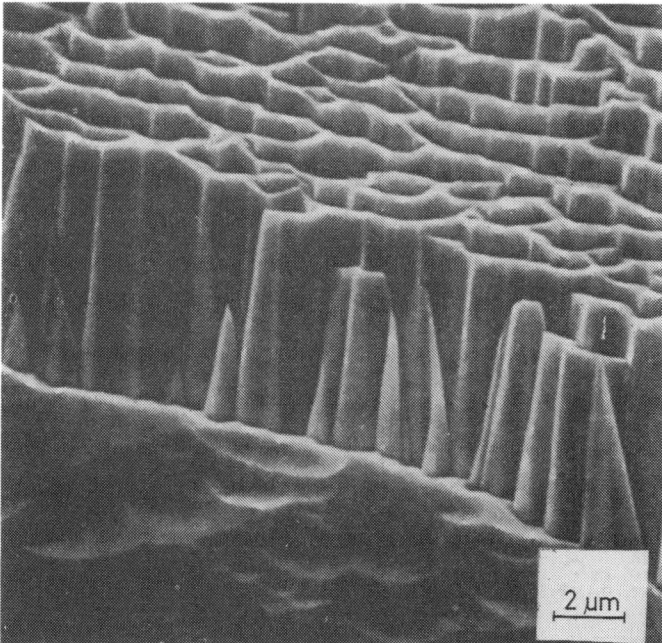

FIG. 10.   Scanning electron micrograph of a polycrystalline tungsten surface after bombardment with
4×10^6 krypton ions per cm^2 at 150 keV (from Staudenmaier [31]).

energy distribution shows a predominance of low-energy particles (mostly < 1 eV) with a fall-off as E^{-2}. Exceedingly few particles emerge with any significant fraction of the incident ion energy.

The theory of sputtering (Sigmund [29]) considers the series of collisions which the impinging ions undergo, creating a cascade of recoiling particles. Both the incident ion and sufficiently energetic cascade atoms have a finite possibility of being scattered back through the surface. The calculation of sputtering yield S involves four steps:

(a)   Determination of the energy deposited by the ion as a function of depth, via the theory of Lindhard et al. [11];

(b)   Conversion of this energy distribution into a number density of low-energy recoil atoms;

(c)   Determination of the number of such particles able to reach the target surface;

(d)   Calculation of the number which are able to overcome surface binding forces and so escape.

Recent experiments by Andersen and Bay [30] on the initial sputtering yield have given results in excellent agreement with Sigmund's theory. The undulations previously observed in Almén and Bruce's data [27] are evidently due to the change in composition brought about by a marked build-up of implanted species creating a surface of low sputtering yield: Almén and Bruce's experiments were lengthy ones in which this process would have reached saturation.

For a similar reason, the sputtered deposition of alloys or compounds is made difficult by the preferential sputtering of one component, leaving an excess of the constituents with low values of S.

Under prolonged bombardment, some interesting changes in surface topography occur on target specimens. Grooves (usually running parallel to the beam direction), mounds and cones may appear, even in single-crystal specimens. In polycrystalline material the differences in sputtering yield of different crystal facets (due largely to different surface binding forces) lead to an etching of the grain structure, which has often been applied for metallography.

The surface topography of ion-bombarded targets is best studied in the scanning electron microscope, which has revealed some complex formations of narrow cones and crevasses (Fig.10) [31]. It is believed that these can be initiated by the presence on the surface of impurities (hydrocarbon or oxide) which resist sputtering. As the surface is eroded around these sites, a structure develops so as to expose those faces having the maximum sputtering yield. Since this is related to the minimum surface cohesive energy, the theory applicable to the growth of crystals under thermodynamic equilibrium has proved relevant (Barker [32]). High cone angles develop because under these conditions the secondary collision cascade lies near to the surface and hence S is large.

6.   ION BACKSCATTERING

This is a means of surface analysis which makes use of an energetic beam of low-mass ions (such as helium nuclei). The energies preferred range from a few hundred keV to several MeV.

The method was used by isolated investigators even twenty years ago, e.g. by nuclear physicists who realised that they could confirm the composition of thin-film targets in this manner. The first application recorded was the analysis of films deposited on to the walls of gun barrels by Sylvan Rubin, who equipped a van de Graaff accelerator laboratory specifically for this purpose around 1954 [33].

In those early days the technique was limited by the need to use tedious magnetic analysis of the scattered particle energies, and it was not until silicon surface-barrier detectors became available in the 1960s that the entire energy spectrum could be recorded simultaneously. This achieved up to a factor of 100 improvement in detection efficiency.

In September 1967 an α-scattering experiment was carried out on the lunar surface by Surveyor 5, by which α particles from a 1-kCi californium source were scattered into large-area silicon detectors. The results of this experiment agreed remarkably well with subsequent chemical analyses of moon samples returned by Apollo 11.

In the laboratory it is customary to use a collimated ion beam of intensity 10^{-9} - 10^{-7} A to bombard the specimen surface, and to measure the energy spectrum of ions scattered through some large angle, typically 165°. This technique allows one to determine the masses of the elements present in the target, to determine more or less quantitatively their relative abundance, and to study the distribution of the composition over depths ranging from about 200 Å to a few μm. Furthermore, the backscattering is sensitive to the crystal structure, via the phenomenon of channelling, and hence it is possible to make observations of crystalline disorder, crystal structure and symmetry.

The ion backscattering technique complements in many respects the electron beam or X-ray techniques (LEED, ESCA, Auger spectroscopy, etc.) which are discussed elsewhere in these Proceedings. Frequently these are sensitive to the surface 20 Å or so but can convey information about the chemical combination of surface atoms. Ion backscattering (IBS) is sensitive to greater depths, but cannot distinguish between atoms of closely spaced masses (e.g. Ni and Cu), nor is it sensitive to chemical binding. IBS yields quantitative results and, while it is particularly sensitive to the heavier species, it can conveniently be supplemented by nuclear reaction spectroscopy (NRS) for the lightest species. The depth probed by IBS (up to a few μm) is often important in the study of semiconductor devices, corrosion films, contaminant layers and electrodeposited films, interdiffusion of coatings, insulator thin-film technology, and so on, and it is in these fields that experiments have so far been carried out.

The theory of ion backscattering is simple as it involves merely an elastic collision process. By classical mechanics we have a relation between the scattered energy E_s and incident energy E_i:

$$E_s = k_s(M_1, M_2, \theta) \, E_i$$

where

$$k_s = \left[\frac{M_1 \cos \theta + (M_2^2 - M_1^2 \sin^2 \theta)^{\frac{1}{2}}}{M_1 + M_2} \right]^2$$

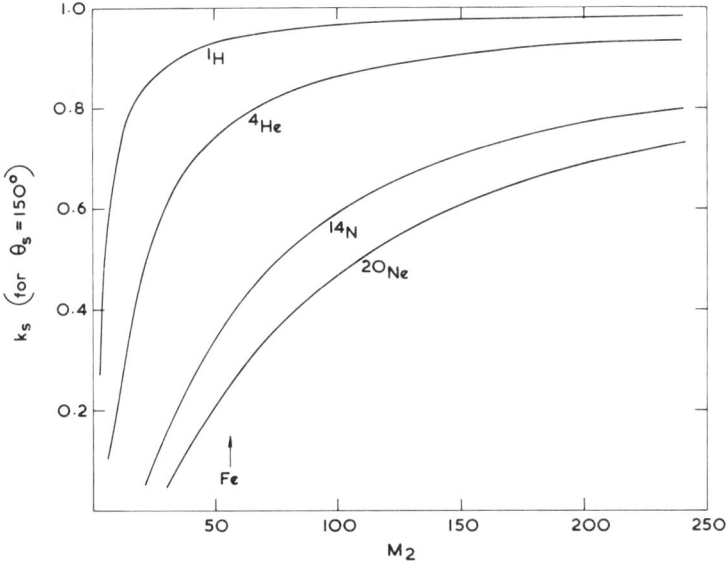

FIG. 11. Ratio of backscattered ion energy to incident energy as a function of ion and target masses.

Here M_1 and M_2 are the ion and target atomic weights and θ is the laboratory angle of scattering:

$$\text{As } \theta \to 2\pi, \quad k_s \to \left[\frac{M_2 - M_1}{M_2 + M_1} \right]^2$$

and this is often a convenient approximation.

The behaviour of k_s as a function of M_2 and for various ion masses M_1 is shown in Fig. 11.

Ions penetrating the surface of the target lose energy (almost entirely by inelastic process) and may make a close collision, resulting in back-scattering, at a depth x. There will be a second energy loss as the scattered particle travels back to the surface on its way to the detector. The detected energy is then

$$E_{det} = \left[E_0 - \int_0^{\frac{x}{\cos \theta_1}} \frac{dE}{dx} \, dx_1 \right] k_s - \int_{\frac{x}{\cos \theta_2}}^0 \frac{dE}{dx} \, dx_2$$

where E_0 is the ion beam energy and θ_1, θ_2 are the angles of incidence and emergence with respect to the surface normal (usually $\theta_1 = 0°$). Thus a solid specimen will give rise to a continuum energy spectrum, with a cut-off at $E = k_s(M_2) E_0$ for each species in the target. Impurities which have a

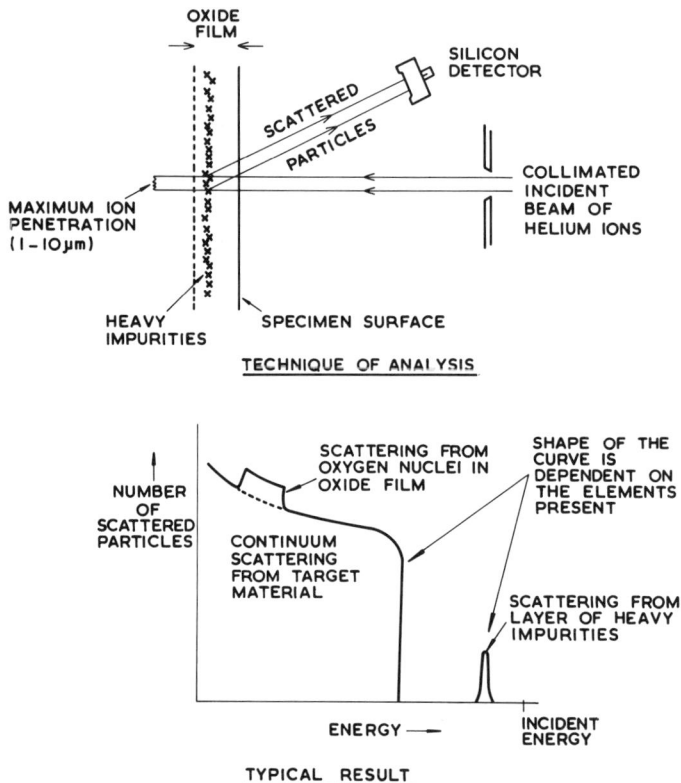

OXIDE
FILM

SILICON
DETECTOR

SCATTERED
PARTICLES

MAXIMUM ION
PENETRATION
(1 - 10 μm)

COLLIMATED
INCIDENT
BEAM OF
HELIUM IONS

HEAVY
IMPURITIES

SPECIMEN SURFACE

TECHNIQUE OF ANALYSIS

NUMBER
OF
SCATTERED
PARTICLES

SCATTERING FROM
OXYGEN NUCLEI IN
OXIDE FILM

SHAPE OF THE
CURVE IS
DEPENDENT ON
THE ELEMENTS
PRESENT

CONTINUUM
SCATTERING
FROM TARGET
MATERIAL

SCATTERING FROM
LAYER OF HEAVY
IMPURITIES

ENERGY ⟶

INCIDENT
ENERGY

TYPICAL RESULT

FIG. 12.   Typical ion backscattering experiment.

restricted depth distribution, e.g. on the surface, or implanted to a uniform
depth, will give rise to a peak in the scattered energy spectrum (Fig.12).
From a single measurement it is not feasible to discriminate between a
light impurity at a given depth and a heavier impurity nearer the surface,
but by the use of different probing ion beams the two can be distinguished.

In most cases the scattering cross-section, which determines the
detection probability, is given simply by the Rutherford scattering formula:

$$\sigma_R \propto \left(\frac{Z_1 Z_2}{E_0} \right)^2 \ \mathrm{cosec}^4 \frac{\theta}{2}$$

and hence the sensitivity increases as the square of the target atomic number.
This means that for heavy species such as gold it is possible to detect as
little as 10^{12} atoms/cm^2 on a surface, i.e. about 10^{-3} monolayers. Light
impurities give far less scattering, and the yield from them will be superimposed on the continuum spectrum from the host material. It is then that
nuclear resonance techniques are valuable (see below). The relative

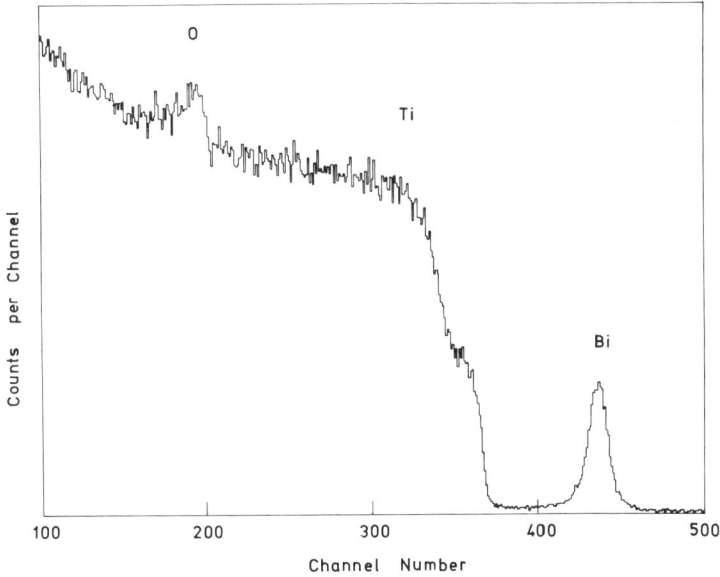

FIG. 13.   Energy spectrum of 1.5-MeV He⁺ ions backscattered from a titanium target, ion-implanted with Bi⁺ ions and subsequently oxidized.

proportions of different elements in the target can be compared by applying the Rutherford scattering formula, but sometimes the fact that this is a centre-of-mass cross-section has been overlooked. The laboratory cross-section is given to a high approximation by

$$\sigma_R \propto \left(\frac{Z_1 Z_2}{E_0}\right)^2 \left[\operatorname{cosec}^4 \frac{\theta}{2} - 2\left(\frac{M_1}{M_2}\right)^2\right]$$

and the correction term is often significant. An additional factor, also frequently overlooked, stems from the fact that the emergent energy loss is a function of k_s and hence the continuum spectrum from a low-mass target atom is spread over a larger energy range than that from a heavier species. This can easily be seen in thin-film scattering, e.g. from self-supporting Al_2O_3, since the peak from ^{16}O is broader than that from ^{27}Al [34]. This effect disappears if the ion energy corresponds to the peak of the curve of dE/dx versus E and is inverted below this energy. As a rule both these correction factors diminish the yield from light target atoms compared with heavier ones.

As an example of ion backscattering, Fig.13 shows the spectrum of 1.5 MeV He⁺ ions scattered from a bismuth-implanted titanium specimen, subsequently oxidized. The spectrum reveals the implanted bismuth, and the 3000 Å oxide film, in which it can be seen that the Ti:O ratio increases with depth. The depth resolution of the technique is determined by the detector energy resolution, and a typical resolution ΔE (full width at half-maximum) of 15 keV corresponds to a depth resolution Δx of about 200-250 Å.

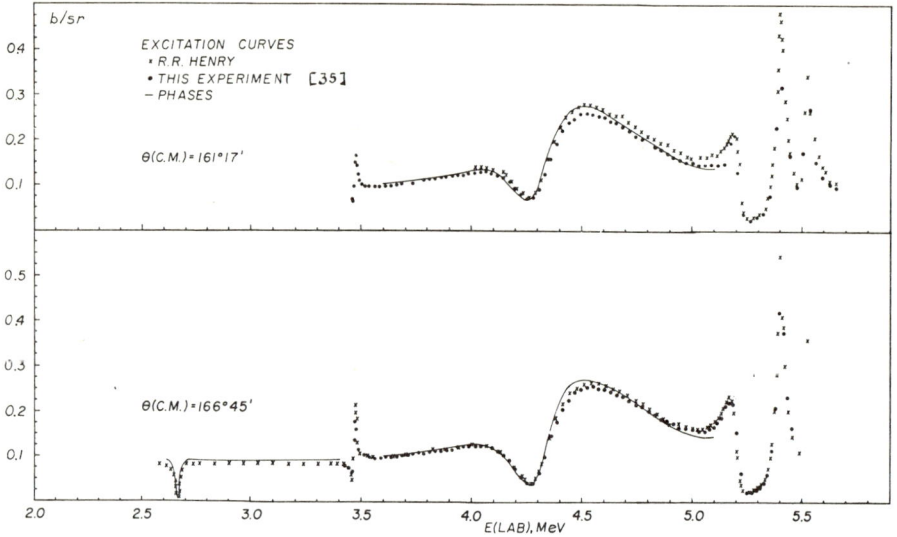

FIG. 14.   Differential elastic scattering cross-section at two backward angles for the ^{16}O(p,p) process (from Harris et al. [35]).

Magnetic analysis, although slow, can provide an energy resolution about ten times better than this, i.e. a depth resolution of 20 Å; this value approaches the depth sensitivity of ESCA and Auger techniques. Some laboratories are installing such magnetic analysers for IBS analysis.

It is clear from Fig.13 that the backscattered yield from oxygen atoms is small, due to the low atomic number compared with most target materials. However, the yield can be increased by taking advantage of nuclear scattering resonances which occur due to the ease with which light bombarding ions can surmount the Coulomb barrier, enter the nucleus, and excite a resonance reaction. Thus at a He$^+$ energy of 2.9 MeV the scattering cross-section is about 30% greater than Rutherford. A much bigger increase is achieved with protons. Figure 14 shows the measured proton elastic scattering cross-section from ^{16}O [35], and at a proton energy of 4.55 MeV the cross-section is about 50 times σ_R over a useful range of energy and scattering angle. This has been utilized by Dearnaley [36] in the examination of thin oxide films on relatively heavy metals such as zirconium; as Fig.15 illustrates, the increased sensitivity more than overcomes the difference in Z^2 for ^{91}Zr and ^{16}O.

These scattering resonances were chosen to be particularly broad so that there would be a uniformly high cross-section for backscattering throughout the films examined. In other cases, it is desirable to be able to probe for a particular atomic species at various depths below the surface. Then a very narrow resonance reaction is preferred, so that the narrow band of energies corresponding to a high cross-section will occur only at a sharply defined depth. Charged-particle resonances such as those which occur in the ^{18}O(p,α), ^{15}N(p,α), ^{27}Al(p,γ) and other reactions have been

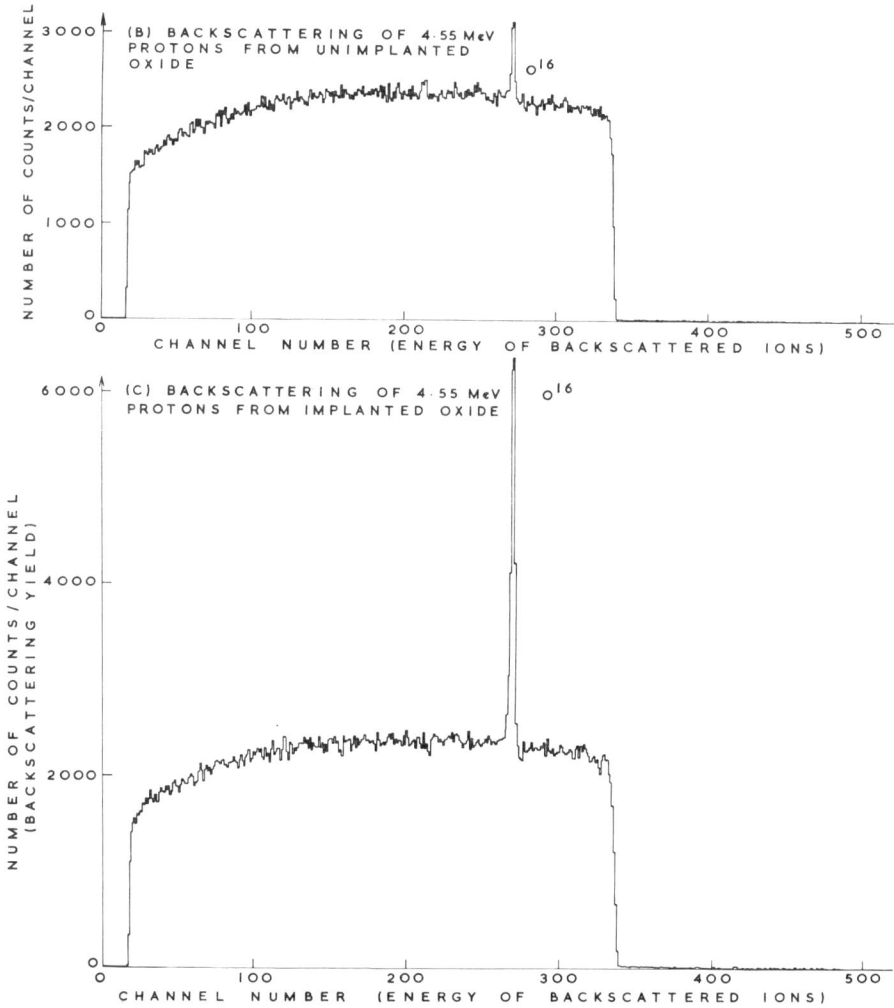

FIG. 15.   Proton backscattering spectrum at 4.55 MeV from a specimen of zirconium, ion-implanted with Cu$^+$ ions and oxidized (from Dearnaley [36]).

used [37] for analytical purposes. Amsel and his colleagues have thus explored the mechanism of anodic oxidation by anodizing materials first in ordinary water and then in water enriched in ^{18}O. By tracing the depth at which the ^{18}O is taken up, by measuring the α-particle yield as a function of proton energy, the anion and cation transport processes can be compared. Similar studies in thermal oxidation have been made by Barnes et al. [38].

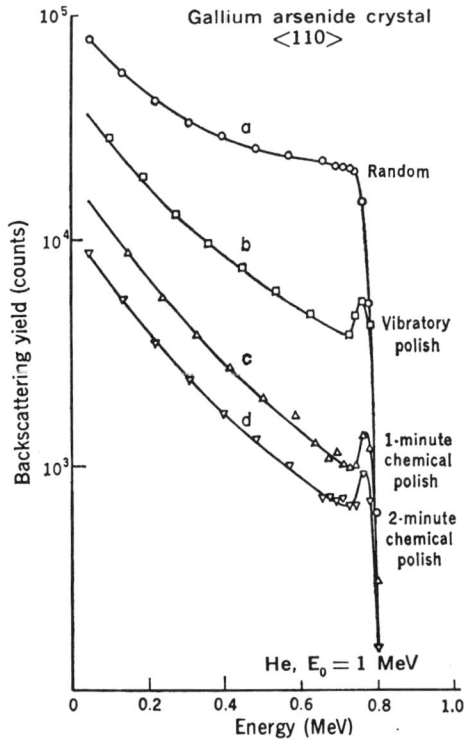

FIG. 16.   Channelled He[+] backscattering spectra from a crystal of gallium arsenide subjected to different surface-preparation techniques (after Whitton [39]).

6.1.   Channelled backscattering

In a crystalline target in which a low-index axis or plane is closely aligned to a well-collimated probing beam, a high proportion of energetic, low-mass bombarding particles will be channelled. We have seen that a consequence of this is that close collisions are suppressed and therefore the probability of ion backscattering through large angles is greatly reduced, in some cases by factors of 50 or more.

Clearly the extent of this reduction in backscattered yield will be dependent upon the crystal perfection. If atoms are displaced from their lattice sites by mechanical or radiation-induced damage, the yield will be higher. Thus channelled backscattering can be used to measure crystal quality and the amount of residual disorder which remains after surface preparation. Figure 16 shows an example, from the article by Whitton [39], of this technique applied to gallium arsenide. If the disorder is introduced by ion bombardment, then the onset and annealing of this damage can very easily be measured by ion backscattering.

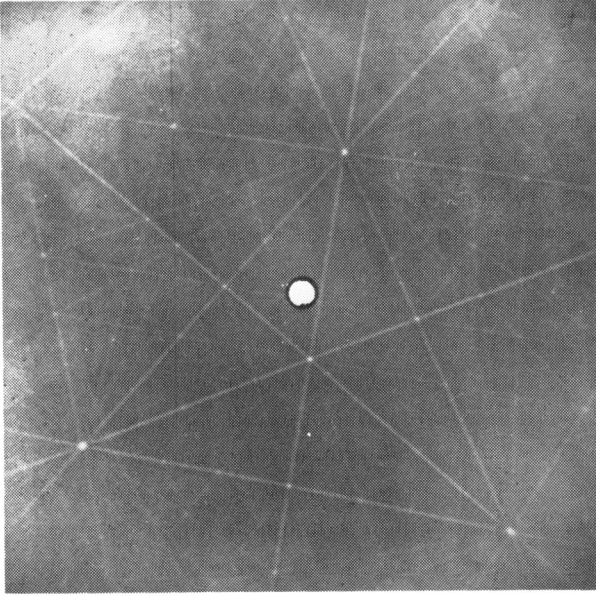

FIG. 17.   Photograph of the distribution of 1.5-MeV He$^+$ ions backscattered from a crystal of germanium, recorded in cellulose nitrate film.

It must be borne in mind that the so-called surface peak that occurs near the high-energy cut-off in the continuum spectrum can arise from three distinct mechanisms:

(a)   The irreducible scattering due to close collisions with atoms at the ends of the atomic rows. The amount of this scattering can be estimated by a cross-section equivalent to that of a circle of radius a_{TF}, the Thomas-Fermi screening radius.

(b)   Scattering by atoms which are displaced by residual mechanical disorder, dislocations, etc.

(c)   Atoms which are bound into a surface corrosion film which is not coherent with the crystal substrate.

It is interesting that the magnitude of the surface peak can be studied even in cases when the oxide is much thinner than the experimental depth resolution.

When an ion that has been backscattered, i.e. has suffered a close encounter with a lattice atom, travels in a direction corresponding to a low-index direction it will very soon arrive at a neighbouring atom site. The resulting collision will deflect the particle, which is unlikely to emerge from the crystal in its original direction of motion. This effect is called 'blocking' and manifests itself particularly well in measurements, using radiation-sensitive films, of the angular distribution of ions backscattered from crystals (Fig.17). Such measurements provide very direct information about crystal symmetry and crystal orientation.

6.2. Lattice location

Channelled ion backscattering (CIBS) has been used to determine the lattice location of impurities in crystals. An impurity atom on a substitutional lattice site will be shielded from a channelled probing beam, while an interstitial will be exposed. There is thus a large difference in aligned yield. Initially, it was thought that the lattice location could be determined merely by two measurements, of the random and aligned yield of ions backscattered from the impurity, along a few major crystallographic directions.

However, it was later realized that the flux of channelled particles is not uniform across the channel but rises to a peak along the channel axis. This 'flux peaking' phenomenon is most easily understood by making the reasonable approximation that channelled trajectories are of a saw-tooth or zigzag form, with the maximum amplitude r_{max} determined by the position and angle of entry into the crystal. Well-channelled ions therefore remain near the channel axis, while other ions spend approximately equal amounts of time over radii extending from zero to r_{max}. The flux falls to a minimum at a radius corresponding to the distance of closest approach ($\sim a_{TF}$) to the atomic rows, and will obviously be a maximum at the channel axis.

This effect greatly complicates the determination of lattice location by CIBS, but at the same time it makes possible a very precise measurement simply because the variation in ion flux across the channel provides a scale by which the position of an impurity site can be measured. To achieve this it now proves necessary to make a series of measurements of backscattered yield as a function of incident beam angle, across a series of major axes and planes. These are termed 'angular scans' and they may each occupy many hours of accelerator time. Moreover, the extent of flux peaking varies with depth over the first few hundred Å, because the oscillatory channelled trajectories are initially correlated, so that the axial flux oscillates in intensity before settling to a steady value. In order to determine the lattice location of relatively shallow impurities the best approach seems to be to compare the backscattered angular scans with those derived by a computer simulation of the ion trajectories and therefore of the flux distribution at all depths [40]. The CIBS technique works best for the case of heavy impurities in a light host lattice, but once again the method can be extended by the use of nuclear reaction techniques, e.g. $^{11}B(p, \alpha)$ and $^{12}C(d, p)$, for light impurities.

In this way the location of about sixteen different species in iron and five in nickel have now been determined [41]. Some of these, such as boron and carbon, are interstitial, as expected, others such as Au, Cu and Pb are substitutional, while many show mixed occupancy of two sites, one of which is usually substitutional. Feldman [41] has shown that if these observations are displayed on a so-called Darken-Gurry plot of atomic size versus electronegativity then the substitutional species cluster in the region of the plot corresponding to the parameters for the host, the mixed-occupancy species lie outside, and the interstitial species are still further removed (Fig.18). This is an impressive example of the way in which the CIBS technique can be used to explore the region of validity of Hume-Rothery's rules for alloy formation.

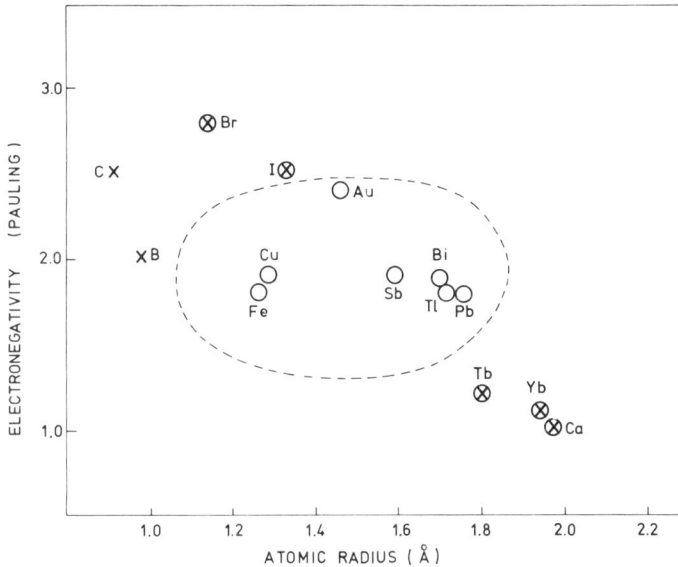

FIG. 18. A Darken–Gurry plot of the lattice site occupancy of various species in iron crystals, determined by the channelling technique (after Feldman [41] and de Waard).

O : 80–100% substitutional;

⊗ : 50–80% substitutional;

x : less than 50% substitutional.

7.   ION IMPLANTATION IN SEMICONDUCTORS

Ion implantation is a process whereby controlled amounts of required foreign species can be introduced into the near-surface regions of a semiconductor for the purpose of producing a device or integrated circuit. In this respect it is similar to the conventional semiconductor processing techniques such as thermal diffusion or epitaxial growth. Since these were already well established as device-manufacturing processes, we must first consider what advantages ion implantation has to offer. Later, we shall see how these advantages are exploited in particular device structures.

The principal advantages of ion implantation are:

(a) The concentration of impurities as a function of depth can be controlled by means of the ion energy and is no longer determined by the mechanism of diffusion. Thus it is possible to implant a buried layer of dopant.

(b) The total amount and purity of material implanted can be accurately controlled. This is not so in the case of diffusion, in which surface phenomena, strain, etc., modify the amount which diffuses in.

(c) It is a comparatively low-temperature process, and therefore the diffusion of unwanted impurities, e.g. Cu, from the surface is minimized.

(d) It is a non-equilibrium process, so that solubility limits can be exceeded, and it is possible to introduce rapidly a material which does not readily diffuse.

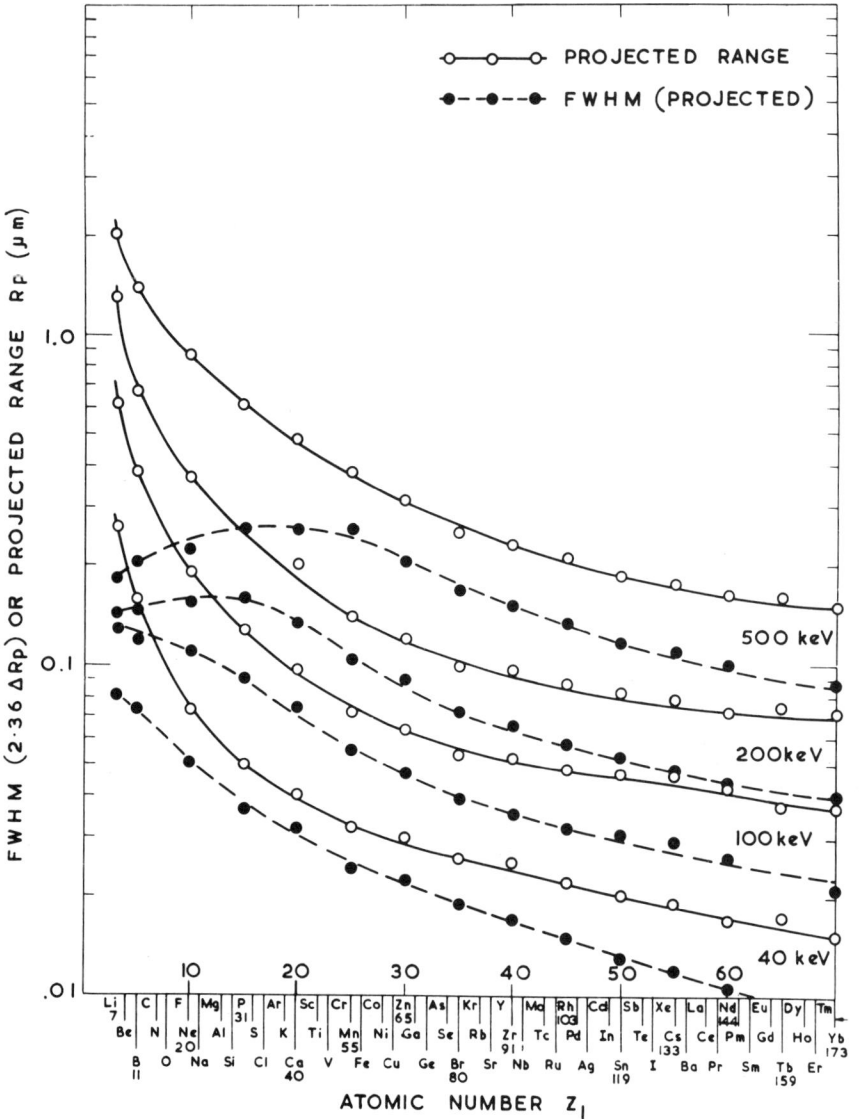

FIG. 19. Ranges and full-widths at half-maximum of the distribution of various ions in silicon, at selected energies, as calculated from the theory of Lindhard, Scharff and Schiøtt.

(e) Ions enter the surface as a directed beam, so that a very high lateral definition of the doped region can be achieved using conventional photo-lithographic masking techniques. It is probably true to say that if ion implantation had not been compatible with photolithographic mask fabrication it would not be much used industrially.

(f) The process is very versatile, in that a single implantation facility can in principle be used for many different ion species. Moreover, the implantation schedule can be varied very easily in a manner which allows the production of short runs of devices.

(g) The process can, in principle, be automatically controlled to a high degree. This may eventually lead to electron beam processing (for mask generation) and ion implantation (for doping) to be used for semi-conductor device manufacture under clean vacuum conditions and without extensive manual operations or wet chemical stages. This, however, is something for the future.

7.1. Ion penetration

The ranges and standard deviations of range in silicon for various ions with energies between 40 and 500 keV are shown in Fig.19. These values are calculated by means of the Lindhard, Scharff and Schiøtt formulas described earlier. They show that, for semiconductor devices of typical dimensions, the lighter doping ions boron and phosphorus must be accelerated to energies between about 50 and 400 keV.

How reproducible is the range distribution, since this will determine the control we have over doping by ion implantation? Figure 20 shows that besides the symmetrical Gaussian distribution expected on LSS theory the experimentally measured profiles show a more or less exponential tail. This can extend, to a significant level of concentration, over nearly half a μm in silicon, and it can be serious in device manufacture because the slope of the tail has been shown [42] to vary by as much as a factor of three between different samples of silicon. Only recently, by radioactive ion transmission experiments, have Blood et al. [43] been able to show conclusively that this tail (in the case of phosphorus implanted into Si) arises entirely by the scattering of ions into channels which are inclined to the beam direction. The variation between crystals is due to differences in crystal perfection. The tail can be eliminated by introducing disorder into the crystal by inert gas bombardment, or by combining implantation with diffusion in a manner which will be discussed.

This knowledge of the factors that determine the penetration of ions into single-crystal semiconductors is of considerable importance in designing a manufacturing process and the machines for performing the implantations.

7.2. Annealing of damage

The ion doses required for semiconductor device fabrication are usually sufficient to introduce considerable disorder, which must be thermally annealed before the chemical doping effects cease to be masked by electrically active defect centres. It is fortunate that this annealing process can be carried out at temperatures which (in silicon) lie below the temperature at which significant diffusion of common dopants occurs. A commonly used anneal temperature is 650°C in silicon, although very high doses may require

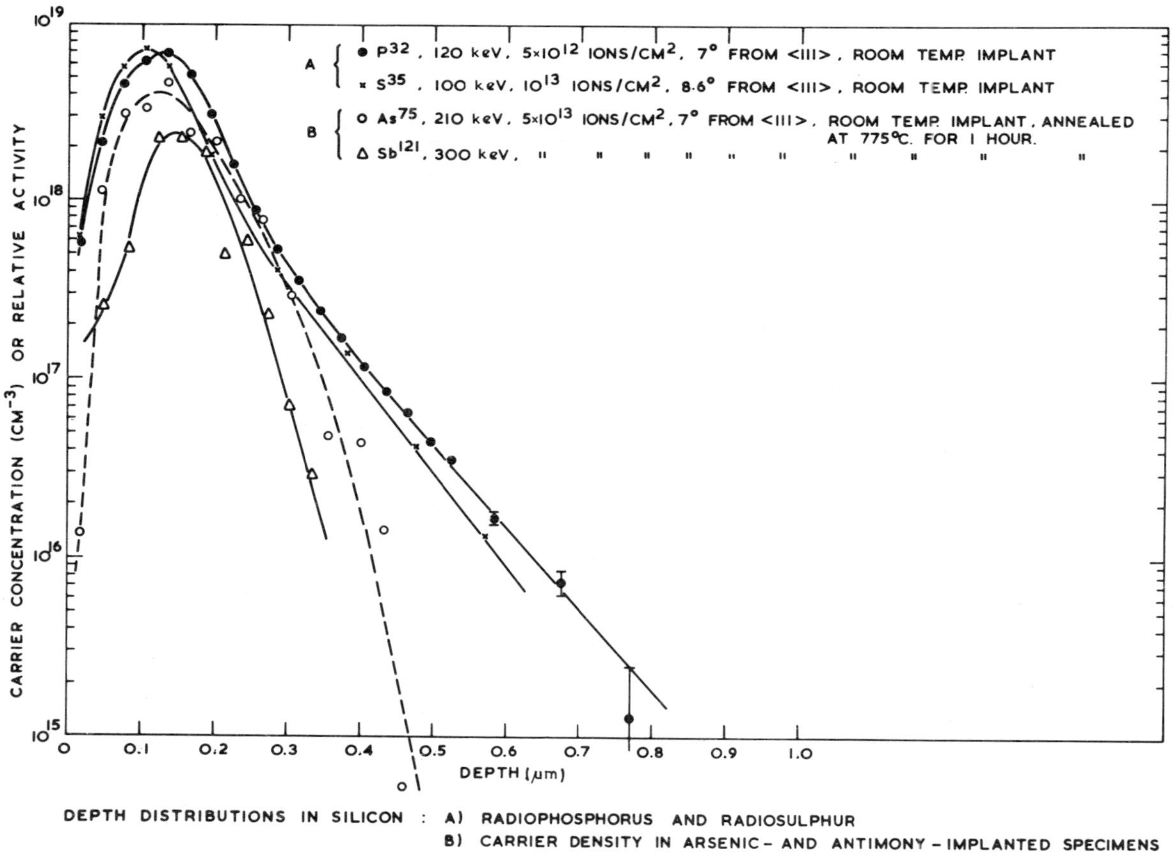

FIG. 20.   Depth distribution of various radioactive ion species implanted into silicon crystals in directions well away from channelling axes or planes. The exponential tail can be seen in each case.

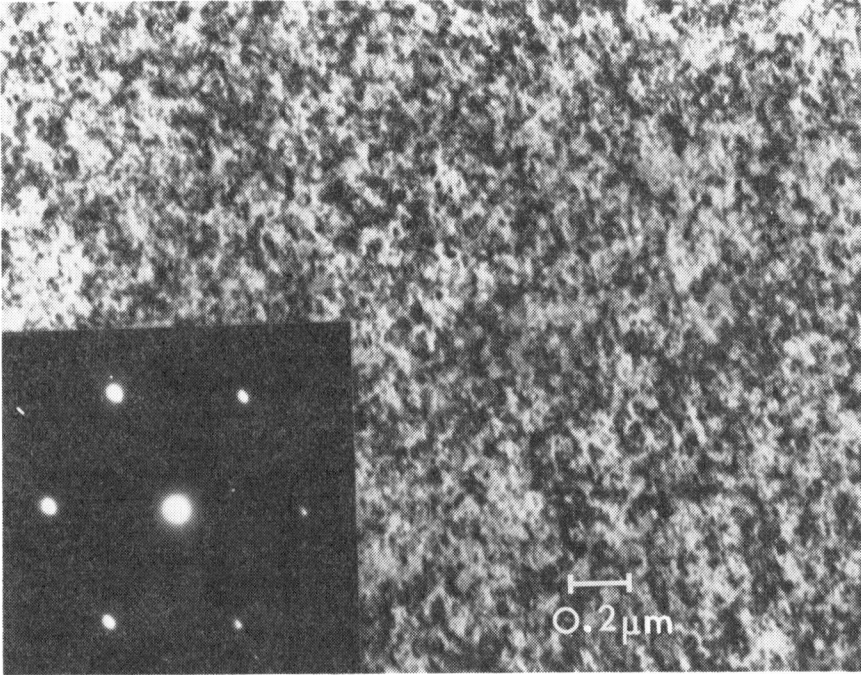

FIG. 21.   Transmission electron micrograph of a phosphorus-implanted silicon crystal that has been annealed at a temperature of 650°C. Inset is the electron diffraction pattern from the same specimen, showing a high degree of order.

higher temperatures, and the time chosen is about 30 minutes. Figure 21 shows a transmission electron micrograph of a piece of phosphorus-implanted silicon after annealing at 700°C; there is still a large amount of localized strain visible, but the electron diffraction pattern superimposed shows that the crystal is relatively perfect. Fortunately, the electron transport properties of the semiconductor are also very good, and the carrier mobility in the implanted region may approach that of bulk, diffused material. Moreover, the carrier density, as determined by measurements of the Hall coefficient, will (in silicon) correspond to the number of phosphorus or arsenic centres introduced; for boron the degree of electrical activity is somewhat less than 100%, due to the formation of thermally stable complexes (e.g. SiB_6). Other group-III dopants (e.g. nitrogen and aluminium) which might be expected to act as acceptors in silicon or germanium have proved useless for this purpose: nitrogen, for example, seems to form silicon nitride within the lattice, and this is a good insulator.

7.3. Gettering

Certain metallic impurities, e.g. copper, nickel and gold, are highly mobile in silicon crystals at elevated temperatures. Since these impurities act as efficient traps for charge carriers, and hence reduce the minority carrier lifetime of a semiconductor, it is frequently desirable to eliminate them from the active region of a device. (In other instances it may be required to introduce them, so as to achieve a particularly low charge storage time.)

Conventional methods of removing these impurities make use of the fact that they will precipitate or 'decorate' dislocations and other defects introduced by mechanical damage. Alternatively, the metals will associate with phosphorus atoms in the heavily doped (and therefore strained) region below a diffused phosphosilicate glass. These techniques of trapping the unwanted species are known as 'gettering'.

During experiments at Bell Laboratories [44] it was discovered, with the aid of ion backscattering, that the damaged zone introduced by ion bombardment can be a relatively effective getter for several metal species. This is important, for it means that the losses and undesirable strain (which leads to bowing of slices) caused by the conventional techniques can be avoided. This is particularly useful for the large-diameter (80 mm) slices which are increasingly used. These observations also explain why, in some ion-implanted specimens, the carrier lifetime is lower than expected: gettering of trapping impurities within the implanted layer could account for this.

8.   ION-IMPLANTED SEMICONDUCTOR DEVICES

The basis of almost all silicon device technology is the 'planar process', in which doping is carried out through 'windows' etched in a thermally grown SiO_2 layer, under a photolithographic mask. As we have seen, ion implantation is entirely compatible with this process, whether the mask generation is achieved with a demagnified u.v. image or electron beam exposure of the 'resist' film. Aluminium has usually been chosen for the 'metallizing' required for contacts and interconnections, and unfortunately the maximum temperature to which it can be subjected (about 525°C) is rather low for the annealing of damage; hence either ion implantation must precede metallizing or the implanted dose must be sufficient to provide an adequate electrical activity at 525°C. Other metals are now being investigated for interconnection films.

A series of important semiconductor device structures will now be considered in turn, in order to illustrate how ion implantation has brought about improvements in yield and performance.

8.1.   MOS transistors

The simplest form of metal oxide transistor consists of a closely spaced pair of heavily doped areas which form respectively the 'source' and 'drain' for charge carriers. Between them lies the 'gate', which is a metallized electrode superimposed on a carefully grown silicon oxide layer. The field effect of this electrode acts so as to produce accumulation or depletion layers in the semiconductor below, in the way discussed in Professor Many's contribution to these Proceedings.

Ion implantation has been used in several ways to improve these devices. Since the ion beam is collimated, the metal gate electrode can itself be used as the mask, to define accurately the boundaries of the source and drain

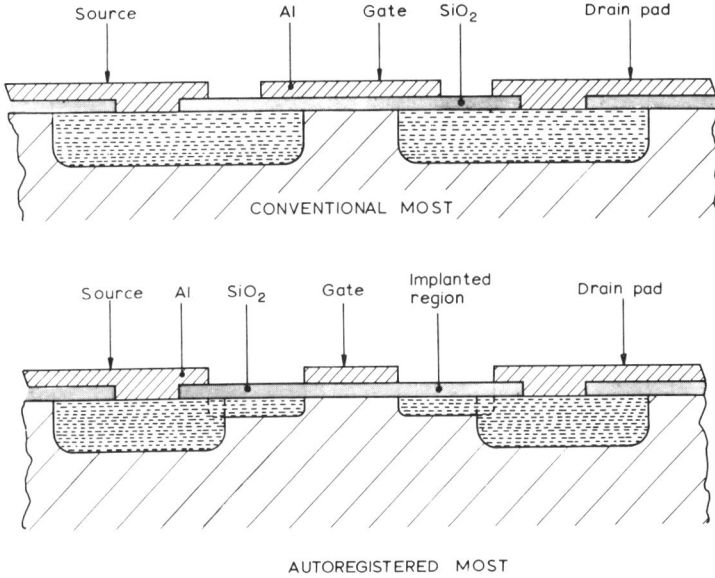

FIG. 22.   An autoregistered ion-implanted metal oxide semiconductor transistor   compared with one prepared by conventional diffusion. The reduced gate overlap is apparent (from Stephen in Ref. [13]).

regions (Fig. 22). By this 'auto-registration' technique the gate width can be made much smaller, since lateral spreading of diffused dopant is avoided. Some but not all of the advantages of this process are also achieved by a doped silicon gate, but this proves to be a relatively difficult method of manufacture. Second, the resistivity in the gate region of the semiconductor can be modified to the optimum level, while choosing a higher value for the substrate. In this way the stray capacitances from source and drain are diminished. Third, the conducting channel can be buried by producing a buried layer of suitable conductivity, and the carrier mobility is thereby increased with respect to the value at the silicon-silicon oxide interface, where it is reduced due to scattering by defects. All these ideas improve the high-frequency performance of discrete MOS transistors, used as switches or for fast telecommunications, and a cut-off frequency of 14 GHz has been achieved in an ion-implanted MOST.

8.2.   MOS integrated circuits

Extraordinarily complex and extensive circuits can be built up from many MOSTs on the same piece or 'chip' of silicon. Ion implantation has proved extremely useful in allowing complementary (or both accumulation-mode and depletion-mode) transistors to be produced side by side. Effectively, n-type and p-type regions of lightly doped semiconductor are created by phosphorus or boron implantation. The purpose is to allow circuits to be designed with the minimum power consumption.

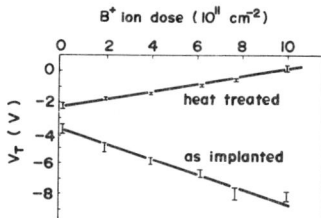

FIG. 23. Threshold voltage of a MOST controlled by boron ion implantation into the gate oxide, followed by annealing.

Another capability is that of controlling the threshold voltage V_T at which an MOS begins to conduct, etc. It is desirable, e.g. in a calculator or electronic wrist-watch, to make this compatible with a low-voltage (1.2 V) battery supply. Unfortunately, typical threshold voltages range up to about 4 V. Ion implantation, e.g. with boron, can lead to an accurately controlled reduction of V_T after a thermal anneal (Fig. 23), and very low ion doses are required for this valuable process.

Perhaps the most striking benefit of ion implantation, deriving from its reproducibility, is the improvement in yield of large-scale integrated (LSI) circuits, typically by an order of magnitude. Since the maximum feasible complexity of a circuit is governed by the yield, it follows that much more complex ICs will become available. Thus the largest present integrated circuit contains 2×10^4 transistors on a single chip, for a versatile pocket calculator. It is no surprise to learn that this, and the smaller MOS ICs used in the highly successful Hewlett-Packard range of calculators, are produced using ion implantation.

The implications of this are quite far-reaching. Rather than build up the circuits which are to form the 'core' of a product, whether it be a calculator, TV set, wrist-watch, etc., from small 'building blocks', it will become economically advantageous to design a specific LSI circuit to provide all the functions required. This has been the trend in the extremely competitive calculator market. The ability of ion implantation to produce short runs of complex circuits with high yield and rapid updating will probably prove of great importance. Manufacturers of advanced products (whether civil or military) will need close access to the relevant design and fabrication facilities.

8.3. Bipolar transistors

Initially, the results with ion-implanted bipolar transistors proved disappointing and non-reproducible: better results could be achieved by diffusion in most cases. The bipolar transistor, with its closely spaced emitter-base base-collector junctions, calls for a high degree of control over junction depths. We have seen that scattering into channels leads to a variable tail on the distribution, which can account for the non-reproducible behaviour. It is important to have a high concentration of dopant in the emitter, a sharply-defined emitter-base junction, and a much lower concentration in the base.

Over the past year or two, much better performance has been achieved in wholly implanted bipolar transistors (following experiments with hybrid transistors, with an ion-implanted base and a diffused emitter). What was realized was that a 'drive-in' diffusion of an implanted arsenic layer could overcome the problem of the exponential tail, since the diffusion coefficient of arsenic in silicon increases with its concentration. Thus the diffusion from the peak of the implanted distribution overtakes the tail and produces an emitter-base junction, typically at a depth of 4000 Å, with greater controllability. Moreover, the highly disordered region created during the high dose ($> 10^{15}$/cm^2) arsenic implantation acts as an effective getter for mobile metallic impurities during the high-temperature (1100-1200°C) drive-in diffusion. This gettered layer, near to the surface, can be etched away if necessary, to leave an emitter with very good electrical properties. This technique provides a more or less flat-topped emitter profile, quite unlike the diffusion profile which has a very high surface concentration, and consequent lattice mismatch and strain (manifested by dislocations running from the emitter boundary).

Thus high-frequency and low-noise bipolar transistors have been produced by ion implantation, with less 'emitter dip' and other troublesome phenomena generally attributed to enhanced diffusion brought about by unnecessary lattice strain.

8.4. High-value resistors

A mundane but important component of an integrated circuit is the resistor. It has long been desirable to employ high-value load resistors, so as to reduce the overall power consumption of a circuit, but it has not been feasible to produce accurately doped regions with a sheet resistivity $>10^3$ ohms/square by diffusion, and therefore it has been necessary to design long meandering resistors sometimes occupying as much as 90% of the total area. The consequent large area of the circuit leads to smaller yields due to the possibility of encountering a crystal defect, dust particle, etc. Ion implantation, on the other hand, enables sheet resistivities of 10^5 ohms/square to be produced, and therefore high-value resistors can be small in size. The benefits of ion implantation are doubled if, besides its ability to introduce small controlled amounts of dopant, use is also made of the reduced carrier mobility in an ion-damaged layer. Thus Ne$^+$ bombardment of a B$^+$-implanted resistor can greatly increase its resistivity and lessen the voltage sensitivity that comes about through partial depletion of a weakly doped layer.

8.5. Diodes

Although the simple forms of diode can be made perfectly well by established diffusion or epitaxy techniques, a variety of specialized types of diode have benefited from ion implantation.

Avalanche photodiodes are operated at a high reverse bias voltage, and their gain and leakage are sensitive to 'microplasma' breakdown occurring at high-field irregularities due to protuberances at the junction. These are believed to be due to anomalous diffusion along dislocations in the semi-conductor. As is to be expected, ion-implanted junctions are essentially free of them.

Large-area diodes, as much as 5 cm^2 or more, are required for charged-particle detection, sometimes with a highly uniform resistive layer for position-sensitive detectors. Ion-implanted structures have proved very successful.

High-frequency IMPATT (impact-avalanche transit time) diodes require a p$^+$ - p - n - n$^+$ structure with the lightly doped regions accurately defined in resistivity and depth over submicron dimensions. The best high-power silicon IMPATTs, working at up to 100 GHz, have been produced at Bell Laboratories by ion implantation.

Variable-capacitance diodes provide a voltage-dependent capacitance for specialized tuning circuits, e.g. with the characteristic $C = kV^{-2}$. The desired function is achieved by control of the dopant profile, which it is difficult to do by diffusion technology. Hence such devices have been produced with a low yield and a high price. A combination of ion implantation together with a small amount of drive-in diffusion to modify the profile has resulted, for one manufacturer, in a yield improvement of two orders of magnitude.

Solar cells are simply large-area shallow-junction devices in which the electron-hole pairs created by illumination separate to provide a small e.m.f. They were one of the first practical devices to benefit from ion implantation, the controlled shallow doping providing an improved blue response and a slightly improved overall efficiency.

8.6.   Compound semiconductors

We have so far dwelt upon the contribution of ion implantation to silicon device technology. There have been relatively few such successes in the compound semiconductors, despite the fact that many observers expected this to prove a more fruitful field of exploitation since there was not a massive entrenched technology as in the case of silicon.

The main obstacles have turned out to be the great variety of crystal defects created by ion bombardment in a compound and the tendency for most materials to decompose at the anneal temperatures necessary to eliminate the defects. In practice, this becomes a problem in the effective encapsulation of devices, as a means of preventing loss of a volatile constituent during heat treatment. Initially, for example, SiO$_2$ films were employed for the encapsulation of GaAs, but although such films are effective for retaining arsenic they allow the out-diffusion of gallium. Silicon nitride forms a better diffusion barrier but has a poor adherence to the semiconductor, so that blisters develop, particularly over the implanted surface, from which it appears easier for arsenic to escape. Dearnaley and D'Cruz [45] have recently developed a technique for improving the adherence by means of a film of evaporated metal between the nitride and GaAs. Such a coating withstands thermal annealing up to above 900°C.

Ion implantation, or rather ion bombardment, has been successfully exploited for producing high-resistivity zones in a variety of compound semiconductors. This is because the defects created during ion bombardment may act as deep-lying levels which shift the Fermi level to the centre of the band gap. This effect is useful for producing semi-insulating zones around individual circuits for isolation purposes, and for this purpose quite small doses of protons are effective (Fig. 24). There is a need, however, for more research to explain the observed depth distribution of defects, since

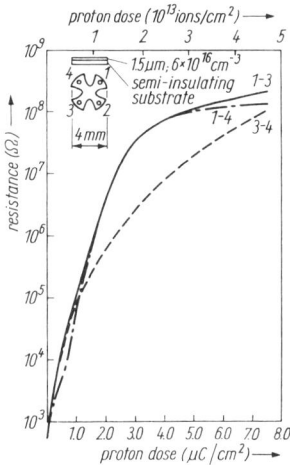

FIG. 24. The change in resistance of an epitaxial GaAs film on a high-resistance substrate brought about by proton bombardment (after R. Allen).

it appears that many of the displacements are caused by ionization processes. Certainly such an effect is to be expected in wide band-gap materials (e. g. the alkali halides), but in a semiconductor the dielectric relaxation times are relatively short and recombination should occur rapidly.

In summary, it appears that ion implantation may well prove as useful in the compound semiconductors as it is in silicon, but that it will take a further few years to achieve this. Once it does come about, certain new requirements will have to be met by the ion implantation equipment. Heavier doping species (e. g. Cd, Te, Zn, Sn) will have to be implanted, and a high degree of mass resolution will be required in order, for example, to implant zinc with the minimum contamination by neighbouring isotopes of copper.

8.7. Ion implantation equipment

This is an appropriate stage at which to summarize briefly the equipment required for ion implantation. It is necessary to stress that, although such apparatus appears unfamiliar, complex, expensive and even dangerous, to a solid-state physicist it is in reality little more so than equipment widely in use for sputtering.

Basically, the requirements are a source of ions, and means of accelerating them towards the target specimen, housed in a clean vacuum system. It has been usual to provide mass analysis between source and target in order to select the required ion species to be implanted, but in those cases in which elemental material may be fed into the source such separation is unnecessary. It is also usual, by a system of magnetic or electric fields, to focus the beam emerging from the source on to the region

FIG. 25.   An 80-keV industrial ion implanter, complete with a high-capacity target chamber (photograph by courtesy of Lintott Engineering Co. Ltd., Horsham, United Kingdom).

occupied by the target. Some means, either by sweeping the beam electrically or (preferably) by mechanically moving the target specimens, is also required to provide a uniform dose or fluence of ions over a surface.

This is all that is necessary, but for research in ion implantation a few additional facilities can be useful: in work on crystals it may be important to orient the specimens in order to control channelling; target temperature control over the range from 77 K up to over 1000 K is desirable if damage processes are to be studied; while a more versatile ion source will be required, with means of converting it readily from one ion species to another.

Some of the more useful ion sources for implantation purposes were developed originally for the separation of isotopes: in each case a simple rugged source is required, capable of delivering large currents over the whole area of a target specimen, using various forms of feedstock material. For research purposes it is advantageous to be able to change from one ion beam to another without elaborate clean-up procedures, while for industrial ion implantation the reliable and efficient production of a few important ion beams is paramount.

There are two possible configurations for the acceleration system: in one case ions are mass-analysed after reaching the full energy required, while, in the other, magnetic analysis is made after acceleration through 20-40 keV and the selected beam is further accelerated. This has the advantages that the high-voltage supply need not be well stabilized; only the specific ions required are accelerated and hence the X-ray hazard from backstreaming secondary electrons is reduced. However, either the magnetic analysis system or the target must be at a high potential, a feature

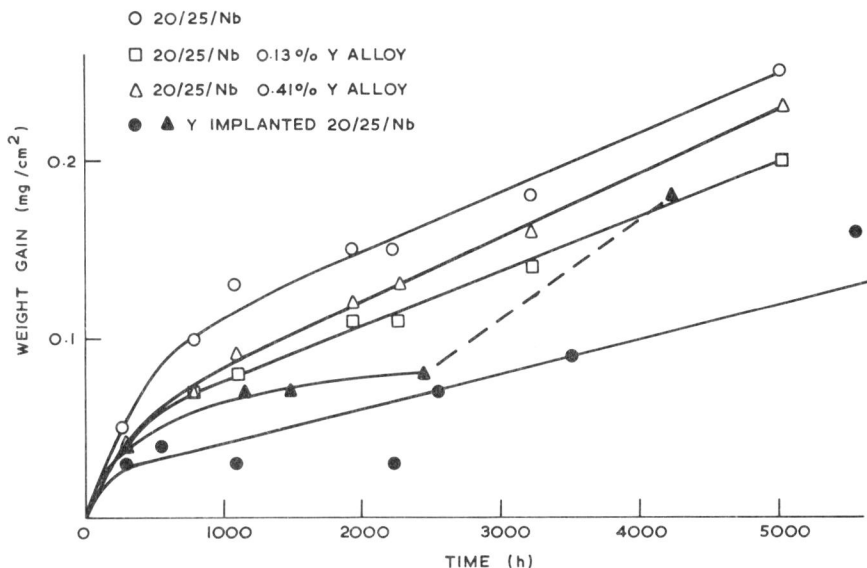

FIG. 26. The influence of yttrium ion implantation ($3.5 \times 10^{15}/cm^2$) on the thermal oxidation of an austenitic stainless steel, in CO_2 at 700°C, compared with alloyed yttrium and untreated steel (from Antill et al. [47]).

which can pose a few design problems. Both approaches have been adopted in systems based upon the Lintott accelerator shown in Fig. 25. In this equipment the target specimens (customarily silicon wafers) are manipulated through the ion beam, which is fixed in space and stabilized in time. There are advantages in this arrangement for production purposes, but in most research facilities the ion beam is swept electrostatically across a fixed specimen. However, it is not so easy to achieve a high degree of uniformity in dose by this procedure.

9.   ION IMPLANTATION OF METALS AND ALLOYS

Now that ion implantation has become an accepted industrial process in the semiconductor industry, attention is being given to its value in other areas of materials technology. Such studies immediately benefit from the amount of experience and equipment developed hitherto for semiconductor applications.

An extensive study of the effects of ion implantation on the surface properties of metals and alloys has been in progress at Harwell since 1971. Preliminary results were reported at the Conference on Ion Implantation held at Yorktown Heights in 1972 [46]. Other laboratories, notably in the United States of America, France and the Federal Republic of Germany, are now also engaged in similar projects.

There are two approaches to any study of ion implantation in metals and alloys. One of these is to regard implantation as a remarkably versatile technique by means of which to investigate the effect of incorporating controlled amounts of specific additives upon surface behaviour, without the difficulties of preparing alloyed specimens which are similar in grain structure and other respects. In the course of such studies a better understanding may emerge of the physical and electrochemical processes which occur near the surface. Some experiments have already succeeded in this respect. Once such knowledge has been gained, consideration may be given to the most effective and economical means of producing the desired surface composition, and this may or may not be by ion implantation.

The second approach is to regard ion implantation as a perfectly feasible metal-finishing process, for example in cases in which the required surface composition has already been established. Here, obviously, conventional surface-coating techniques will have been considered and rejected and the areas to be treated will be limited, while the components themselves are likely to be either intrinsically costly or crucial to the performance or safety of an expensive system. Such systems are to be found in the aerospace, nuclear energy, undersea and military fields.

The following is a list (not exhaustive) of some surface properties which may in most cases have already been shown to be influenced by ion implantation:

> Corrosion resistance
> Electrochemical behaviour
> Wear resistance
> Coefficient of friction
> Bonding ability

All these properties are controlled by the composition within a depth of a few μm below the surface, and this is the region accessible to ion implantation. Other considerations, such as cheapness, strength, thermal, electrical, nuclear or other bulk properties, may determine the choice of material for a component. Often, conflicting surface requirements are met by a coating, applied by painting, electroplating, diffusion, spray-coating, hot-dipping, or cladding, but there are instances where such techniques prove inadequate, usually because of interfacial corrosion or bonding failure. It is not easy to detect potential trouble of this kind. Ion implantation shares with diffusion the ability to produce a surface which is coherent with the substrate, with no interfacial weakness and negligible dimensional change. However, it is in principle more versatile than diffusion, although (as in the semiconductor devices) there are cases where ion implantation is best coupled with diffusion in order to obtain an adequate depth of protection. Ion implantation brings about scarcely any change in dimensions, and so a component may be given a fine mechanical finish and then be implanted as a final step to control its surface composition.

We shall now summarize a number of applications of ion implantation to corrosion, friction and wear.

Yttrium ion implantation has been applied by Antill et al. [47] to the high-temperature oxidation of a stainless steel used for cladding nuclear fuel elements. It has been known for some years that fractions of 1% of yttrium alloyed into 20% Cr/25% Ni/Nb-stabilized austenitic steel would inhibit oxidation in CO_2 gas. However, there are drawbacks of high cost and reduced tensile strength and ductility due to segregation of yttria at grain boundaries. The work demonstrated that a shallow layer of ion-implanted yttrium can be just as effective as yttrium alloyed throughout the steel

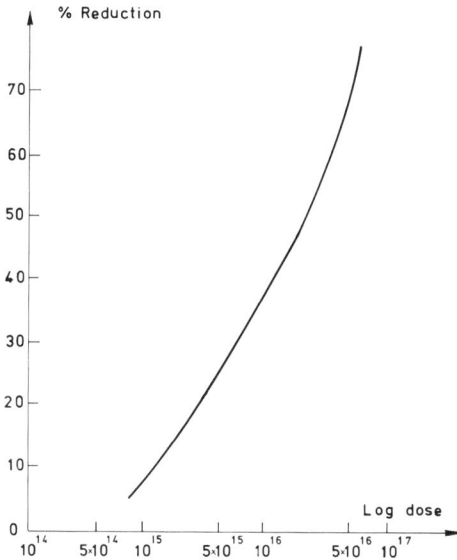

FIG. 27.   The percentage reduction in oxidation of titanium brought about by the ion implantation of calcium, as a function of dose (from Dearnaley [36]).

(Fig. 26), even though the total implanted dose (3.5×10^{15} ions/cm^2) is equivalent to only a single monolayer in terms of surface density. This surface treatment is effective over eight months of oxidation, during which time the steel consumed in oxide growth is many times the thickness of the treated layer. In fact, examination by ion microanalysis after oxidation reveals that most of the yttrium is still located at the metal-oxide interface, where it must progressively advance as oxidation proceeds. The yttrium furthermore improves the adhesion of the oxide, and so reduces spalling.

These experiments, making use as they do of an exceedingly thin layer of yttrium-doped steel, have invalidated the most widely held views concerning the method by which yttrium conveys protection. It is not yet possible to state precisely how it does inhibit corrosion, but the explanation may involve the formation of an yttrium-rich barrier oxide near the metal interface: perovskites are known to be relatively impermeable to ion transport.

In a parallel study of the more exploratory kind Dearnaley et al. [48] have carried out a study of the effects of ion implantation of various impurities into titanium and 18/8/1 stainless steel upon their initial phase of high-temperature oxidation in dry O_2 (Fig. 27). The results indicate a close correlation between the electronegativity of the impurity ions and their effects upon oxidation. Equally novel was the finding that the effects were reversed in the case of the two metals: every ion that inhibited oxidation in titanium would enhance it in stainless steel, and vice versa. These findings, coupled with measurements of the effect of similar implantations on the electronic conductivity of anodic titanium dioxide, strongly suggest that, in the early stage at least, the rate-determining mechanism is electron transport through the growing oxide film. The parabolic time dependence of oxide thickness is readily explained and the behaviour of the additive would be one of compensation of an n-type semiconductor (in the case of TiO_2) and of a p-type semiconductor (in the oxide grown on 18/8/1 steel). This model represents an extension of the classic Wagner-Hauffe rules, based upon the valence of impurities in the oxide: it would seem that the electronegativity, measuring as it does the power of an atom to attract electrons, is a better basis for discussion.

Dearnaley et al. [49] have made similar experiments in polycrystalline zirconium, implanted with a very wide variety of ions. In almost every case the oxidation, carried out in dry O_2 at 400°C, was enhanced, the only exceptions being found after implantation of Fe^+ or Ni^+. This is in marked contrast to the behaviour of the neighbouring group-IVB metal, titanium. Ca^+ ions, which inhibit oxidation of titanium, produced a large enhancement of the oxidation of zirconium, and there was no simple correlation with electronegativity. The behaviour is further complicated by the fact that some impurities will inhibit oxidation if present in low concentration (of the order 0.1 - 0.3%) but will enhance it in high concentration (of the order 1 - 5%).

The difference in behaviour of these two metals, Ti and Zr, may be attributed to the difference in crystal structure of their oxides. Titanium dioxide can undergo crystallographic shear [50] and thereby eliminates anion defects, effectively by wide variations in stoichiometry; ZrO_2 exists in various modified forms of the fluorite (CaF_2) structure but does not exhibit crystallographic shear. The stress caused by the volume change which accompanies oxidation in zirconium is relieved by mechanical cracking of the oxide, which reduces the effective barrier thickness. Under these conditions the ionic size of impurity atoms can therefore be an important factor, so long as the conditions for solid solubility in the ZrO_2 lattice are maintained. Iron and nickel have small ionic radii, while Cu^+, Tl^+, Eu^{++} and other species which enhance oxidation have large ionic radii. Large concentrations of a stabilizing impurity may, however, exceed the solubility limit and then quite possibly degrade the mechanical properties of the oxide film. It is therefore necessary to choose an optimum concentration in order to achieve the optimum inhibition: ion implantation can be an effective way of determining this.

Copper is another metal which shows interesting oxidation effects following ion implantation. Boron implantation [51], aluminium implantation and other species all inhibit thermal oxidation and atmospheric tarnishing. However, preliminary results [52] indicate that bombardment of single-crystal copper with Cu^+ ions can greatly enhance the subsequent thermal oxidation. These effects deserve more investigation.

Aqueous corrosion following ion implantation has been studied in aluminium by Street et al. [53] and strong effects were observed as a result of argon bombardment. The use of a potentiostat and Luggin probe allowed detailed measurements to be made without removal of the shallow ion-implanted surface. This study opens up many interesting possibilities in the use of implantation for the control of electrochemical behaviour, e.g. in electroplating.

Turning next to mechanical properties, the work of Hartley et al. [53] has demonstrated striking alterations in the coefficient of friction under nonlubricated conditions between ion-implanted steel and a tungsten carbide

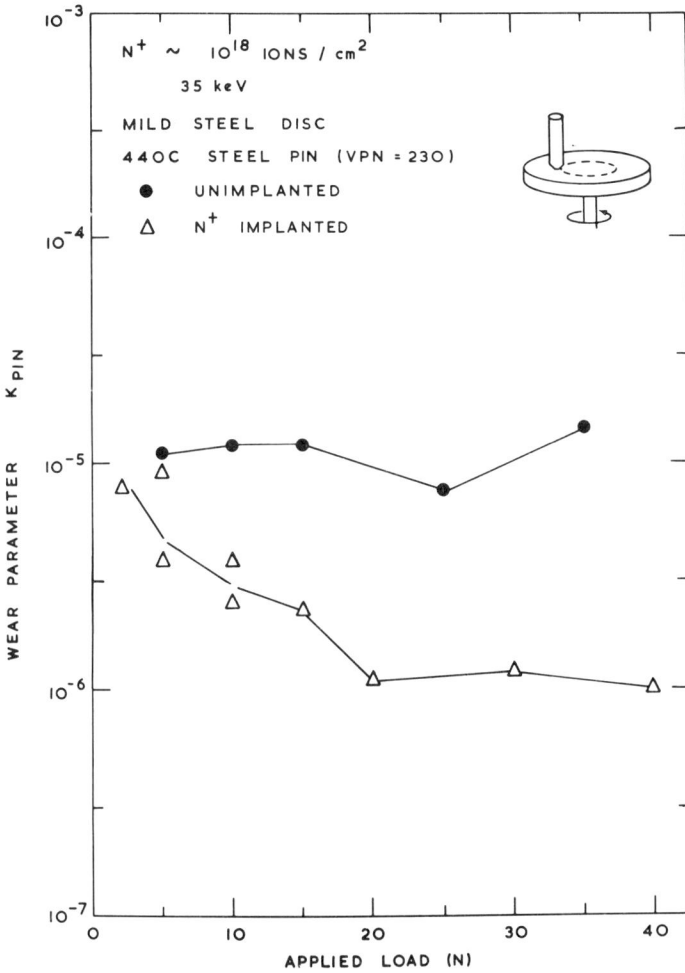

FIG. 28.   The influence of N^+ ion implantation on the wear rate of mild steel, as measured in a standard pin-and-disc wear tester (from Hartley et al. [53]).

test ball. The effects observed are probably linked with processes of adhesion between asperities, which are raised to high temperatures during friction under high loads. In the cases, such as in Pb^+-implanted steel, in which large and fluctuating coefficients of friction were measured, electron microscopy revealed transverse shear cracks in the base of the wear groove, while in cases, e.g. following Sn^+ implantation, in which the friction was reduced, no such tearing and stick-slip adhesion occurred. It seems therefore that ion implantation can modify the tendency for surfaces to adhere during friction, and so provides a means of dry lubrication.

Conventional methods of dry lubrication, e.g. for space applications, involve coatings of bonded MoS_2 or plated lead. Soft films of lubricant and binder sometimes give trouble owing to the accumulation of wear debris. Ion-implanted lead, on the other hand, is likely to be dispersed as fine precipitates near the surface of a hard bearing steel: as the steel slowly wears, lead will be released continuously to be smeared across the bearing surfaces. Such a thin film of soft metal remains in good thermal contact with its substrate so that hot-spot temperatures are reduced. (It is to be noted that lead is far less suitable for atmospheric lubrication due to its ready oxidation.)

In many cases wear is more important than friction. Wear is a more complex phenomenon and it is often related to corrosion in that high temperatures accelerate the production of a corrosion film, which itself wears and can produce abrasive debris. Hartley [53] has extended his measurements to studies of the wear rate of ion-implanted surfaces, using a standard pin-and-disc machine. The results have proved remarkable, for ion implantation of steel by a variety of different species leads to reductions in wear rate by factors of up to 25 (Fig.28). The tentative explanation is that injection of ions leads to an increase in the compressive stress near the surface, and this lessens the tendency for loose particles to be torn away from the material, e.g. following local adhesion to the sliding component. Effectively, the surface grains are held as if in a vice and tend to be polished rather than dislodged. The visual appearance of the surface, which appears burnished, bears out this argument, and the wear-resistance lasts for periods which are long compared with the time taken to abrade the ion-implanted layer. It is as if an initial period of mild wear ensures subsequent low rates of wear, a situation similar perhaps to the 'running-in' of an automobile engine.

There is growing interest in the effects of ion implantation on superconducting properties in metal and alloys. Bett and Howlett [54] in early work at Harwell implanted Sn^+ ions into niobium so as to create a thin surface layer of Nb_3Sn. It was thought that a sufficient number of fine precipitates of this compound could screen the underlying Nb from the penetration of a magnetic field by pinning fluxons at the surface, so enhancing the superconducting properties. Alternatively, the surface could itself carry a superconducting current by a tunnelling mechanism between precipitates. Niobium tape, 3 mm wide by 10 μm thick, was implanted with tin at 100 keV to a dose sufficient to produce about 25 at.% Sn to a depth of about 1000 Å. The tape was annealed in argon at 950°C for 20 minutes. There was a significant increase in critical current and critical field compared with unimplanted niobium. More recent work on niobium has been reported by Crozat et al. [55] and Freyhardt et al. [56]. The latter used Ni^+ ions to create a large number of small voids, about 30 Å in diameter, which together with some larger voids act as strong pinning centres for fluxons. Increases in critical field and critical current were again observed.

Buckel and Stritzker [57] have used proton bombardment of palladium alloys to determine the concentration which yields the optimum superconducting transition temperature, and values as high as 16 K were achieved in hydrogen-implanted Pd-Ag alloy. This work helped to elucidate why palladium alloys become superconducting.

In a similar vein, Chang and Rose-Innes [58] used ion implantation to study the mechanisms controlling the critical currents in very pure type-II superconductors. By implanting Mo$^+$ ions into a Ni-Mo alloy it was found possible to enhance or diminish the critical current, and the effect was dominated by that part of a cylindrical specimen which lay parallel to the applied magnetic field.

Thus ion implantation can be used, as in corrosion science, both to produce alloys with desirable properties and to carry out experiments aimed at a better understanding of physical mechanisms.

In conclusion it may be said that the possibilities lying ahead for ion implantation in metals represent as interesting a challenge and as fruitful an area for exploitation as the semiconductor field is now recognized to be.

REFERENCES

[1]   SCHIFF, L.I., Quantum Mechanics, McGraw-Hill, New York (1949) 271.

[2]   GOMBAS, P., Handbuch der Physik 36, Springer, Berlin (1956) 109.

[3]   MOLIERE, G., Z. Naturforsch. A2 (1947) 133.

[4]   BOHR, N., K. Dan. Vidensk. Selsk., Mat.-Fys. Medd. 18 8 (1948).

[5]   BETHE, H.A., Ann. Physik 5 (1930) 325.

[6]   BLOCH, F., Ann. Physik 16 (1933) 285.

[7]   FERMI, E., TELLER, E., Phys. Rev. 72 (1947) 399.

[8]   LINDHARD, J., SCHARFF, M., K. Dan. Vidensk. Selsk., Mat.-Fys. Medd. 27 15 (1953).

[9]   LINDHARD, J., SCHARFF, M., Phys. Rev. 124 (1961) 128.

[10] FIRSOV, O.B., Sov. Phys.-JETP 9 (1959) 1076.

[11] LINDHARD, J., SCHARFF, M., SCHIØTT, H.E., K. Dan. Vidensk. Selsk., Mat.-Fys. Medd. 33 (1963) 14.

[12] LINDHARD, J., NIELSEN, V., SCHARFF, M., THOMSEN, P.V., K. Dan. Vidensk. Selsk., Mat.-Fys. Medd. 33 (1963) 10.

[13] DEARNALEY, G., FREEMAN, J.H., NELSON, R.S., STEPHEN, J., Ion Implantation, North-Holland, Amsterdam (1973).

[14] JESPERSGÅRD, P., DAVIES, J.A., Can. J. Phys. 45 (1967) 2983.

[15] NEILSON, G., THOMPSON, M.W., Phys. Lett. 46A (1973) 45.

[16] ERIKSSON, L., DAVIES, J.A., JESPERSGÅRD, P., Phys. Rev. 161 (1967) 219.

[17] LINDHARD, J., K. Dan. Vidensk. Selsk., Mat.-Fys. Medd. 34 14 (1965).

[18] BLOOD, P., DEARNALEY, G., WILKINS, M.A., J. Appl. Phys. 45 (1974) 5123.

[19] CHESHIRE, I., DEARNALEY, G., POATE, J.M., Phys. Lett. 27A (1968) 304; Proc. R. Soc. (London) Ser. A., 311 (1969) 47.

[20] BRIGGS, J.S., PATHAK, A.P., UKAEA Rep. AERE TP 522 (1973) and J. Phys. C 7 (1974) 1929.

[21] KINCHIN, G.H., PEASE, R.S., Rep. Prog. Phys. 18 (1955) 1.

[22] BRICE, D.K., Radiat. Eff. 6 (1970) 77.

[23] NELSON, R.S., MAZEY, D.J., Can. J. Phys. 46 (1968) 689.

[24] WATKINS, G.D., J. Phys. Soc. Jap. 18 Suppl. III (1963) 22.

[25] EERNISSE, E.P., Appl. Phys. Lett. 18 (1971) 581.

[26] HIPPEL, A. von, Ann. Physik 80 (1926) 672.

[27] ALMÉN, O., BRUCE, G., Nucl. Instrum. Methods 11 (1961) 257.

[28] FARMERY, B.W., THOMPSON, M.W., Philos.Mag. 18 (1968) 415.

[29] SIGMUND, P., Phys. Rev. 184 (1969) 383.

[30] ANDERSEN, H.H., BAY, H.L., Radiat. Eff. 19 (1973) 139.

[31] STAUDENMAIER, G., Radiat. Eff. 13 (1972) 87.

[32] BARKER, P., private communication (1973).

[33] RUBIN, S., PASSELL, T.O., BAILEY, L.E., Anal. Chem. 29 (1957) 736.

[34] NICOLET, M.A., MAYER, J.W., MITCHELL, I.V., Science 177 (1972) 841.

[35] HARRIS, R.W., PHILLIPS, G.C., MILLER JONES, C., Nucl. Phys. 38 (1962) 259.

[36] DEARNALEY, G., in Proc. Int. Conf. on Applications of Ion Beams to Metals, Albuquerque, 1973. Plenum Press, New York (1974) 63.

[37] AMSEL, G., et al., Nucl. Instrum. Methods $\underline{92}$ (1971) 481.
[38] BARNES, D.G., CALVERT, J.M., LEES, D.G., private communication (1974).
[39] WHITTON, J.L., quoted in Ref. [34].
[40] ALEXANDER, R.B., UKAEA, Harwell, Rep. AERE R-6849 (1971).
[41] FELDMAN, L., in Proc. Int. Conf. on Applications of Ion Beams to Metals, Albuquerque, 1973, Plenum Press, New York (1974) 317.
[42] BLOOD, P., DEARNALEY, G., WILKINS, M.A., in Proc. Int. Conf. on Ion Implantation in Semiconductors and Other Materials, Yorktown Heights, 1972, Plenum Press, New York (1973).
[43] BLOOD, P., DEARNALEY, G., WILKINS, M.A., in Proc. Conf. on Lattice Defects in Semi-conductors, 1974, Inst. Phys., London (1975).
[44] SEIDEL, T.E., MEEK, R.L., Proc. Conf. on Ion Implantation in Semiconductors and Other Materials, Yorktown Heights, 1972, Plenum Press, New York (1973).
[45] DEARNALEY, G., D'CRUZ, A.D.E. (to be published).
[46] Proc. Int. Conf. on Ion Implantation in Semiconductors and Other Materials, Yorktown Heights, 1972, Plenum Press, New York (1973).
[47] ANTILL, J.E., et al., ibid., p. 45.
[48] DEARNALEY, G., GOODE, P.D., MILLER, W.S., TURNER, J.F., ibid., p. 405.
[49] DEARNALEY, G., WEIDMAN, L., GOODE, P.D., in Proc. Int. Conf. on Applications of Ion Beams to Metals, Albuquerque, 1973, Plenum Press, New York (1974).
[50] ANDERSSON, S., WADSLEY, A.D., Nature $\underline{211}$ (1966) 581.
[51] CROWDER, B.L., TAN, S.I., IBM Tech. Disclosure Bull. $\underline{14}$ (1971) 198.
[52] RICKARDS, J., DEARNALEY, G., in Proc. Conf. on Applications of Ion Beams to Metals, Albuquerque, 1973, Plenum Press, New York (1974) 101.
[53] HARTLEY, N.E.W., DEARNALEY, G., TURNER, J.F., SAUNDERS, J., ibid.
[54] BETT, R., HOWLETT, B.W., unpublished data described in Chap. 6 of Ref. [13].
[55] CROZAT, P., et al., in Proc. Conf. on Applications of Ion Beams to Metals, Albuquerque, 1973, Plenum Press, New York (1974) 27.
[56] FREYHARDT, H.C., LOOMIS, B.A., TAYLOR, A., ibid.
[57] BUCKEL, W., STRITZKER, B., ibid.
[58] CHANG, C.C., ROSE-INNES, A.C., Proc. Int. Conf. on Low-Temperature Physics, Kyoto, Jap. Phys. Soc., Tokyo (1970).

APPLIED CATALYSIS

D. A. DOWDEN
Agricultural Division,
Imperial Chemical Industries Ltd,
Billingham, Cleveland,
United Kingdom

Abstract

APPLIED CATALYSIS.
   A short account of applied heterogeneous catalysis is given against the background of surface physics and solid-state physics. It is shown from the results of pure research and the facts of industrial practice that the catalytic properties of solids depend upon their electronic structure and that much of recent solid-state physics and chemistry has relevance to applied catalysis. Active catalysts are solids of high area and small particle size which function in reactive gases at elevated temperatures and pressures, so that solid-state reactions and diffusion processes leading to change of phase and to loss of surface area by sintering ensue unless inhibited. The development and maintenance of both appropriate catalytic specificity and high area in porous solids of adequate strength present the major problems of applied catalysis; the solutions to the problems depend, inter alia, upon the use and the progress of solid-state science. The maintenance of standards of life depends upon technologies which involve catalysis so that the improvement of the efficiency of catalytic processes and the development of new processes are of great importance to both industrialized and developing countries.

INTRODUCTION

   It cannot be doubted that heterogeneous catalysis is a phenomenon associated with the surfaces of solids which is of the greatest importance to the chemical industries of all developed and developing countries. Catalysis makes a major contribution not only to productivity but also to the control of the environment, especially to the abatement of noxious and toxic effluents. In its applied aspects it is largely centred in chemistry but chemical physicists have for some time been working to establish sounder foundations by the development of new experimental methods and new theories which conjoin catalysis to solid-state physics.

   The object of the present work is to show the relevance of the physics and chemistry of the solid state to the whole field of applied catalysis. The following diagram suggests the interdependence of pure science, applied science and technology:

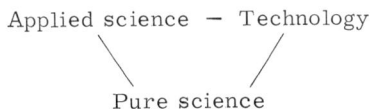

Applied science — Technology
\   /
 \   /
Pure science

Briefly, if pure science is knowledge, and technology is science in action for profit, then applied science teaches knowledge of science for effective action. The aims of applied science are to analyse, amplify and to add

specific precision to the knowledge upon which the principles and prescrip-
tions of technology depend, so that a technology becomes more effective
technically and economically. Thus applied science stands squarely with
a footing in pure science and in technology.

   This paper deals with the catalytic properties of metals, semiconductors
and insulators (to use a classification of solids familiar to the physicist), and
the use of this information in the development of catalysts and in the solution
of the many problems facing the technologist.

1.   THE CHEMICAL REACTION

1.1. Stoichiometry

   The transformation by a single chemical reaction of reactants (R_i)
into product (P_j) is represented [1] by the stoichiometric equation:

$$\nu_1 R_1 + \nu_2 R_2 + \ldots + \nu_j R_j \rightleftharpoons \nu_{j+1} P_{j+1} + \ldots + \nu_c P_c \tag{1}$$

in which the ν's are the stoichiometric coefficients (negative for reactants
and positive for products).

   But many chemical changes of industrial importance are complex
because of the occurrence of many reactions even under fixed conditions;
with r simultaneous independent reactions there is a set of r equations:

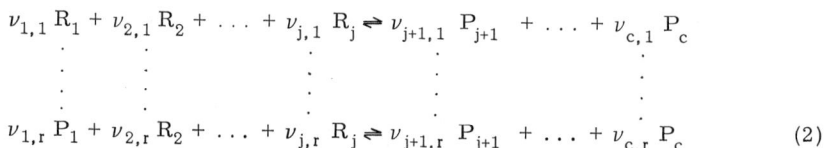

$$\nu_{1,1} R_1 + \nu_{2,1} R_2 + \ldots + \nu_{j,1} R_j \rightleftharpoons \nu_{j+1,1} P_{j+1} + \ldots + \nu_{c,1} P_c$$

$$\nu_{1,r} P_1 + \nu_{2,r} R_2 + \ldots + \nu_{j,r} R_j \rightleftharpoons \nu_{j+1,r} P_{j+1} + \ldots + \nu_{c,r} P_c \tag{2}$$

The initial state (I) of the system at time t_0 alters to the final state (II) at
time t after a period $t-t_0$ and the reaction is completely specified if the
temperature (T K), the pressure (p) or the volume (V), and the identity and
masses of all the chemical species present, are known at all times from t_0 to t.

   Reactions seldom proceed to the degree that state II comprises essentially
only products so that the composition of the system in state II depends also
upon the extent of reaction xi(ξ) [1]. In a single reaction the increase in
mass ($m_i - m_i^0$) of the i-th component is proportional to its molecular weight
M_i and to ν_i:

$$m_1 - m_1^0 = \nu_1 M_1 \xi$$

$$m_i - m_i^0 = \nu_i M_i \xi$$

$$m_c - m_c^0 = \nu_c M_c \xi \tag{3}$$

or in numbers (n) of moles:

$$n_1 - n_1^0 = \nu_1 \xi$$
$$\vdots \qquad \vdots \qquad \vdots$$
$$n_i - n_i^0 = \nu_i \xi$$
$$\vdots \qquad \vdots \qquad \vdots$$
$$n_c - n_c^0 = \nu_c \xi \qquad\qquad\qquad (4)$$

with a set of similar relations when r simultaneous independent reactions occur.

The conversion (C) of a reactant is the fraction of the initial quantity (mass, number of moles) which has reacted. For the i-th reactant of the single reaction therefore:

$$_nC_i = \frac{n_i^0 - n_i}{n_i^0} = \frac{\nu_i \xi}{n^0} \qquad\qquad\qquad (5)$$

and for the r reactions:

$$_nC_i = (\nu_{i,1}\, \xi_{i,1} + \nu_{i,2}\, \xi_{i,2} + \dots + \nu_{i,r}\, \xi_{i,r})/n_i^0 \qquad\qquad\qquad (6)$$

The yield (Y) of a product (c-th component) with respect to reactant i is that fraction of converted reactant i which is equivalent to the amount of product P_c. Thus for the single reaction:

$$_nY_c = \frac{n_c - n_c^0}{n_i^0 - n_i} = -\frac{\nu_c \xi}{\nu_i \xi} = -\frac{\nu_c}{\nu_i} \qquad\qquad\qquad (7)$$

as it must be, but for the complex reaction:

$$_nY_c = \frac{\nu_{c,1}\, \xi_{c,1} + \nu_{c,2}\, \xi_{c,2} + \dots + \nu_{c,r}\, \xi_{c,r}}{\nu_{i,1}\, \xi_{i,1} + \nu_{i,2}\, \xi_{i,2} + \dots + \nu_{i,r}\, \xi_{i,r}} \qquad\qquad\qquad (8)$$

The product of C and Y in consistent units is a measure of the fraction of a given reactant converted to a given product and is called the pass-conversion.

The masses of the reactants decrease monotonically with increasing $t-t_0$ but the mass $m_{n,r}$ of product $P_{n,r}$ may pass through an extremum or increase continuously. If $m_{n,r}$ passes through a maximum then $P_{n,r}$ is an intermediate, otherwise it is a terminal product. Sometimes it is found that specific components (i, j ...) appear as intermediates with maxima of m_i, m_j ... occurring sequentially with increasing $t-t_0$. Such primary, secondary, etc., products can sometimes be identified by plotting $_nY_{j+1}, \dots, _nY_c$ against $_nC_i$.

With the masses constant at time t_0, from Eq. (3) one gets:

$$dm_1/dt = \nu_1 M_1 d\xi/dt$$
$$\vdots \qquad\qquad \vdots$$
$$dm_c/dt = \nu_c M_c d\xi/dt \qquad\qquad\qquad (9)$$

i.e.

$$d\xi = \frac{dm_1}{\nu_1 M_1} = \frac{dm_2}{\nu_2 M_2} = \dots = \frac{dm_c}{\nu_c M_c} \qquad\qquad\qquad (10)$$

Also from (4):

$$d\xi = \frac{dn_1}{\nu_1} = \frac{dn_2}{\nu_2} = \ldots = \frac{dn_c}{\nu_c} \tag{11}$$

and similarly for the ρ-th reaction of the r simultaneous reactions:

$$d\xi_\rho = \frac{d_\rho m_1}{\nu_{1,\rho} M_1} = \frac{d_\rho m_2}{\nu_{2,\rho} M_2} = \ldots = \frac{d_\rho m_c}{\nu_{c,\rho} M_c} \tag{12}$$

$$= \frac{d_\rho n_1}{\nu_{1,\rho}} = \frac{d_\rho n_2}{\nu_{2,\rho}} = \ldots = \frac{d_\rho n_c}{\nu_{c,\rho}} \tag{13}$$

The composition of a system evidently depends upon the values of $m_1^0 \ldots m^0$, T, p and $\xi_1 \ldots \xi_r$, and adjustment of these parameters may result in maximum values of m_{j+1} and small values of all others after the interval t-t_0. There is now selectivity for the (j+1)th component and if high yields, $_nY_{j+1}$, are obtained, one has a selective reaction.

1.2.   Reaction rate and affinity

The corresponding rates of reaction are:

$$v = \frac{d\xi(t)}{dt}, \quad \frac{dn_i}{dt} = \nu_i v \text{ or } \frac{dm_i}{dt} = \nu_i M_i v \tag{14}$$

for the single reaction and

$$v_\rho = \frac{d\xi_\rho}{dt} \text{ and } \frac{dn_i}{dt} = \sum_\rho \nu_{i,\rho} v_\rho \tag{15}$$

for the ρ-th reaction of the r simultaneous reactions. Prigogine and Defay [1] point out that for a uniform closed system

$$\frac{d\xi}{dt} = v(T, p, \xi) \text{ or } v(t, \xi) \tag{16}$$

and the velocity of reaction v may be regarded as a function of state. Also from the second law of thermodynamics the entropy change (dS) accompanying a change in a closed system at temperature T is

$$dS = \frac{dQ}{T} + \frac{dQ'}{T} \tag{17}$$

in which dQ/T is the entropy change for a reversible change (i.e. $dQ_1 = 0$) and dQ + dQ' (i.e. dQ' > 0) is the heat change for the irreversible change. The entropy change dQ'/T is created within the system during an irreversible process and it is always greater than zero.

For systems in partial equilibrium, in which only one chemical change can occur of extent ξ at time t, ξ changes by dξ in the interval. Because this is the only irreversible process in the system, the production of entropy must be determined solely by ξ, so that

$$dQ' = \overline{A}d\xi \geq 0 \qquad\qquad (18)$$

and $dQ' = 0$ at equilibrium but $dQ' > 0$ for spontaneous reaction. De Donder's function of state $\overline{A}$ is the affinity of the reaction. Thus

$$\frac{dQ'}{dt} = \frac{\overline{A}d\xi}{dt} = \overline{A}v \geq 0 \qquad\qquad (19)$$

and if $\overline{A} > 0$ then $v \geq 0$, if $\overline{A} < 0$, $v \leq 0$ and if $\overline{A} = 0$ then $v = 0$, so that the affinity has always the same sign as the rate of reaction, and if the affinity is zero the rate of reaction is zero (the system is in equilibrium).

For the case where $v = 0$ and $\overline{A}v = 0$, one may have either $v = 0$ and $\overline{A} = 0$ (i.e. true equilibrium) or $v = 0$ and $\overline{A} \neq 0$ (i.e. false equilibrium).

A catalyst increases the rate of reaction in the forward direction for a single reaction with $\overline{A} > 0, v \geq 0$, but in the reverse direction for $\overline{A} < 0, v \leq 0$. (The affinity can be calculated from the chemical potentials, μ_i, of the components of the reaction:

$$\overline{A} = -\sum_i \nu_{i,\rho}\mu_i) \qquad\qquad (20)$$

These conclusions can be generalized for the r simultaneous reactions with similar results for the rate of the ρ-th reaction:

$$\frac{dQ'}{dt} = \sum_\rho \overline{A}_\rho V_\rho \geq 0$$

Because dQ'/dt must be positive and appears as the sum of the entropy production, positive or negative, of the various simultaneous reactions, it is possible for a system in which two reactions take place to have

$$\overline{A}_1 v_1 < 0 \text{ and } \overline{A}_2 v_2 > 0$$

as long as

$$\overline{A}_1 v_1 + \overline{A}_2 v_2 > 0$$

This thermodynamic coupling allows the first coupled reaction 1 to advance because of its relation to the coupling reaction 2. Oxidative dehydrogenation exemplifies this situation in its simplest form:

$$CH_3OH \rightleftharpoons HCHO + H_2$$

$$H_2 + \tfrac{1}{2}O_2 \rightarrow H_2O$$

Note that the rate (v_1) of the coupled reaction

$$\leq \frac{\overline{A}_2 v_2}{\overline{A}_1}$$

No reaction can proceed beyond a value of ξ corresponding to the equilibrium constant (K) given by $\overline{A}^* = RT \ln K(T,p)$ where the asterisk denotes the standard affinity of the reaction. Because $v = 0$ at equilibrium it is obvious that the rates of the forward and reverse reactions are then equal. Close to equilibrium $v \propto \overline{A}$ [2].

Reaction rates are always found by measurement and are often given as a power law expressed in the form:

$$v = k[R_1]^a \ [R_2]^b \ \dots \ [P_{j+1}]^m \ \dots \ [P_c]^z = kf(C)$$

in which k is the rate constant and the square brackets indicate concentrations (C). The exponents a, b...z are seldom equal to the stoichiometric coefficients but are small positive or negative numbers or zero. Other forms may be used derived from some type of adsorption isotherm. The rate constant of a simple reaction varies with temperature (T K) according to the equation:

$$k = f(T) \exp\left[-\frac{E}{RT}\right]$$

in which the function of T is the so-called frequency factor and E the activation energy. When the frequency factor is independent of temperature

$$k = B \exp\left[-\frac{E}{RT}\right]$$

and the equation represents the Arrhenius law. It is often found that $\log B \propto E$, corresponding to the 'compensation effect' [3].

The interactions between chemical species can be described in principle by a function that gives the energy of the system, represented as a surface in hyperspace, for all positions of the component atoms (nuclei and electrons). The surface will generally possess extrema among which the minima will correspond to more or less stable compositions of matter. The course of a reaction can then be represented by a line connecting the points corresponding to the system in states I and II (reactants and products respectively). There are many such lines and they may pass through more than one extremum. However, the reaction co-ordinate follows the line which contains not only the points of lowest energy but also the saddle-point of least energy increment (saddle-point minus adjacent minimum). The configuration at the traverse corresponds to the activated complex and the energy increment is the activation energy; there may be a shallow well for the activated complex and the path of the reaction co-ordinate through the well affects the transmission coefficient. In a simple reaction the co-ordinate passes through no significant minima, except that of the activated complex, but in a complicated reaction several such may be traversed.

The configurations at the intervening minima correspond to intermediate compounds and complexes which can be relatively stable or transient.

The kinetic rate equations can be elaborated in the context of the theory of absolute reaction rates but this will not be included here as most of the reactions of concern are too complex to be handled by such methods.

1.3. Sorption [4]

The physisorption of inert gases by solids is extensively used to provide a measure of the total area exposed by solids but it will not be treated explicitly here. On the other hand, chemisorption and absorption are chemical processes (confined respectively to the surface and the bulk of the solid) which play an important part in chemical reactions occurring on and in solids.

When absorption cannot occur, the number of moles of component i adsorbed at the surface is defined by

$$n_i^o = n_i - n_i' - n_i''$$

in which n_i is the total number of moles of i and n_i' and n_i'' are the numbers in the gas and liquid phases respectively.

The adsorption of component i is then

$$\Gamma_i = \frac{n_i^o}{S} \quad \text{(Gibbs)}$$

with S the area of the surface (not to be confused with entropy). The chemisorption reaction cannot usually be written as a stoichiometric equation because the number of active centres on the catalyst surface forming a specific chemisorbed complex with individual reactant species is almost always unknown. Instead, the Gibbs' adsorption of component i is used which for the gas-solid surface is essentially the amount chemisorbed. The adsorption is relatively easily measured for a single adsorptive and expressed as an isotherm and an isobar, but little is known of the chemisorption of adsorptives from multicomponent systems or of adsorption during catalytic reactions.

Nowadays it is clear that surfaces may attract adsorptives into more than one chemisorbed state and that the rate of chemisorption may be very fast with a negligible activation energy (unactivated chemisorption) or slower with an activation energy (activated chemisorption). A modified Lennard-Jones diagram can still be used to represent, purely diagrammatically, the overlapping potential energy curves (Fig.1). Although the details of the interplay between chemisorption and catalysis are largely obscure, it remains an important tool in the investigation of the chemistry of solid surfaces and catalysts and of the processes between chemically adsorbed species.

There are marked correlations between the chemisorptive and catalytic properties of solid surfaces as well as between the electronic and geometric structure of solids.

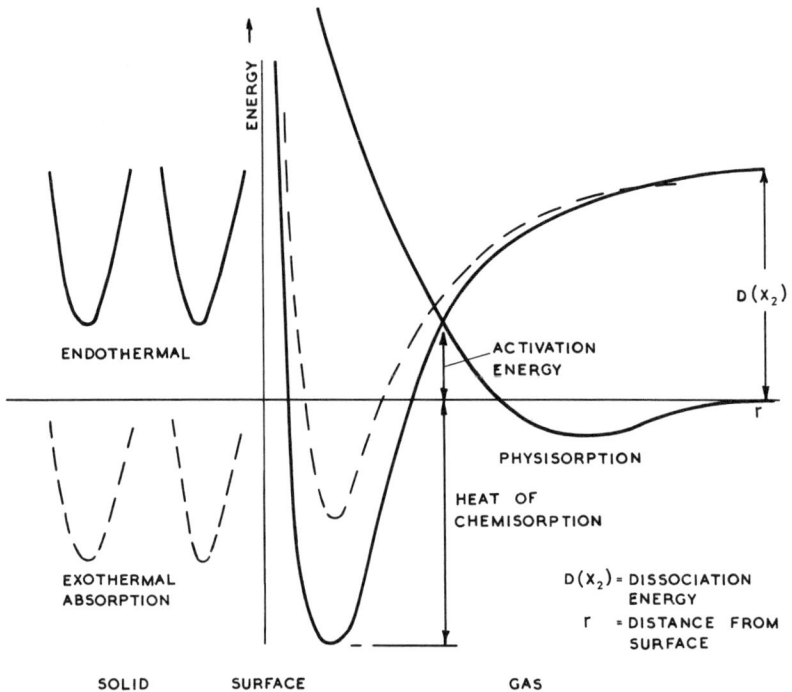

FIG.1.   Lennard-Jones diagram.

1.4. Catalysis

A system is said to be 'catalysed' when the rate of change from state I
to state II is increased by contact with a specific material agent which is not
a component of the system in either state and which is not consumed in the
reaction.   The agent itself is a 'catalyst'; it possesses 'catalytic activity'
and the phenomenon is called 'catalysis'.

If both system and catalyst exist in the same phase, the catalysis is
'homogeneous' but where each exists in a different phase, so that interaction
can take place only at the interface between the phases, the catalysis is
'heterogeneous'.

A catalytic effect can be demonstrated, investigated or used either
intermittently or continuously with the aid of suitable apparatus or plant.
In the discontinuous ('batch', 'static') arrangement the catalyst is placed
in a vessel (reactor) and a quantity of reactants added thereto by temporarily
connecting the vessel to a reservoir of the system in state I.   (Alternatively,
the catalyst may be added after the reactants.)   After a period of time,
under appropriate reaction conditions, the reaction vessel is emptied of
products, which are transferred to a receiver for the system in state II.
This cycle of operations is repeated for as many times as the persistence

of the catalytic effect allows and the total reaction time measures the 'life' of the catalyst.

In the continuous arrangement, the vessel containing the catalyst is connected simultaneously to both the reservoir (state I) and the receiver (state II), in such a way that the fluids flow from one to the other over the catalyst retained in the reactor. By means of suitable devices the rate of flow and the reaction conditions can be adjusted to bring the fluids leaving the reactor into state II; when this can no longer be achieved the catalyst has lost activity and the useful time on line is the catalyst life for the particular conditions.

A catalyst for reactions beginning from state I is a chemical species whose presence affects the energy surface as follows:

(a) Unchanged at points corresponding to physical separation of the added species from the states I, II, III, etc.

(b) Changed at other points so that chemical interaction has taken place between chemical components and the added species.

(c) The changes provide a new reaction co-ordinate between states I and II which passes via an activated complex of lower free energy than that occurring in the absence of the added species.

(d) The changes result in a more favoured reaction co-ordinate between state I and state III rather than between state I and state II. The added species is then called a 'selective' catalyst for the reaction I to III. If the new surface offers a number of preferred equivalent reaction co-ordinates between state I and other states II, III, IV, etc., the catalyst is 'non-selective'. Evidently, such surfaces can be constructed, in principle, for nuclei and electrons so that reactions involving excited and ionized states can be taken into account as in photochemical reactions, photocatalysis, electrocatalysis, etc.

The construction of free energy surfaces of this complexity is at present impracticable so that possible reaction co-ordinates are represented in more qualitative terms. For instance, points along the reaction co-ordinate may be represented by known or suspected appropriate compounds, complexes, radicals, ions, etc., and where possible use can be made of simple quantum-chemical models and theories.

It is generally conceded that chemical interactions may be understood wholly or in part in terms of the interdependent properties of atoms and their nearer neighbours, e.g. valency, co-ordination number, symmetry, atomic (molecular) orbitals, electronic configuration and electronic energy. Therefore, if a particular species is present in a fluid or solid phase in situations where the above characteristics are very similar, then its characteristics and reactivity are expected to be similar also. In this way the similarity in the patterns of behaviour of homogeneous, heterogeneous and enzyme catalysis can be understood, provided that mass transfer effects are absent. It also follows, for heterogeneous systems, in which the solid can expose 'active centres', stable or metastable with various characteristics, that the activity will depend upon local electronic and geometric factors of the type mentioned above.

It is then evident that the content of all chemistry can be used to help in the understanding and the practical appreciation of heterogeneous catalysis (Table I).

TABLE I. RELEVANT PHYSICS, CHEMISTRY AND ENGINEERING OF CATALYSIS

Catalyst	Catalysis phase	Science and data
1. Scale = atomic Atoms: normal, excited Ions: cations, anions	Homogeneous gas phase; photochemical	Atomic physics and parameters; electronic structure, orbitals, spectroscopy, ionization potentials, electron affinities, etc.
2. Scale = molecular Small molecules, covalent, ionic, normal, excited. Acidic: basic		As above + molecular physics and chemistry
3. Scale = molecular (polyatomic) radicals, solvated ions, complexes (inorganic, organometallic); polynuclear complexes	Homogeneous gas and liquid phase	As above + content of radical and complex chemistry
4. Scale = colloidal Sols: metals, oxides, sulphides, etc. Micelles Co-acervates		As above + colloid science + surface science
5. Scale = solid particles Metals, transitional and non-transitional Semiconductors: N, P-types Insulators, acids, bases	Heterogeneous gas and liquid phase	As above + solid-state physics and chemistry; X-ray crystallography, work functions, ESCA, band structure, magnetism, semiconductivity, micromeritics, etc.
6. Scale = porous aggregates All types of solid		As above + hydraulics, heat and mass flow, strength of composites

2.   FACTORS CONTROLLING CATALYST ACTIVITY

To effect the chemisorption of molecules at a solid surface, the solid must be immersed in a bath of the fluid reactants. Then, if the paths of the atoms comprising the reactants are traced from the fluid phase through the chemisorbed state(s) to their final states in the molecules of product, the following processes may be recognized:

(1) Mass transfer from the bulk of the fluid to the surface of the boundary layer.
(2) Diffusion of reactants through the boundary layer.
(3) Physical adsorption onto the solid surfaces.
(4) Surface diffusion in the physisorbed layer.
(5) Chemisorption into one or more states.
(6) Transitions between chemisorbed states.
(7) Surface diffusion of species in the same chemisorbed state.

(8) Reaction of a single chemisorbate alone, or with other chemisorbates or physisorbates of the same or of other kinds to give different chemisorbed species.

(9) Passage of chemisorbed species into the solid.

(10) Processes (7), (6), and (5) for the new chemisorbed species.

(11) Surface diffusion of molecules of product in the physisorbed layer.

(12) Desorption of product molecules from the physisorbed layer.

(13) Diffusion of products through the boundary layer.

(14) Mass transfer of product into the bulk of the fluid.

The slowest of these is the 'rate-controlling' step. None of steps (3), (4), (11) and (12) is likely to be rate-controlling and we shall not be directly concerned here explicitly with steps (1), (2), (12) and (14). Thus consideration is given only to catalysed reactions proceeding in the 'chemical regime' (i.e. controlled by the rate of a chemical process) and not to those proceeding in a 'diffusion regime' (i.e. controlled by rates of physical mass transfer). The detailed information available on stages (5) to (10) is scant and limited to special systems observed under conditions not usual in applied catalysis. Nevertheless such results as have been obtained provide some quantitative foundation upon which the more general concepts can be based.

2.1. Surface area

Because a catalysed reaction proceeds through chemisorbed species or complexes, the volume in which reaction occurs must be contained within the surface or interfacial layer [4]. Then for a given reaction, proceeding under fixed conditions in a steady state on a given surface, the rate of reaction must be proportional to that volume and also to the area of solid surface accessible to the reactants and products because the thickness of the interfacial layer is constant. Ideally the accessible area of the solid A_s is determined by measuring the volume of the interfacial layer using standard adsorption techniques and adsorptives of appropriate dimensions and properties. More often the total interfacial area (S) is found from the physisorption of an inert gas or nitrogen. The rate constant (or some related function) divided by the total area provides a measure of the activity per unit area, i.e. the specific activity. The total area divided by the mass of the catalyst gives its specific area (s).

In technology it is therefore usual to make catalysts with the highest possible specific area by methods which produce either finely divided or porous solids.

For a mass (m) of a single solid phase the total area is the product of the specific area, and the mass:

$$S = ms$$

but for a catalyst containing more than one phase

$$S = \sum_{\rho} m_\rho s_\rho$$

wherein the subscript ρ refers to the ρ-th solid phase.

2.2. Intrinsic activity

The catalytic activity of a single solid phase must depend upon the
nature of its constituents, whose properties may be deduced from the
more or less complete compilation of the phenomena with which the substance
and its leptons can be associated. In modern quantum-chemical terms the
chemistry of the solid may be related to the electronic configuration of each
lepton in its ground and excited states, the site co-ordination number and
symmetry, the interaction between nearest and next-nearest neighbours,
and the interleptonic distances. In past years these contributions have been
grouped into 'electronic' and 'geometric' factors.

When the activity of a catalyst is due entirely to a single solid phase,
the specific activity of that phase is its intrinsic activity, but for a multiphase catalyst whose activity depends upon the properties of more than one
solid phase, the specific activity may be a complicated function of the
intrinsic activities and the specific areas of the components.

2.2.1. Single phases

The electronic and geometric factors affecting the properties of leptons
in the surfaces of single crystals depend not only upon the Miller indices of
the exposed faces and thus upon the crystal shape but also upon the amounts
of 'good' and 'bad' surface. Good surface comprises those extended areas
having a regular geometry equivalent or near to that expected from the
crystal structure. Because many surfaces are reconstructed, on cleavage
or in use, truly good surfaces appear to be of rare occurrence. Bad surface
embraces point defects, line defects and combinations of these:

Point defects:	Corners, emergent edge dislocations, the cores of screw dislocations in the surface.
Line defects:	Edges, steps, narrow vicinal faces, the surface components of screw dislocations, etc.
Point aggregates:	Surface equivalents of vacancy clusters, crowdions, F, F', F_2, F_3, V_1, V_2 centres, etc.
Line aggregates:	Exposed shear structures, grain boundaries, etc.

Because the preparation of a solid phase in the presence of another,
even the mere juxtaposition of two phases, may inter alia modify the
respective morphologies, the properties of single phases must in general
be examined without admixture. A close approximation to this ideal situation
is obtained when the active phase, in the form of crystals or particles which
are not too small, is disposed upon the surface of an inert support of negligible area; academic research with films deposited upon glass or silica
provides examples.

Clearly the method of preparation of even a single-phase catalyst may
affect the nature and the extent of good and bad surface. However, as will
be suggested later, some of these defects are recognizable only in relatively
large crystals and may not exist as such in small ones. If, indeed, the
character of the surface of a particle varies with its size, then the phenomenon
of mitohedry is encountered [5] and the intrinsic activity of a single-phase
catalyst may vary with its specific area.

Little is available concerning the activity of defects of various kinds but the problem will be dealt with later when the properties of the different classes of solid are summarized. Such knowledge is in any case of little practical value because, although it is possible with care to reproduce catalyst activity, it is not yet possible to produce small particles of predetermined morphology.

2.2.2. Multiple phases

The specific activity of a multiphase solid catalyst depends upon the proportions by weight of the components, their specific areas and their intrinsic activities. But at the points, lines and areas of contact between the more bulky phases, new interfacial phases (interphases) may be formed by reaction, as in the chemisorption of species from one phase onto another or by more extensive incursions to give interfacial compounds or new bulk phases. The new phases may sustain new classes of defects.

It is not possible yet, for many types of industrial multiphase catalyst, to measure the specific areas or the intrinsic activities of the several phases but it is valuable, even essential, to identify the solid phases which are present in situ.

2.3. Catalyst strength

The majority of applications of catalysts require that the small particles providing the interfacial area shall be agglomerated into porous granules or compacts which can be packed into and retained in the reactor. The aggregates must be sufficiently strong to withstand crushing, corrosion and erosion for a long time. Any weakness results in pressure drop, loss of catalyst, and premature discharge of the catalyst with all its attendant costs.

2.4. Catalyst life

From the moment the catalyst is contacted with the reactants ('brought on line') it begins to change and in the course of time the alterations are such as to diminish the activity of the charge of catalyst to values which are unacceptable. The changes are both physical and chemical and result in loss of accessible surface area, loss or change in catalytic activity and, frequently, loss of strength. The length of time for which the catalyst performs in a manner meeting the specifications of the plant design is the catalyst life; it depends upon the factors given in this section and upon the overall economics of plant operation.

It follows from the above description of the principal parameters affecting catalyst activity that they are not independent and furthermore that they are uniquely dependent upon the nature of the solid substances comprising the active catalyst in situ. Nevertheless it is possible, as in chemistry as a whole, to classify the quality and the quantity of activity in various ways, e.g. according to the position of the leptons in the periodic table of the elements, the character of the bonding in the solid, the electro-magnetic properties of the solid, etc., and thus to approach the fundamentals. The correlations between the parameters are often such as to set limits to the usefulness of some phases but for the purposes of this paper it is convenient to treat them, at first, as though they are independent.

3. FACTORS AFFECTING SURFACE AREA

Again, it is necessary to distinguish between single-phase and multi-phase catalysts because of the many processes which may ensue at interfaces.

3.1. Single-component phases

3.1.1. Particle size and shape

It is easily shown [6] that the relationships between surface area and size, for particles of different shapes, are as follows:

(a) Cubes and spheres

$$S = \frac{6 \sum (\eta_i \lambda_i)^2}{L\rho (\eta_i \lambda_i)^3}$$

where L is the most frequently occurring length (the cube edge, diameter of sphere) in the distribution of particle sizes; η_i is the fraction of particles of the i-th size; λ_i is the ratio of the dimension of the same particles to the characteristic length L; and ρ is the density of the particles. The parameters within the brackets are given by the distribution of particle sizes, so that if the distribution round L remains unchanged as L changes $S \propto 1/L\rho$ or $S = fc/L\rho$ and $fc \to 6$ as the quotient of the summations tends to unity. When all particles have the same size $S = 6/L\rho$.

(b) Square or cylindrical rods

$$S = \frac{4 \sum (\eta_i \lambda_i)^2}{D\rho \sum (\eta_i \lambda_i)^3}$$

D is again a characteristic linear dimension, say the most frequently occurring thickness or diameter, and η_i, λ_i and ρ have the same significance as before. The area $S \propto 1/D\rho$ or $S = fc/D\rho$ and $fc \to 4$ as the quotient of summations $\to 1$.

(c) Thin plates

$$S = \frac{2 \sum (\eta_i \lambda_i)^2}{T\rho \sum (\eta_i \lambda_i)^3}$$

so that $S \propto 2/T$ and for equidimensional plates $S = 2/T\rho$ where T is the characteristic thickness.

The inverse relationship between surface area and particle size for all shapes of particles shows why most of the methods for making solids of high area involve at some stage the preparation of the phase (or a precursor) in the form of small particles. It is also evident that processes leading to particle growth cause loss of area. Little can be done to control or modify the shape of very small particles.

As fine particles cannot be conveniently handled on the large scale, they are usually compacted by various devices to give porous aggregates (granules, pills, pellets, briquettes, etc.). The aggregates can be regarded as being composed of porous particles (of a range of sizes) which latter are themselves coherent assemblages of non-porous microcrystals or amorphous bodies, also of different sizes. Then, unless the ultimate micro-entities are zeolitic, the pore structure of the aggregates is formed only by the interstices between the various particles, and the pore size distribution depends upon the particle size distribution and the geometry of the packing. A porous catalyst is then to be viewed both as an aggregate of small particles and as a foraminiferous body with pores of sizes in accord with some distribution function. Lately it has become customary to group pores according to their diameters (d) as follows [6]:

(1) Macropores $d > 50$ nm (500 Å)
(2) Mesores, 2nm $d < 50$ nm
(3) Micropores $d < 2$ nm (20 Å)

Non-porous solids usually have rugose surfaces, and the ratio of the total area to that calculated from the macroscopic dimensions of the surface is the roughness factor. The pore sizes given above correspond roughly to the diameters of the smaller interstices formed by the simple cubic packing of equal spheres of diameters 120 nm (1200 Å) and 5 nm (50 Å). Industrial catalysts may be non-porous like the Pt-Rh wire gauzes used in ammonia oxidation, microporous like active carbons, or microcrystalline, ~ 5 nm, as in some metallic hydrogenating catalysts.

3.1.2. Thermodynamics of small particles and pores

In a small cube 10 atoms on edge, about one half of the atoms lie on the surface and, as is well known for small particles, surface energy becomes an appreciable fraction of the total energy [4]. In consequence the small particles and pores will tend by change of size and shape to states which minimize the free energy, and the mechanism of the change will depend, inter alia, upon the properties of both the solid and the fluid phases.

The free energy of a solitary crystal, not at equilibrium with the ambient fluid and unable to exchange leptons with another solid phase, can be diminished only by change to the equilibrium shape. The equilibrium shape is the Wulff form but the Wulff relations cannot be applied to the small crystallites typical of most catalysts. The Gibbs-Curie proposal [4], that the equilibrium form of a crystal of constant volume is that for which the function:

$$Q = \sum_{\gamma} \sigma^{\gamma} A^{\gamma}$$

is a minimum (σ^{γ} is the surface tension and A^{γ} the area of the face designated by γ), is more qualitatively useful. At constant temperature and crystal

TABLE II. PROPERTIES OF SMALL CRYSTALS OF PURE
COMPONENT 1

Vapour pressure, p'_1	$\ln(p'_1)/(p'_1)_0 = (2\sigma^{\gamma}/r^{\gamma})(v''_1/RT)$
Melting point, T_m	$\ln(T_m/T_{m,0}) = -(2\sigma^{s,1}/r^{s,1})(v^s/\Delta H_f)$
Solubility, (x'_1 = mole fraction)	$\ln(x'_1/x'_1)_0 = (2\sigma/r)(v^0_1/RT)$
Chemical potential, μ'_1	$\mu'_1 - (\mu'_1)_0 = (2\sigma^{\gamma}/r^{\gamma})v''_1$
Conc. of vacancies, C', under a curved surface [7]	$\ln(C'/C_0) = (2\sigma^{\gamma}/r^{\gamma})(v/RT)$

The subscript zero everywhere refers to the values over a plane surface; the v's are
molar volumes; ΔH_f is the heat of fusion.

volume, and if adsorptions can be neglected, the surface tension σ equals
the surface free energy g^o and Q can be put equal to the total surface free
energy, $G^o = \sum_j g^o$, γA^{γ}. The Gibbs-Curie proposal is therefore a reasonable hypothesis and isolated crystals can be expected to tend to minimize
G^o by changes in form which minimize σA, as by exposing faces of low
Miller index.

The driving force for alteration of shape is also expressed in the various
analogies (Table II) with the Kelvin equation [4, 7], which gives the vapour
pressure (p') of a droplet of radius r of a liquid of molar volume v", surface
tension σ and normal vapour pressure p_0:

$$\ln\left(\frac{p'}{p_0}\right) = \frac{2\sigma}{r}\frac{v''}{RT}$$

The chemical potentials, vapour pressures and solubilities of small particles
are greater but the melting points lower than those of large particles. It is
also probable that small crystals have smaller heats of fusion and of subli-
mation than large crystals. The net effect in an assembly of particles with
a range of sizes is that the larger particles grow at the expense of the
smaller. The larger free energy of the smaller crystals also means that
the equilibrium constants of reaction involving them are correspondingly
affected.

The equilibrium shapes of very small particles containing ~10 leptons
have not been determined but can hardly be considered without taking into
account the effects of chemisorbed species. Presumably such 'solids'
are similar in structure to the corresponding polynuclear complexes with
metal-metal bonds as in the rhodium complex $Rh_6(CO)_{18}$ where the metal
atoms form an octahedron. In larger crystals, singular surfaces correspond
to minima in the Wulff plot and are usually low-index planes of the crystal
with the leptons in normal lattice sites. As the temperature is raised,
some roughening may occur due to the formation of adatom-vacancy pairs

stabilized by the increased configurational entropy. However, no large-scale disordering is expected for such surfaces even near the melting point; the concentration of single vacancies at the melting point is: Cu, 2×10^{-4}; Ag, 1.7×10^{-4}; Au, 7.2×10^{-4}; HC, 9.0×10^{-4} [8]. However, the vicinal planes according to Burton and Cabrera [9] and Herring [10] possess a certain density of monatomic ledges which roughen with increasing temperature due to the formation of kinks and jogs and the dissociation of leptons through a series of sites of decreasing co-ordination number; thus surface diffusion ensues [11]. Impurities which are chemisorbed on some of these sites can affect the rates of diffusion, either increasing or decreasing them depending upon the sites which chemisorb most strongly.

Herring has deduced, for crystals large enough to comply with the thermodynamic models and at temperatures above 0 K, that the equilibrium shapes can comprise plane regions of finite extent joined by smoothly curved regions or by curved regions with sharp edges and corners; alternatively, the shape may be polyhedral. In very small particles where edges and corners account for an appreciable fraction of the atoms of the surface, these conclusions cease to be well based. Many catalyst particles are large enough to conform to the model (e.g. in catalysts operating at high temperatures) but many are very small (e.g. in platforming) and the question must be reopened. In very small crystals the interior is under a pressure $\sim 2\sigma^{\gamma}/r$, which for a surface tension of 1000 erg/cm^2 and a radius r = 20 Å amounts to $\sim 4 \times 10^4$ atm; the pressure is substantial even if adsorption reduces σ^{γ} by a factor of 3. Such a particle would be expected to be non-porous and approximately spherical at temperatures at which the lattice leptons may diffuse. Under these pressures, solids which are normally amorphous may crystallize.

Dutch workers [12] have proposed that small particles of metals expose low-index planes in a near spherical shape. The surfaces of compounds may minimize surface energy by change of valency, co-ordination number and symmetry.

Related reasoning shows that the physical and chemical properties of material within or bounding small pores must also be modified by surface energy effects; e.g. in the attainment of an equilibrium distribution of pore sizes small pores will receive leptons from those of larger radius.

3.1.3. Mechanisms of size changes [7]

The rates of the changes to the equilibrium shapes and size distributions depend upon the nature of the processes; matter can be transported between sites through the volume of the solid, across the surface or through the ambient fluid (evaporation/condensation, solution/deposition).

Although experimental work bearing upon mechanism has been for the most part confined to non-porous particles much larger ($\sim 1\mu$ m) than those frequently found in active catalysts, yet the results are worthy of extrapolation if only to provide a conceptual framework. The following stages of growth will be recognised:

(i) Formation of primary particles; the leptons combine to give metastable, non-porous entities.

(iia) Surface modification; isolated primary particles change their shape to accord with the Gibbs-Curie rule. If adsorption effects are small, planes of high index should disappear and the shape of the particle should approach a sphere [12], but if there are strong chemisorptions from the fluid phase then facetting can occur and higher index faces may be stabilized.

(iib) Necking; osculating particles are bridged by accommodation of material in the concavities of the fillet. The particles may have been formed in contact or brought together by Brownian-like movements.

(iii) Assimilation; the larger particles grow at the expense of smaller neighbours.

(iv) Aggregate consolidation; the pores of necked aggregates are filled in to form larger particles with further smoothing.

(v) Metastabilization; the crystallites become so large that further change is slow in the absence of additional energy inputs (thermal or from the application of pressure).

Mechanisms have been investigated by following the sintering of solids of simple geometry in contact (e.g. polycrystalline spheres in contact with other spheres, flat plates or grooves). For instance, the equation connecting the radius (x) of the neck formed between two spheres of radius r to the time of sintering (t) for matter transport by diffusion is

$$x^5/r^2 = \text{a constant } x(\sigma^\gamma vD/RT)t$$

The constant (~ 10) depends upon the geometry of the system; σ and v as before are the surface tension and molar volume of the substance, respectively, and D is the self-diffusion coefficient.

The experimental studies on the early (necking) stages of sintering, in particles of size $> 1\,\mu m$, support bulk diffusion as the principal mechanism of transport. In smaller particles it was supposed that surface diffusion might be dominant. The data also support the assumption that the grain boundaries are vacancy sites.

However, it is not possible to conclude for any given single-component catalyst, existing as very small particles, which is the principal mechanism of transport. It is to be noted especially that porous catalysts have a pore volume $\sim 50\%$ and that shrinkage of the compacts is often observed.

Effect of temperature

As all the transport processes contributing to sintering require the movement of atoms or vacancies and an increase in energy, they are all speeded by increase of temperature which must in consequence be carefully controlled. Unfortunately, the rates of the catalysed reactions also depend exponentially upon the temperature and it is usually found that the reaction temperature can be varied within only a very narrow range. Other methods must be found for controlling sintering. Long ago Tammann suggested that the onset of sintering occurs at a temperature near to half the melting point (°K). Later Hüttig proposed that surface movement becomes fast at a temperature of 0.25-0.3 T_m. These parameters are now known as the Tammann and Hüttig temperatures.

Particle size

No matter what the mechanism of material transport, the distance is smaller and the rate faster the smaller the particle size of particles in contact. Coble and Burke [7] point out that for the bulk diffusion model, according to which rates are proportional to the third power of the diameter, there can be enhancement of rates by factors of many thousands, necessitating much lower operating temperatures (by hundreds of degrees) to maintain the lower sintering rates.

3.2. Multicomponent phases

When the catalyst particles comprise more than one solid phase the process of sintering may be accompanied by the chemisorption of leptons from one phase onto another solid solution of one phase in another or compound formation. Many years ago Jander and Hüttig postulated that this sequence of events occurred in all reactions between solids [13]. Very little work has been done in recent times on the chemisorption stage but the others have been given much attention.

3.2.1. Solid solution

It is generally conceded that chemical diffusion in oxides of the sodium chloride structure (e.g. the MgO-NiO couple[1]) occurs primarily in a rigid anion lattice by a cation exchange mechanism involving a cation vacancy. The activation energies for other mechanisms, such as ring rotation or migration of interstitial ions are too large. The vacancies in the cation lattice are Schottky defects or are due to the presence of aleovalent cations in substitutional solid solution (e.g. $3Mg^{2+} = 2Al^{3+}$). For the MgO-NiO couple in air the aleovalent ion is Ni^{3+} the concentration of which appears to depend exponentially on the concentration of Ni^{2+}.

3.2.2. Formation of compounds at interfaces

It has been seen that small particles take part in solid-solid reactions more quickly (or at lower temperatures) than large particles. The mechanism

---

[1] See M. APPEL and J.A. PASK, J. Am. Ceram. Soc. 54 (1971) 152.

depends upon the type of solid but can be exemplified by those leading to the formation of spinels (in particular of ferrites) which have undergone investigation over many years.

Spinels [14]

If the rate of reaction at the interface between the two solid phases is faster than the diffusion of the leptons, and the boundaries of the phases are at equilibrium, if the layer of spinel product is not holed and diffusion along line defects is small, and if the ion, electron and hole currents are independent of one another and coupled only through the requirement of electrical neutrality, one can present three mechanisms:

(1) Counterdiffusion

The oxygen anions do not move but the cations migrate in opposite directions, thus:

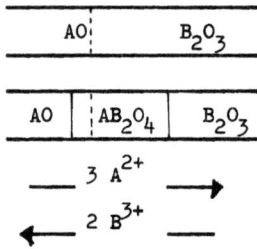

According to this mechanism the position of the inert marker (broken line) does not change and the amounts of spinel on either side of the marker are in the ratio 1 to 3; magnesium aluminium spinel forms thus. The layer grows in thickness (Δx) with time t following a parabolic law $(\Delta x)^2 = (8/3) (D/RT) \cdot (\mu_{AO'} - \mu_{AO''}) t$ where $\mu_{AO'}$ and $\mu_{AO''}$ are the chemical potentials per equivalent of AO on the two sides of the interphase and

$$D = \frac{6[A^{2+}] D_A [B^{3+}] D_B}{2[A^{2+}] D_A + 3[B^{3+}] D_B}$$

where $[A^{2+}]$ and $[B^{3+}]$ are the molar fractions and D_A and D_B are the molar diffusivities of the two ions.

(2) Bipolar diffusion

If the anions also diffuse then the following situation is reached:

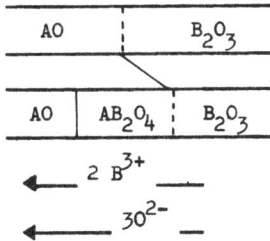

Here only B^{3+} moves together with its counter ion O^{2-}. The relation between x and t is similar to that in case 1 but with

$$D = \frac{6[B^{3+}]D_B [O^{2-}]D_O}{3[B^{3+}]D_B + 2[O^{2-}]D_O}$$

If the trivalent ion diffuses faster than the divalent then the inert marker remains in the AB_2O_4/B_2O_3 interface. If the divalent ion is the faster, the marker remains in the AO/AB_2O_4 interface as in $CoO-Al_2O_3$ and $CoO-Cr_2O_3$ couples.

(3) Biphase diffusion

When B^{3+} is reducible, say to B^{2+} (like Fe^{2+} in Fe_2O_3) the B-ion may diffuse through the spinel interphase while oxygen, evolved at the ABO_4/B_2O_3 interface, is transported through the gas phase and recombines at the AO/ABO_4 interface:

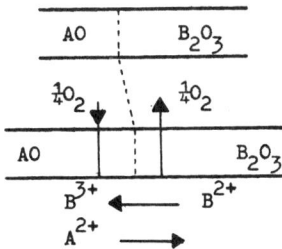

The third mechanism is common in the formation of ferrites; in $MgFe_2O_3$, for which the phase diagram ($MgO-FeO-Fe_2O_3$) shows that Fe_2O_3 dissolves in the ferrite phase for the most part as $\sim Fe_3O_4$, there must be counter-diffusion of Mg^{2+} and Fe^{2+} ions and a strong dependence of the rate on the partial pressure of oxygen. A further example is $CdO-Fe_2O_3$ [15].

Kuczynski's results on the sintering of polycrystalline spheres of MgO, NiO and Fe_2O_3 in contact with plates of the same substances show clearly the stresses which are set up initially in single-crystal plates (Fe_2O_3 on MgO) by the reaction at the interface and the subsequent neck formation and growth. All these results have been obtained with particles which are much

larger than those usually encountered in catalysis. Neck growth at an
advanced stage between MgO and Fe_2O_3, NiO and Fe_2O_3 and ZnO and Fe_2O_3,
follows the parabolic relation noted above, but a second mechanism involving
evaporation and condensation occurs at higher temperatures for $ZnO-Fe_2O_3$.
Kooy [16] shows how voids may be formed when the area of contact in a
couple is restricted. Even more complex situations may arise at the points
of contact between phases which are themselves ternary oxides, etc.

3.3. Active phases in situ

It can now be perceived why the proper characterization of the phases
in many catalysts containing very small particles in situ still presents
intractable problems which hinder the extension of knowledge of catalysis.
Because of the reactivity of small particles it appears to be fairly generally
true that the phases which are thermodynamically allowed will, in practice,
be formed by solid-solid and solid-gas reactions. The solid phases found
in catalysts in steady states are usually those which would be in equilibrium
with the reactants in the fluid phase, under the reaction conditions, if equilibrium were in fact established. For example, the important shift-reaction,
which produces hydrogen and carbon dioxide from carbon monoxide and water,
is catalysed by magnetite (Fe_3O_4), i.e. the equilibrium solid phase in the
Fe-C-O and Fe-H-O systems under the conditions of the reaction:

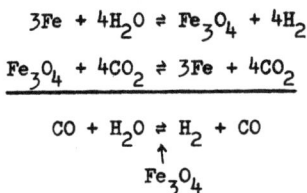

$$3Fe + 4H_2O \rightleftharpoons Fe_3O_4 + 4H_2$$
$$\underline{Fe_3O_4 + 4CO_2 \rightleftharpoons 3Fe + 4CO_2}$$
$$CO + H_2O \underset{\underset{Fe_3O_4}{\uparrow}}{\rightleftharpoons} H_2 + CO$$

Because the reactions with and within the solid state occur in a short time
(hours) compared with the life (months) of a catalyst, and because activity
is determined by the composition of a thin interphase, the steady-state activity
is attained quickly and in the best catalysts changes only slowly with time.

4.   CLASSIFICATION OF CHEMICAL REACTIONS

An account of applied catalysis is facilitated if some of the many kinds
of chemical reactions are grouped together and exemplified (Table III).
Often the reactions can be reversed, as in hydrogenation and dehydrogenation,
but the back reactions are not shown separately unless they are of great
industrial importance, e.g. cracking can be considered to be a type of
depolymerization or 'de-condensation'. The name and the equation of the
reaction describes only its stoichiometry and implies little concerning
the mechanism.

Many complex industrial processes can be considered to be, indeed
often are, sequences of linked, simpler reactions. The hydrogenolysis of

TABLE III. TYPES OF CHEMICAL REACTIONS

	Name	Description	Examples
1.	Carbonylation (De-)	Addition (elimination) of CO	$CH_3OH + CO \rightarrow CH_3COOH$: $CH_3CHO \rightarrow CH_4 + CO$
2.	Carboxylation (De-)	Addition (elimination) of CO_2	$C_6H_5ONa + CO_2 \rightarrow C_6H_4{\nearrow}^{OH}_{\searrow COONa}$; $CH_3COOH \rightarrow CH_4 + CO_2$
3.	Condensation	Union of 2 molecules; no eliminations	$2CH_3CHO \rightleftharpoons CH_3CH(OH)CH_2CHO$ (Aldol); $3C_2H_2 \rightarrow C_6H_6$; $RH + C_2H_4 \rightleftharpoons R.C_2H_5$
4.	Cracking	Scission of a molecule, usually a hydrocarbon	$RC_2H_5 \rightleftharpoons RH + C_2H_4$
5.	Hydration (De-)	Add. (elim.) of H_2O to (from) a molecule	$C_2H_2 + H_2O \rightleftharpoons CH_3CHO$; $C_2H_4 \rightleftharpoons C_2H_5OH$; $CO \rightleftharpoons H.COOH$; $RCN \rightleftharpoons RCONH_2$
6.	Hydrolysis	Add. of H_2O to a molecule $\rightarrow$ 2 products	$CH_3COOR + H_2O \rightleftharpoons CH_3COOH + ROH$; $RCONH_2 \rightleftharpoons RCOOH + NH_3$; $R_2O \rightleftharpoons 2ROH$
7.	Hydrogenation (De-)	Add. (elim.) of H_2 to (from) a molecule	$N_2 + 3H_2 \rightleftharpoons 2NH_3$; $C_2H_2 \rightleftharpoons C_2H_4$; $C_4H_8 \rightleftharpoons C_4H_{10}$; $C_6H_6 \rightleftharpoons C_6H_{12}$; $RCN \rightleftharpoons RCH_2NH_2$
8.	Hydrogenolysis	Add. of H_2 to 1 molecule $\rightarrow$ 2 products	$ROH + H_2 \rightarrow RH + H_2O$; $RC_2H_5 \rightarrow RCH_3 + CH_4$; $RNH_2 \rightarrow RH + NH_3$; $RCl \rightarrow RH + HCl$
9.	Hydrohalogenation (De-)	Add. (elim.) of H halides to (from) 1 molecule	$C_2H_2 + HCl \rightleftharpoons CH_2 = CHCl$
10.	Isomerization	Structure change of 1 molecule	$CH_3CH_2CH = CH_2 \rightleftharpoons CH_3CH = CH.CH_3 \rightleftharpoons (CH_3)_2C = CH_2$; $C_6H_{12} \rightleftharpoons C_5H_9(CH_3)$
11.	Oxidation	Add. $\frac{1}{2}O_2$ or O_2 to 1 mole and elim. of H_2O, CO_2, CO	$CH_3OH + \frac{1}{2}O_2 \rightarrow HCHO + H_2O$; $CH_3CH = CH_2 \rightarrow CH_2 = CH.CHO$ $+ H_2O$; $CH_4 \rightarrow CO_2 + 2H_2O$
12.	Oxygenation	Add. $\frac{1}{2}O_2$ to 1 unsat. mole; no elim.	$C_2H_4 + \frac{1}{2}O_2 \rightarrow C_2H_4O$; $CO \rightarrow CO_2$; $SO_2 \rightarrow SO_3$
13.	O-insertion	Add. $\frac{1}{2}O_2$ to 1 sat. mole; no elim.	$RH + \frac{1}{2}O_2 \rightarrow ROH$; $H_2 \rightarrow H_2O$; $NH_3 \rightarrow NH_2OH$
14.	Peroxidation	Add. O_2 to 1 mole; no elim.	$H_2 + O_2 \rightarrow H_2O_2$; $-CH = CH- \rightarrow -\overset{O-O}{C\!H-C\!H}-$; $CH_3CHO \rightarrow CH_3COOH$
15.	O-transfer	Transfer of O between molecules	$CO + H_2O \rightleftharpoons CO_2 + H_2$; $CH_4 + H_2O \rightleftharpoons CO + 3H_2$
16.	Polymerization (De-)	Multiple condensation (and reverse)	$nC_2H_4 \rightarrow C_{2n}H_{4n}$; $nHCHO \rightleftharpoons (CH_2O)_n$

some alcohols may be equivalent to dehydration to olefine followed by hydrogenation of the double bond:

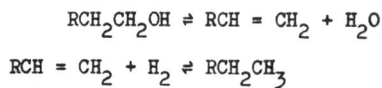

$$RCH_2CH_2OH \rightleftharpoons RCH = CH_2 + H_2O$$

$$\underline{RCH = CH_2 + H_2 \rightleftharpoons RCH_2CH_3}$$

i.e. $RCH_2CH_2OH + H_2 \rightleftharpoons RCH_2CH_3 + H_2O$

Similarly, the hydrocracking reaction, essential to many petroleum refineries, can be represented as the cracking of a hydrocarbon into a smaller hydrocarbon and an olefine followed by saturation of the olefine.

Table III shows that some types of reaction are reductions (e.g. hydrogenation, hydrogenolysis) and proceed in a reducing atmosphere while others are oxidations and may take place in an oxidizing atmosphere. The natures of the appropriate catalysts in the steady state may therefore be significantly different from one another and from those which effect the changes, like hydration, which occur under more or less neutral conditions.

5.   ACTIVITY OF SOLIDS

The chemistry of the elements is codified in the periodic table which exposes the special properties of the transitional elements and their compounds as a function of the presence of unfilled d-shells of electrons. Chemisorption is a chemical reaction with the leptons of the surface of a solid, and catalysis depends upon chemisorption, so that a relation between the patterns of behaviour of solids towards chemisorption and catalysis and the periodic table is expected and found.

The classification of chemical reactions suggests that a broad group of reactions can be recognized as redox in character. Following the same chemical theme, it can be inferred that these are associated with a relatively facile transfer and exchange of electrons with suitable catalysts which should then themselves be redox systems. The redox catalysts will be elements or compounds with leptons of variable valency, capable of sustaining a steady state in which the valency varies locally by small integers around some average value. The transitional elements are noted for their variable valencies and they feature prominently as redox catalysts. Furthermore, the electronic conductivity of a solid frequently arises from the presence of leptons of variable valency and we are provided with a link between catalysis and solid-state physics because the typical catalyst for a redox reaction is a typical conductor, i.e. a metal, semimetal or semiconductor. The associated physical properties (colour, magnetism, etc.) have also been correlated with catalytic activity by early workers [17] and now form the basis of powerful techniques for the investigation of surface chemistry.

In contrast, solids which are typical insulators are poor redox catalysts. Instead, they induce reactions which proceed through polarized or ionically dissociated species of the types encountered in general catalysis by acids and bases. The typical insulators are composed of leptons which are derived from the non-transitional elements.

The connection with solid-state physics is evident if catalysts are broadly classified as metals, semiconductors and insulators and further subdivided into transitional and non-transitional species.

6.   METALS

The metals of the periodic table of the elements (excluding the lanthanides and the actinides) are given in Table IV. The pure scientist is widely interested in the metals and their alloys in his quest for omniscience but the applied scientist and the technologist are principally concerned with systems which can be used easily and cheaply at temperatures and pressures less than $\sim 1000°C$ and ~ 500 atm respectively. It must be possible to prepare the metals readily in high area and they must be stable; in particular,

TABLE IV. METALS OF THE PERIODIC TABLE

1a	2a	3a	4a	5a	6a	7a	8			1b	2b	3b	4b
Li	Be												
Na	Mg											Al	
K	Ca	Sc	Ti	V	Cr	Mn	Fe	Co	Ni	Cu	Zn	Ga	
Rb	Sr	Y	Zr	Nb	Mo	Tc	Ru	Rh	Pd	Ag	Cd	In	Sn
Cs	Ba	La	Hf	Ta	W	Re	Os	Ir	Pt	Au	Hg	Tl	Pb

(Metals underlined are candidate catalysts.)

the metals must resist oxidation in the presence of small concentrations of oxidants under the reaction conditions. For such metals the oxides are readily reduced (see Table IV) and the equilibrium constants for the reducing reactions are large without the need to use either reducing agents of high chemical potential with respect to oxygen (C, Al) or extreme conditions of temperature and pressure. Hydrogen is the preferred reductant:

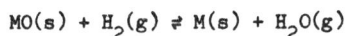

$$MO(s) + H_2(g) \rightleftharpoons M(s) + H_2O(g)$$

because it introduces the minimum of undesirable side reactions. Carbon monoxide might, for instance, yield a stable carbided surface instead of the required clean metal:

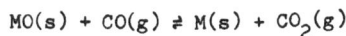

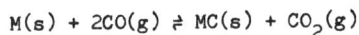

$$MO(s) + CO(g) \rightleftharpoons M(s) + CO_2(g)$$

$$M(s) + 2CO(g) \rightleftharpoons MC(s) + CO_2(g)$$

Metals possessing stable oxides (e.g. Na, Mg, Al) are used only for special purposes and require anhydrous reactants and de-oxygenated atmospheres.

Relatively few of the metals of the periodic table have reducible oxides and a smaller number, the noble metals, form no stable oxides, but it happens that most of these possess outstanding or unique catalytic activity. Under reducing atmospheres, e.g. in the hydrogenation of a substance containing no oxygen or sulphur atoms, the partial pressure of oxidants is negligible and all the metals underlined in Table IV are candidate catalysts. In atmospheres containing an oxidant, for example the oxidation of ammonia to nitric oxide or of ethylene to ethylene oxide, the selection is much smaller and comprises essentially the noble metals. If sulphur is present, the number of metals resistant to sulphidation is smaller still because of the smaller equilibrium constants of reactions of the type:

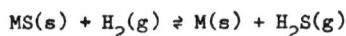

$$MS(s) + H_2(g) \rightleftharpoons M(s) + H_2S(g)$$

A transitional metal or alloy is defined as one comprised of atoms having an average number $\bar{\jmath}$ (the electron to atom ratio) of d-electrons in a 3d, 4d, or 6d-shell, with $0 < \bar{\jmath} < 10$. Such 'd-metals' possess a surface chemistry which is markedly different from that of the non-transitional

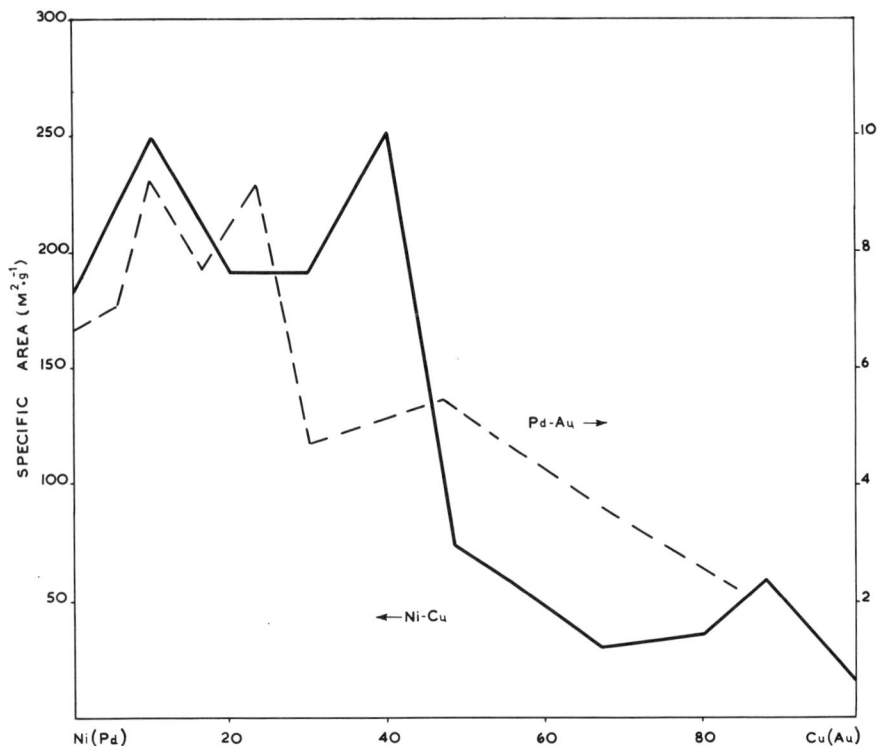

FIG.2.   Variation of specific area of alloy catalysts.

('s,p-') metals. The d-metals have unfilled d-orbitals (holes in the d-band) contributing characteristic physical and chemical properties which suggest many interesting correlations, e.g. with crystal structure, cohesive energy, work function, magnetic properties, etc.

6.1. Specific area

The melting points of the metals range from 234 K (Hg) to 3683 K (W) with corresponding Tammann temperatures of 117 K (Hg) to 1842 K (W), and Hüttig temperatures of 78 K (Hg) to 1228 K (W). From the earlier general discussion of specific area it is evident that mercury will sinter easily and would be difficult to make and use in high area whereas tungsten will be resistant to particle growth.

The transitional metals, with the exception of lanthanum, have higher melting points (1493 K (Mn) to 3683 K (W)) than all the non-transitional metals (234 K (Hg) to 1356 K (Cu)) except beryllium. Because catalytic activity per unit mass is proportional to the specific area, the transitional metals may already be expected to show the higher specific activities.

The broad correlation of specific area with melting points is shown in a marked manner in those parts of the long periods where the electronic structure of the metals shows relatively sharp changes with γ, as in the Ni-Cu and Pd-Ag, face-centred cubic, series of solid solutions (Fig.2) near the equiatomic composition.

6.2. Intrinsic activity

The review will be simplified by reference to reactions in (a) reducing atmospheres, and (b) oxidizing atmospheres, and by proceeding from chemisorption to catalysis.

6.2.1. Reducing atmospheres

The most important reactions are numbers 1, 4, 7, 8 and 10 of Table III (involving the chemisorption of almost all the simple molecules and hydrocarbons) but especially the interactions with hydrogen about which much is known.

Chemisorption

In earlier days the chemical reaction of adsorption with solid surfaces was recognized and distinguished from physisorption by the high heat of reaction (chemisorption). For small molecules a heat of chemisorption ~ 10 kcal $\cdot$ mole^{-1} would have been taken as indicative of chemisorption. If an adsorbate in that mode also took part in a catalytic reaction, the diagnosis was complete. In recent times the discovery of a number of different chemisorbed states on a given surface, including some weakly held forms, has greatly complicated the issue. The observations are now too many and too complex to be detailed; instead, the correlations of the strong chemisorptions with the electronic structure of the alloys are summarized:

(a) Dioxygen is chemisorbed by all metals as adatoms;
(b) Dihydrogen is taken up by all d-metals as adatoms; the chemisorptions by s,p-metals are much less strong;

$$H_2(g) + 2* \rightarrow 2\ \overset{H}{\underset{*}{|}}\quad (* = \text{surface atom});$$

(c) Dinitrogen adsorbs as adatoms on all the d-metals except nickel and the precious metals of group 8. There is no chemisorption by s,p-metals;
(d) All other molecules show the same pattern of behaviour as hydrogen. Saturated molecules are dissociatively adsorbed but unsaturated adsorptives can give both dissociated and undissociated adsorbates:

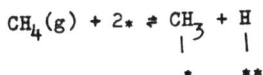

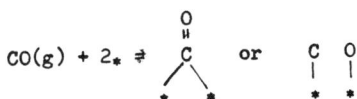

$$CH_4(g) + 2* \rightleftharpoons \underset{*}{\overset{}{CH_3}} + \underset{**}{\overset{}{H}}$$

$$CO(g) + 2* \rightleftharpoons \overset{O}{\underset{* \quad *}{\overset{\|}{C}}} \quad \text{or} \quad \underset{* \quad *}{\overset{}{C \quad O}}$$

It is evident that the bonding with metals having unfilled d-orbitals is much the strongest but there is some evidence [18] that the chemisorption on manganese is relatively weak. The difference between the d-metals and the s, p-metals is shown dramatically by the chemisorption of dihydrogen on the Pd-Au series of solid solutions [19]; the palladium-rich alloys adsorb strongly but the gold-rich alloys, with filled 4d-bands, are inert. Accommodation coefficients seem to show similar trends — high on Pt, lower on Cu [20].

Heats of chemisorption (ΔH_θ) always decline with increasing coverage (θ) by adsorbate up to $\theta = 1$ and the qualitative statements given above reflect roughly the variation of the heats with the nature of the adsorbent. Initial heats ($\theta \sim 0$) represent more precisely the binding energy uncomplicated by adsorbate interactions. ΔH_0 is much larger for adsorption on the group-8 metals than on group 1B metals but increases still more for the d-metals further to the left in groups 7, 6, 5 and 4; ΔH_0 for hydrogen on manganese is anomalously small. Many relations have been suggested between ΔH_0 and the physical properties of the metals: e.g. with the cohesive energy, the work function, the maximum energy of the electrons, Pauling's empirical metal valencies and with various combinations of these [21]. The relations describe only the broad trends and the calculations give at best qualitative results; little can be said about the variation of ΔH_0 with the geometry of the solid surface or the particular type of chemisorbed complex.

The rates of chemisorption at small values of θ on the different metals give the same pattern as the extents and the heats of chemisorption, just as would be expected from the Lennard-Jones diagram (Fig.1). Adsorptives of low ionization potential or large electron affinity are strongly adsorbed. If the adsorption is very strong and the adsorptive is a reactant, the reaction rate may be small and the reaction is 'inhibited' or 'self-poisoned'; if the adsorptive is an impurity, the reaction may be 'poisoned'. The adsorptives of low ionization potential usually have regions of high electron density (unsaturated molecules), non-bonding ('lone pairs') or anti-bonding electrons or negative charges. The electronegative elements (O, S, halogens), with their derivatives, are among the most common poisons and probably form surface structures on metal surfaces not unlike the surfaces of the corresponding oxides, sulphides and halides.

Absorption (solution)

Only small adatoms (H, C, N, O) can migrate from the surface to interstitial positions in the bulk where they form solid solutions or compounds (hydrides, carbides, nitrides, oxides, hydrocarbides, carbonitrides, etc.) [22].

The existence ranges of the solutions and compounds form a pattern which, as has been recognized for a long time, is quite similar to that for the strong chemisorption of the gases. The heats of formation increase from right to left in the long periods, and heats of chemisorption for diatomic molecules (X_2) have often been related to the corresponding bond (M-X) enthalpies calculated from the heat of formation of the compound MX.

Hydrogen: In parallel with chemisorption, solubility and hydride formation increase as the d-bands begin to empty. For the Cu-Ni, Ag-Pd, Au-Pt systems the solubility is much greater in the d-metal-rich alloys. In palladium alloys the increase is spectacular and accompanied by hydride formation. Indeed

the uptake of hydrogen has been associated specifically with the presence
of unfilled orbitals in the 4d-band of the alloys [23]. The Cu-Ni alloys behave
similarly if the hydrogen is supplied at a high chemical potential, as by
electrolytic deposition [23]. Apart from palladium in group 8, solution is
small and endothermal and remains so until group 5 when it becomes large
and exothermal and accompanied by hydride formation. A closer look at
the solubility of hydrogen in metals reveals very interesting changes with
γ. At manganese ($\gamma = 7$) solution is exothermal at temperatures below
$\sim 650°C$ but endothermal at higher temperatures [24] and the reversal has
been associated with the complex crystal structure of α-Mn and in turn with
its electronic structure [25]. The boundary between the endothermal and
exothermal absorbers in the vicinity of Mn and Cr in the first long period
does not appear so sharp as that in the other long periods between groups 5
and 6. It has been shown [26] that the heats of solution of hydrogen in Ti-Nb,
Ti-Mo, Nb-Re and Mo-Re alloys become endothermic at values of γ between
5.5 and 6.1 in the region of composition where there is a minimum in the
density of states at the Fermi surface.

Carbon: The sorption of carbon atoms and molecules is to be expected
but is a rare occurrence in practice. The formation of carbides from organic
molecules is, on the other hand, very common and is more often a nuisance
than a help. Again the same pattern emerges — strong bonding in groups 4,
5 and 6, decreasing to instability in group 8; in the higher periods the
decrease in stability is very marked between groups 6 and 7.

Nitrogen: The solubility of nitrogen and the stability of the nitrides
follow those of the carbides. The chemisorption of dinitrogen on the series
Cu-Ni-Co-Fe becomes marked at iron where the nitride Fe_4N is formed
exothermally. The strengths of the bonds M-X all attain critical values in
the regions between groups 5 and 7 and groups 8 and 1B, i.e. near the middle
and at the end of the series of transitional metals. The factors which affect
these bond strengths appear to influence the strengths of chemisorption
also.

Catalysis

The rates of the redox reactions give the same pattern of activity as
the chemisorptions but, following from the foregoing discussion, the number
of metals investigated is much smaller, and smaller in industrial than in
academic work. Only the simplest reactions, e.g. atom recombination,
ortho- to para-hydrogen conversion and hydrogen-deuterium exchange
($H_2 + D_2 \rightleftharpoons 2HD$) have been examined over a wide range of metals including
those with irreducible oxides.

The rate constants for H_2/D_2 exchange on the metals of the first long
period from Ti to Cu [27] show the characteristic sharp decline from Ni
to Cu and a small minimum at manganese. In the binary solid solutions
Ni-Cu, Pd-Cu, Pd-Ag, Pd-Au and the Pt-Cu alloys, the d-metals are always
more active than the s,p-metals for H_2/D_2 exchange, CH_4/D_2 exchange, the
hydrogenation of multiple bonds, hydrogenolysis of saturated hydrocarbons,
decarbonylation, etc. Over the palladium alloys there is often a quite marked
decrease in the rate of hydrogenation near the composition at which the 4d-
band is just full. Presaturation of the palladium alloys with hydrogen (which

fills the 4d-orbitals in somewhat the same way as Cu, Ag, or Au) lowers the activity but does not change the pattern. As $\bar{z}$ decreases from 10 in the long periods, hydrogenation activity tends to decline but for given $\bar{z}$ it increases with atomic number. Because the strong chemisorption of nitrogen is marked only at iron, and because metals of lower $\bar{z}$ (in that period) have stable oxides, the synthesis of ammonia is limited to Fe, Mo, Ru, W, Re and Os.

Polymerization of unsaturated molecules does not take place selectively on metals but it occurs parasitically in many of the reactions which are catalysed by them. The hydrogenation of unsaturated hydrocarbons is always accompanied by the formation of small amounts of polymer which can adversely affect catalyst activity and introduce difficulties in both research and technology. Intermediates may oligomerize to give a desired product; thus the hydrogenation of carbon dioxide can take a number of paths as follows:

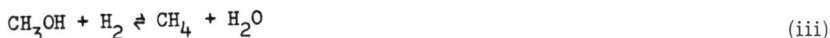

$$CO_2 + H_2 \rightleftharpoons CO + H_2O \tag{i}$$

$$CO + 2H_2 \rightleftharpoons CH_3OH \tag{ii}$$

$$CH_3OH + H_2 \rightleftharpoons CH_4 + H_2O \tag{iii}$$

Over the Cu-Ni system at atmospheric pressure the 'reverse-shift' reaction (i) proceeds selectively on the copper-rich alloys whereas the reaction through to methane ($CO + 3H_2 = CH_4 + H_2O$) dominates on the nickel-rich alloys and on Co and Fe also. At elevated pressures (20 - 110 atm) copper catalyses (ii) to yield methanol and the parasitic production of methane increases as nickel is added. Reactions leading to higher alcohols, olefines and paraffins begin to take place at modest pressures over catalysts based on nickel, cobalt and iron:

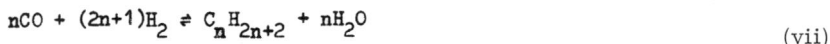

$$3CO + 6H_2 \rightleftharpoons C_3H_7OH + 2H_2O \tag{iv}$$

$$nCO + 2nH_2 \rightleftharpoons C_nH_{2n+1}OH + (n-1)H_2O \tag{v}$$

$$\rightleftharpoons C_nH_{2n} + nH_2O \tag{vi}$$

$$nCO + (2n+1)H_2 \rightleftharpoons C_nH_{2n+2} + nH_2O \tag{vii}$$

as though some chemisorbed intermediate remains longer on the metals of smaller $\bar{z}$ allowing condensation and polymerization to molecules of greater size. Ruthenium metal yields waxy paraffins of very high molecular weight [28].

The sharp differences between the d-metals and the s,p-metals are shown in the selective hydrogenation of unsaturated molecules. Substituted benzenes with unsaturated side-chains are easily saturated over nickel catalysts, whereas the aromatic ring remains unaffected over copper:

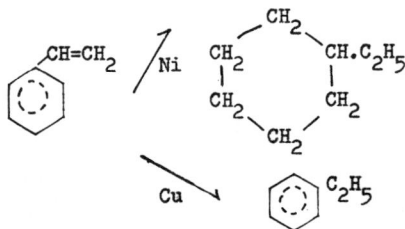

Because selectivity tends to increase as activity decreases, the unique activity of palladium in selective hydrogenation (acetylenes to olefines, dienes to mono-olefines) is more difficult to understand.

The hydrogenolysis of C-X bonds:

$$RX + H_2 \rightleftharpoons RH + HX$$

where X = OH, NH_2, SH, halogen, etc., is much slower over s, p-metals than over d-metals, as is demonstrated by the hydrogenation and hydrogenolysis of unsaturated fatty acids:

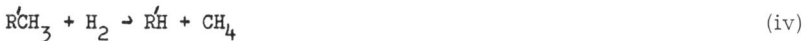

$$RCH_2C \overset{O}{\underset{OH}{\diagup}} + H_2 \rightarrow RCH_2C \overset{O}{\underset{H}{\diagup}} + H_2O \qquad (i)$$

$$RCH_2C \overset{O}{\underset{H}{\diagup}} + H_2 \rightarrow R\dot{C}H_2OH \qquad (ii)$$

$$R\dot{C}H_2OH + H_2 \rightarrow R\dot{C}H_3 + H_2O \qquad (iii)$$

$$R\dot{C}H_3 + H_2 \rightarrow R\dot{H} + CH_4 \qquad (iv)$$

Metallic lead catalyses reactions (i) and (ii) without saturating multiple bonds in the hydrocarbon chain; copper takes the reaction as far as (ii) but saturates the chain and may produce a little hydrocarbon (iii), but nickel gives saturated hydrocarbon (iii) and may cause some further hydrogenolysis (demethanation (iv)). Only the noble metals of group 8 can be used for the hydrogenolysis of carbon-halogen bonds.

Many complex processes in the petroleum refining industry, such as catalytic reforming ('platforming') depend upon the hydrogenating-dehydrogenating and associated properties of the most active metals (Pt, Ir, Pd, Rh). In more recent years the selectivity and the stability of single metals have been improved on one hand by addition of s, p-metals or semimetals and on the other by combinations with more refractory metals. These so-called bi-metallic catalysts probably contain alloys but the name indicates the difficulties inherent in catalyst characterization.

6.2.2. Oxidizing atmospheres

Oxidation occurs in reactions 11 to 15 of Table III with dioxygen as the oxidant but corresponding reactions with sulphur are also included. With dioxygen as the oxidant, only Rh, Pd, Ir, Pt, Au and, under certain conditions, Ag can persist as metals; in the presence of sulphur, silver metal must be excluded.

Chemisorption

It has already been noted that dioxygen is chemisorbed by all metals, even by gold which was once thought to be an exception. The ease with which oxygen forms subsurface oxide layers makes the measurement of heats of chemisorption difficult but it is clear that initial heats are always large and increase from group 8 to group 4 in the long periods. Unlike the other gases there is some evidence, especially for silver, that the heats increase in going from group 8 to group 1B although this effect is not much in evidence between nickel and copper. Present data show no singularities around manganese.

Most of the information concerning the chemisorption of sulphur has been derived from experiments on the system $M + H_2S \rightleftharpoons MS + H_2$ where M is a single crystal of Ni, Cu or Ag [29]. The adsorption isotherms so derived show that the heats depend upon the surface exposed and that the sulphur tends to accumulate at or near steps in vicinal faces. Low-energy electron diffraction (LEED) reveals that with increasing coverage the adsorbed species form a number of different surface arrays, some simply related to the symmetry of the metal atoms of the surface plane and others more complicated.

Absorption

The extent of solid solution of dioxygen in metals is small except in silver but all metals except gold form more or less stable oxides. The heats of formation of the oxides and sulphides increase from group 1B to group 8 and sharply to the left of group 8.

It is interesting that the heats of solution of oxygen in the Ti-Nb, Ti-Mo alloys fall abruptly when $\gamma \geq 5.6$ just as do the heats for hydrogen [23].

Catalysis

Again the d-metals are more active than the s,p-metals in the oxidation of hydrogen, carbon monoxide, hydrocarbons, ammonia, sulphur dioxide, etc. The intrinsic oxidation activity of platinum is, in general, greater than that of all other solids and it is widely used to catalyse non-selective oxidations of all types as well as a few selective oxidations under special conditions (e.g. $CH_4 + NH_3 \rightarrow HCN + 3H_2$ as $CH_4 + NH_3 + \frac{3}{2}O_2 \rightarrow HCN + 3H_2O$). Many of these oxidations form basic processes in the chemical industry ($SO_2 \rightarrow SO_3 \rightarrow H_2SO_4$; $NH_3 \rightarrow NO \rightarrow NO_2 \rightarrow HNO_3$) and in the abatement of pollution (h. carbs $\rightarrow CO_2$, H_2O; $CO \rightarrow CO_2$; NO, $H_2 \rightarrow N_2$, H_2O).

The alloys of Pd with Ag and Au show characteristic variations in catalytic properties as the 4d-band fills [30] with sharp changes in activity and activation energy near the composition 0.5 Pd:0.5 Ag (or Au). As in hydrogenation, the selectivity increases as the activity decreases, a remarkable example being provided by the oxidation of ethylene to carbon dioxide and ethylene oxide over evaporated Pd-Ag alloy films [31]. Whereas the yield of carbon dioxide is very high over pure palladium and falls sharply to a low value (at 0.6 Pd), which remains roughly constant up to pure silver, the yield of ethylene oxide rises almost linearly from zero near the composition (0.4 Pd) at which the 4d-band of Pd is just full to a much higher value at pure silver. Silver is an essential component of catalysts for the production of ethylene oxide from ethylene, one of the most important petrochemical processes. Alloys from the Pd-Au system have been claimed in the patent literature for the oxidative dehydrogenation of butene to butadiene but are not yet used in industry.

6.3. Conclusions

The chemisorptive, absorptive, catalytic and electron-structure properties of metals are strongly correlated. The patterns of behaviour show that
d-metals are markedly different from and, with few exceptions, more active
than s, p-metals. Among the d-metals themselves there is evidence of
anomalous behaviour for electron to atom ratios in the range 5.5 to 7 with
the suggestion of a minimum of activity in the same region. Thus there is
a hint of a twin-peaked pattern of intrinsic activity, for reactions involving
hydrogen, with maxima in group 8 and groups 4 to 6.
    The same pattern of activity is found (for metals with reducible oxides)
for massive solids (single crystals, plates, foils), evaporated films and
polycrystalline metals prepared by various reductions whether supported
on inert solids or not. Consequently the same patterns are observed in
industrial practice and can be used in the design and the modification of
metal-based catalysts.

7. SEMICONDUCTORS

Inasmuch as all solids conduct at a sufficiently high temperature and
inasmuch as the 'typical' semiconductor is not composed of the very small
particles characteristic of most catalysts, the classification 'semiconductors'
can introduce difficulties — but it is suggestive to the solid-state physicist.
    The typical non-metallic conductors can be further subdivided as
follows [32]:

(a) Elements: C, Si, Ge, α-Sn, P, As, β-Sb, Bi, Se, Te
(b) Compounds: e.g. 2-4, Mg_2Ge, Ca_2Pb
                     2-5, $ZnO(S, Se, Te)$, $CdO(S, Se, Te)$
                     3-5, AlP, $GaAs$, $InSb$
                     3-6, In_2Te_3
                     4-4, SiC
                     4-6, $PbS(Se, Te)$, SnO_2, PbO_2
                     5-6, Sb_2Te_3, Bi_2Se_3, Sb_2O_3, Bi_2O_3
(c) d-band semiconductors: compounds containing leptons with unfilled,
                     low-lying d-orbitals

The elemental semiconductors are 's, p'-solids and from the discussion
of metals they are expected to be less active than d-metals. Only the 2-6,
4-6 and 5-6 semiconductors and the transitional metal compounds, mostly
oxides, sulphides and halides, have attained wide use in catalytic processes.
The stability of the oxides of the alkaline earths, Si, Al, In and P, preclude
the use of their semiconducting compounds as catalysts.

7.1. Specific area

Most of the elemental semiconductors are of relatively low melting
point (Sn, As, Sb, Bi, Se) and of high volatility (P, As, Sb, Se); they cannot
easily be made or maintained in high area. Moreover some tend to form
volatile hydrides (As, Sb, Se) or low melting oxides (P, As, Sb, Bi, Se, Te)
under appropriate atmospheres. The compound semiconductors, except

SiC and the 2-6 oxides and sulphides, have similar deficiencies. Carbon (melting point (MP) > 3500°C) alone among the elements, as amorphous carbon and graphite, possesses useful properties in applied catalysis, largely because of the very high areas in which more or less pure turbo-stratic carbons can be produced.

The oxides of the transitional metals are of high melting point and very low volatility with the exception of CrO_3 (196°C), V_2O_5 (MP 690°C), MoO_3 (sublimes 1155°C) and Re_2O_7 (sublimes 250°C). The sulphides have adequately high melting points although there exist low-melting (700-800°C) non-stoichiometric phases in the Co-S and Ni-S systems. The halides are generally of low melting point ($TiCl_3$, $CuCl_2$) or volatile ($TiCl_4$, WCl_{5-6}).

Phases of low melting point can be used in catalysts but they must be supported in stable porous solids.

7.2. Intrinsic activity

As with metals, reactions will first be considered in reducing atmospheres and then in oxidizing atmospheres.

7.2.1. Reducing atmospheres (Reactions 4, 7, 8, 10 and 16 in Table III)

Because of the variable valency of the leptons in the several classes of compounds, non-stoichiometry and shear-structures are encountered. Despite the indications given in Section 3.3 and some crystallographic characterization, the composition of the phases in situ cannot usually be deduced unequivocally. Indeed, the nature of the leptons in some catalysts, as determined by modern physical methods, has been the subject of extended polemics and remains one of the most important of the difficult problems of catalysis. It can be said that under reducing atmospheres the leptons will exhibit average valencies below the maximum for the element.

Chemisorption

The subject is vast and no attempt will be made to review it, only the main features are outlined with special reference to the solids of particular interest as catalysts. Solid-state physicists have reported many results concerning the surface states of semiconductors used by the electronics industry but although chemisorption and catalysis are important in the fabrication and functioning of electronic devices such effects will be omitted.

Dioxygen is strongly chemisorbed by all the semiconductors to give adsorbed species of the types O_2, O_2^-, O_2^{2-}, O^- and O^{2-}, depending upon the conditions and the nature of the solid [18, 33]. Dinitrogen is apparently not chemisorbed by any semiconductor although there is a suggestion that it may be taken up by some oxides of transitional metals in their lower valency states [34].

The elemental semiconductors and the compound semiconductors (with the exception of the oxides and sulphides) chemisorb hydrogen only slowly and at high temperatures. (Carbon and graphite take up hydrogen slowly at temperatures >600 K whereas nickel chemisorbs rapidly at 70 K). Methane and the oxides of carbon also chemisorb slowly at high temperatures but halogens, hydrogen halides, other hydrides (H_2S, NH_3, PH_3) and more polar molecules are chemisorbed (e.g. on GaP) [35] at room temperature.

The chemisorption of hydrogen on the oxides of the transitional metals of the first long period gives some indication of a twin-peaked pattern resembling that found on the metals. The major features are a strong chemisorption at Co_3O_4 and Cr_2O_3 with weak chemisorptions at MnO and Fe_2O_3 in which the cations possess half-filled 3d-shells; the insulator CaO chemisorbs negligible quantities of hydrogen [18].

Absorption

Hydrogen alone under reducing conditions has the ability to enter the lattice from the surface. The effect is very dependent upon the nature of the solid and the presence of lattice defects; hydroxyl groups are formed in small concentration in some oxides but the process has received little attention.

Catalysis

The semiconductors, other than the oxides and the sulphides, have not been carefully investigated. For those elements for which the few results are available there must be serious doubt concerning the state of oxidation of the surface. They are poor hydrogenation catalysts, as the negligible chemisorption of dihydrogen already indicates. Although the easy hydrogenation of nitrobenzene to aniline is catalysed by lead sulphide, no industrial chemist would choose to use a catalyst of such low activity. The reverse, endothermal reaction of dehydrogenation requires higher temperatures for thermodynamic reasons, and because of the exponential relationship between reaction rate and temperature even the less active solids can be effective in dehydrogenation at higher temperatures. Krylov [36] gives a series of activities for the dehydrogenation of isopropanol (GaSb> InAs > Ga_2Te_3> CdS> Ge> InSb> GaAs> Ga_2Se_3) and finds correlations, albeit with much scatter, with the band-gaps of the semiconductors, the spacings between the atoms, the differences in the electronegativities of the constituent atoms and the work function. The closer the solid approaches the metallic state the more active it becomes. Because of its instability this type of semiconductor is of negligible importance compared with the oxides and sulphides.

For the oxides and the sulphides of the metals of the long periods between titanium and gallium, both pure research and industrial practice provide activities which show again the twin-peaked pattern of activity already noticed for the metals. In the series of phases Sc_2O_3, TiO_2, V_2O_5, Cr_2O_3, MnO, Fe_2O_3, Co_3O_4, NiO, Cu_2O, CuO, ZnO and Ga_2O_3, each shown to have its proper crystal structure, there are distinct maxima of activity at Cr_2O_3 and Co_3O_4 with much smaller values elsewhere (Fig. 3). Even for the reducible oxides it is possible by careful experimentation to obtain their activities at moderate temperatures but these must be much less reliable than those of the irreducible oxides. The pattern seems to apply for the H_2/D_2 exchange, for the hydrogenation of ethylene and cyclohexene, and, over the irreducible oxides, for the dehydrogenation of

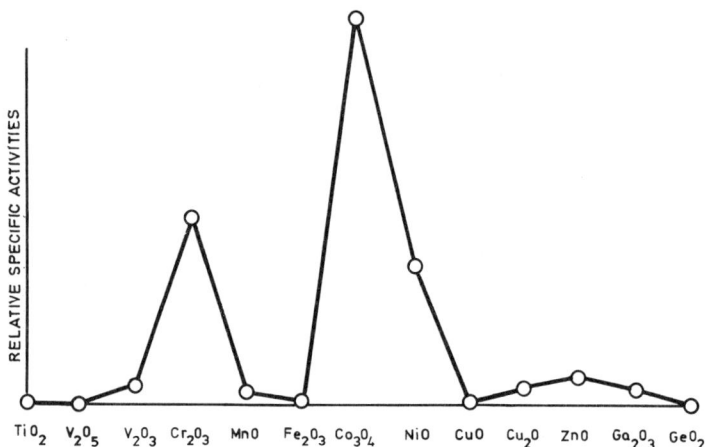

FIG.3.   Hydrogen-deuterium exchange over oxides of first long period.

paraffins. The high dehydrogenation activity of Cr_2O_3 is confirmed by its extensive application in industrial dehydrogenation catalysts at 550 - 650°C as the principal active phase or as a desirable additive (a 'promoter'). The results for the second and third long periods are incomplete but it is well known in industrial practice that 'MoO$_2$' and 'WO$_2$' in group 6 also have appreciable dehydrogenation activity; indeed the oxides and the sulphides of group 6 are outstanding in many respects.   In fact, industrial processes are seldom as simple as the type reactions chosen by the academic researcher dehydrogenation may be accompanied by demethanation (cracking) and by cyclization (aromatization):

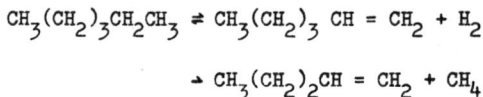

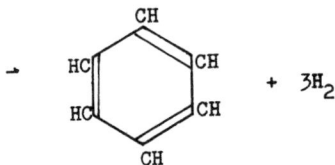

$$CH_3(CH_2)_3CH_2CH_3 \rightleftharpoons CH_3(CH_2)_3 CH = CH_2 + H_2$$

$$\rightarrow CH_3(CH_2)_2CH = CH_2 + CH_4$$

Demethanation is a parasitic reaction but aromatization is desirable in the petroleum refining and petrochemical industries. Chromia and molybdenum oxide are the most effective ('selective') oxides for dehydrogenation and cyclization. The oxides of other transitional metals possess activity for the reactions of hydrogen but the ultimate choice of a catalyst depends upon factors besides those which lead to high activity and selectivity.

Two further oxide catalysts frequently employed in hydrogenation–
dehydrogenation   must be mentioned – Fe_3O_4 and ZnO.   Magnetite and zinc
oxide have both been used as the principal active phases for the dehydrogenation of ethyl benzene to styrene (a monomer of great importance to the
plastics industry):

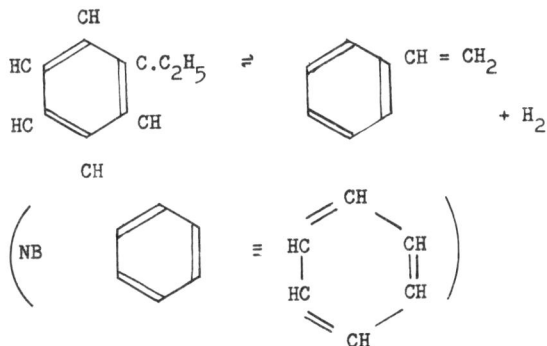

However, magnetite is a reducible oxide, so it must be used in the presence
of steam ($3Fe + 4H_2O = Fe_3O_4 + 4H_2$) [37]. The same situation arises in
the carbon monoxide shift-reaction used in hydrogen manufacture:

$$CO + H_2O \rightleftharpoons CO_2 + H_2$$

which can be viewed as a dehydrogenation of water assisted by the formation
of carbon dioxide. Stoichiometric zinc oxide has a small activity in the
reactions of hydrogen but the oxide is much more active when non-
stoichiometric, i.e. when the valency of the zinc leptons is less than two;
it is a characteristic of many leptons (e.g. Zn, Cd) that they lose activity
if placed in a matrix which inhibits such valency changes [38]. Combinations of zinc oxide with chromia are the most usual catalysts for the high-
pressure, high-temperature hydrogenation of carbon monoxide to methanol
and higher alcohols (Section 6.2.1 (Catalysis)). Generally the intrinsic activity
of oxides is much less than that of the most active metals and they are therefore expected to be more selective, at least in the sense that they introduce
fewer parasitic reactions. As we have seen, carbon monoxide must be
hydrogenated to methanol over a less active metal (Cu) in order to prevent
the formation of methane even at moderate temperatures, but zinc cannot
be used because of its low melting point: zinc oxide as a degenerate semiconductor may be considered to be an approximation to zinc metal. Hydrogenolysis occurs under more extreme pressures and temperatures, as in
the conversion of organic acids to alcohols.
    The intrinsic isomerization activities of the oxides, as exemplified by
the rate constants per unit area in the transformations of but-1-ene to cis-
and trans-but-2-ene, again display, with some differences, the twin-peaked
pattern of activity [39]; the high activity of zinc oxide is again to be associated with a reduced state of the lattice.
    Because olefines are reactive hydrocarbons they are vital industrial
intermediates and the processes for making them cheaply are in great

demand, e.g. cracking and dehydrogenation of saturated hydrocarbons
(above). The catalysed reactions of olefines themselves form the basis of
a large part of the petroleum refining and petrochemicals industries and
also confront us with some fundamental problems of mechanism. Olefines
can dimerize, trimerize, oligomerize to oily liquids or greases, or polymerize to give hydrocarbons of very high molecular weight which may be
amorphous or crystalline:

$$nC_mH_{2m} \rightleftarrows (CH_2)_{nm}$$

Disproportionation (or dismutation) of olefines to form two different olefines
also occurs [40]:

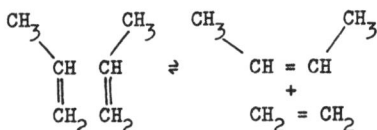

It is interesting that the polymerization of the smaller olefines to high
polymers and the disproportionation reactions are catalysed by oxides of
the transitional metals, particularly those of group 6 (Cr, Mo, W). Dimeri-
zation and oligomerization, on the other hand, are catalysed by a wide range
of oxides in complex combinations usually including an acidic component.
      Sulphide phases are stable in reactions with hydrogen only if the
equilibrium partial pressure of hydrogen sulphide, or its equivalent, is
exceeded so that the reactants must contain sulphur compounds. Sulphur
compounds are among the most powerful poisons for metallic catalysts;
consequently sulphide catalysts are most often used in the treatment of
sulphurous reactants. The reactions catalysed are of the same types as
those accelerated by metals and oxides but the activities fall in the series:
metals > oxides $\gtrless$ sulphides. Sulphides can be used to hydrogenate unsaturated
molecules, either selectively ($CH \equiv CH + H_2 \rightleftharpoons CH_2 = CH_2$) or completely
(cyclopentadiene to cyclopentane), to hydrogenolyse C-O, C-H, C-S bonds
and to effect dehydrogenations ($CO + H_2O \rightleftharpoons CO_2 + H_2$). Hydrogenolysis is
widely used to remove sulphur from the toxic impurities in hydrocarbon
feedstocks; thus:

$$CH_3SH + H_2 \rightleftarrows CH_4 + H_2S$$

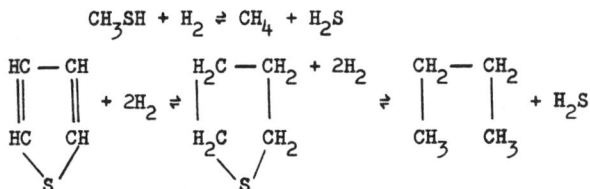

The hydrogen sulphide is subsequently removed by absorption in alkaline
solutions or in solid absorbents (ZnO, Fe_2O_3). The method is one of the
most effective for the abatement of pollution from combusted fuels ($H_2S \rightarrow SO_2$).
The pattern of intrinsic activity of sulphides of the first long period for the

hydrogenolysis of carbon disulphide ($CS_2 + 4H_2 \rightarrow CH_4 + 2H_2S$) is again twin-peaked with maxima in the vicinity of Cr_2S_3 and Co_4S_3 or NiS [41]. Data from industrial literature support this conclusion because the most active catalysts used in hydrodesulphurization are composed of at least one of the pair MoS_2, WS_2 together with at least one of the pair CoS_x, NiS_x. Again the compounds of the d-metals of group 6 exhibit special efficacy; even in the polymerization of olefines these sulphides show appreciable activity [36].

The halides are inactive in the reactions of hydrogen until they decompose and expose non-halide phases, but for the polymerization of olefines to poly-olefines of very high molecular weight they possess remarkable properties when associated with metal alkyls [36].

7.2.2. Oxidizing atmospheres (Reactions 11-15 in Table III)

Applied catalysis employs only the oxides (which now tend in situ to states of higher valency than are found under the conditions of 7.2.1) and some halides. It is helpful to distinguish between non-selective oxidation in which a hydrocarbon, for instance, is oxidized with oxygen to carbon dioxide and water and a selective oxidation wherein another valuable compound is produced (e.g. $CH_2 = CH.CH_3 + O_2 \rightarrow CH_2 = CH.CHO + H_2O$). The non-selective catalysts are variously used to remove toxic, noxious or hazardous pollutants from exhaust systems. Carbon monoxide is oxidized to carbon dioxide in gas-mask canisters and trace amount of evil-smelling organic compounds or quantities of combustible gases are oxidized to harmless CO_2 and H_2O; ammonia is converted to nitrogen and water. The complete oxidation of sulphur dioxide to the trioxide is an essential step in the production of sulphuric acid.

Again, regularities in intrinsic activity are found but they differ in some respects from those observed for metals and for oxides in reducing atmospheres. For the gas phase, oxidation of a number of compounds maxima have been reported at the oxides of Mn, Co and Cu [33] in the first long period: Ti < V < Cr < Mn > Fe < Co > Ni < Cu > Zn. Generally oxidation activity decreases with increasing atomic number in each group. Some variations in the detail of the pattern are found among the different reactants but there can be no doubt that the oxides of Mn, Co and Cu provide the most active oxide catalysts in all circumstances although their intrinsic activity is less than that of the most active metals: $Pt \geq Pd > Ag \geq MnO_x$, CoO_x, CuO_x. The distortions may arise from the intrusion of heat and mass transfer effects and uncertainties concerning the nature, stoichiometry and exposure of the solid phases under reaction conditions.

Very early in the history of catalysis it was noticed that the more active oxides are the less stable; the pattern of activity given above is reproduced by the series of inverse metal-oxygen bond strengths obtained from thermo-chemical data [33]: $TiO_2 < V_2O_4 < Cr_2O_3 < MnO_2 > Fe_2O_4 < Co_3O_4 > NiO < CuO > ZnO$. Direct measurement of the equilibrium pressures of oxygen over the oxides at various temperatures provides a direct measure of the M-O bond strength and good agreement with the series given above. It can also be seen that the typical p-type semiconductors are more active than the n-type although such a simple classification is inadequate for complex systems.

The exchanges of $^{18}O_2$ with $^{16}O_2$, either homomolecular:

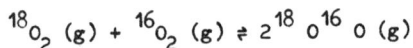

$$^{18}O_2 \text{ (g)} + {}^{16}O_2 \text{ (g)} \rightleftharpoons 2\,^{18}O\,^{16}O \text{ (g)}$$

or heteromolecular:

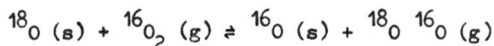

$$^{18}O \text{ (s)} + {}^{16}O_2 \text{ (g)} \rightleftharpoons {}^{16}O \text{ (s)} + {}^{18}O\,^{16}O \text{ (g)}$$

under certain conditions, proceed at the same rate (hence through the same oxygen species) with activation energies which parallel the M-O bond strengths. At moderate temperatures all these reactions appear to involve the following steps using surface anions of the solid:

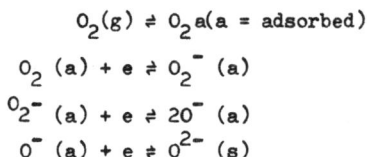

$$O_2(g) \rightleftharpoons O_2 a(a = \text{adsorbed})$$
$$O_2 \text{ (a)} + e \rightleftharpoons O_2^- \text{ (a)}$$
$$O_2^- \text{ (a)} + e \rightleftharpoons 2O^- \text{ (a)}$$
$$O^- \text{ (a)} + e \rightleftharpoons O^{2-} \text{ (s)}$$

The large electron affinity of oxygen and the change in the conductivity and the surface potential of the semiconductors during adsorption show that oxygen is an electron-acceptor. Then application of barrier layer theory provides an explanation of the small coverage with adsorbed gas at saturation and the fall of the heat of chemisorption with average. Oxide lattices with mobile leptons, because of low melting points or for other reasons, exchange oxygen with the gas phase to a considerable depth, even throughout the bulk in small particles.

Germain relates the formation of O^-, taken to be the active intermediate at moderate temperatures, to the redox properties of the solid as follows:

$$M^{n+} + O^{-2} \rightleftharpoons M^{(n-1)} + O^-$$

Whereas the reaction between CO and oxygen pre-adsorbed on a gallium-doped nickel oxide appears to proceed according to the equation:

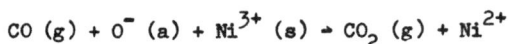

$$CO \text{ (g)} + O^- \text{ (a)} + Ni^{3+} \text{ (s)} \rightarrow CO_2 \text{ (g)} + Ni^{2+}$$

the mechanism of oxidation in lithium-doped nickel oxide is different [42]:

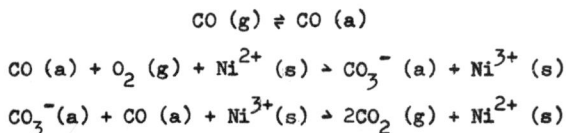

$$CO \text{ (g)} \rightleftharpoons CO \text{ (a)}$$
$$CO \text{ (a)} + O_2 \text{ (g)} + Ni^{2+} \text{ (s)} \rightarrow CO_3^- \text{ (a)} + Ni^{3+} \text{ (s)}$$
$$CO_3^- \text{(a)} + CO \text{ (a)} + Ni^{3+} \text{(s)} \rightarrow 2CO_2 \text{ (g)} + Ni^{2+} \text{ (s)}$$

In some vanadium oxide catalysts the oxidation goes through O^- but this species is not lattice oxygen but adsorbed oxygen [43]:

$$2V^{5+}(s) + O^{2-}(s) + CO\ (g) \rightarrow 2V^{4+}\ (s) + CO_2\ (g)$$

$$V^{4+}\ (s) + O_2\ (g) \rightarrow V^{5+} \ldots\ O_2^-\ (a)$$

$$V^{5+} \ldots\ O_2^-(a) + V^{4+} \rightarrow 2\left[V^{5+} \ldots\ O^-\right]$$

$$\left[V^{5+} \ldots\ O^-\right] + CO\ (g) \rightarrow V^{4+}\ (s) + CO_2\ (g)$$

$$\left[V^{5+} \ldots\ O^{2-}\right] + V^{4+}(s) \rightarrow 2V^{5+}\ (s) + O^{2-}\ (s)$$

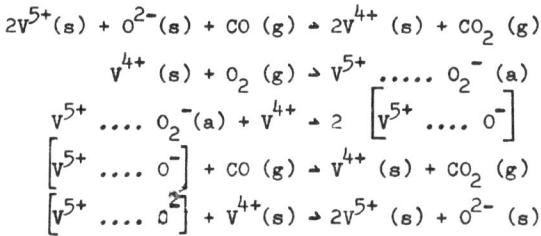

The oxidation of sulphur dioxide catalysed by vanadium species illustrates the progression from homogeneous to heterogeneous catalysis. The reaction is catalysed by vanadium sulphates dissolved in aqueous solutions of sulphuric acid or potassium sulphate, in molten potassium polysulphates and in crystalline polysulphates. Evidently long range order is not necessary for catalysis to occur and the industrial 'vanadium pentoxide catalysts indeed operate as molten complex potassium-vanadium sulphates held in the pores of porous silica. Selective oxidations include many of the most important modern catalytic processes and pose difficult problems because of the ubiquity of the parasitic non-selective reactions. In accord with the general rule, it is found that the less active oxides form the more selective catalysts. Among the oxides of the transitional metals those of vanadium, molybdenum, tungsten and uranium are much more selective than those of chromium, manganese, cobalt, nickel and Cu(2). The non-transitional oxides are much less active and, apart from Cu(1), tin and antimony, not outstandingly selective. However, selectivity cannot be considered independently of the reactants and products; the only successful selective oxidations are those which give a product having some resistance to further oxidation. The conversions of naphthalene and orthoxylene to phthalic anhydride:

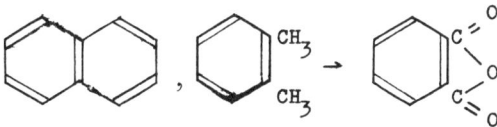

and of benzene, butenes, butanes, etc., to maleic anhydride:

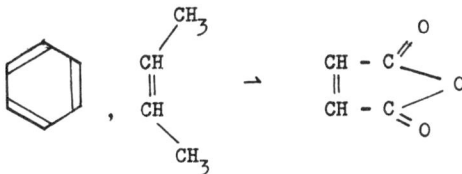

yield relatively stable ring systems. The oxidative dehydrogenation of butene to butadiene, the oxidation of propylene to acrolein and the ammoxidation of propylene to acrylonitrile all form molecules with some stabilization due to resonance:

$$CH_2 = CH-CH_2-CH_3 + \tfrac{1}{2}O_2 \rightarrow CH_2 \cdots CH \cdots CH \cdots CH_2 + H_2O$$

$$CH_2 = CH-CH_3 + O_2 \rightarrow CH_2 \cdots CH \cdots CH \cdots O + H_2O$$

$$CH_2 = CH-CH_3 + NH_3 + 1\tfrac{1}{2}O_2 \rightarrow CH_2 \cdots CH \cdots O \cdots N + 3H_2O$$

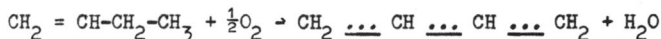

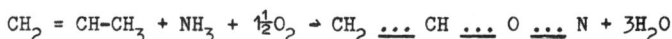

The oxidation of methane to the less stable formaldehyde molecule ($CH_4 \rightarrow HCHO$) is very difficult to achieve in an economic space-time yield.

Cuprous oxide is the only binary oxide-catalyst which has both high activity and high selectivity but specifically for the oxidation of olefines, e.g. propylene to acrolein; copper metal and cupric oxide are active and non-selective. Among the transitional oxides those of V and Mo are essential components of the most selective oxides; V_2O_5 itself is the precursor of the active phase in naphthalene oxidation. In some systems $V_{12}O_{26}$ is probably the actual catalyst and is one of the few oxides which have been investigated as single crystals by LEED; it exposes an unrecon-structed face composed of vanadium-centred square pyramids [44].

The prominence of oxides which also possess a strong dehydrogenation function is probably not accidental as research has shown that most selective oxidations are multistep reactions and that many include the abstraction of a hydrogen atom as the first step, e.g.:

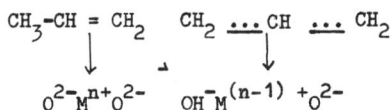

To proceed from the allyl species to acrolein, water is eliminated from the surface:

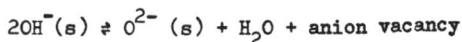

$$2OH^-(s) \rightleftharpoons O^{2-}(s) + H_2O + \text{anion vacancy}$$

and another hydrogen is lost from the organic intermediate:

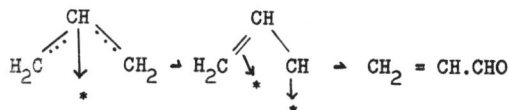

which then takes up oxygen to give the aldehyde. The best selective oxidation catalysts contain at least one additional component which in olefine oxidation is presumed to maintain the basis oxide in its highest valency state:

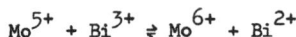

$$Mo^{5+} + Bi^{3+} \rightleftharpoons Mo^{6+} + Bi^{2+}$$

but the details are not known with certainty. Germain has noted that the selective catalysts are those formed from the elements with d^0 or d^{10} cations in their stable oxides and finds that selectivity varies directly as the ionic potential within the A and B subgroups [33].

8.   INSULATORS

The catalytic properties of the insulators are markedly different from those of the conductors but as the catalysts are used in high area, small particle size and, usually, in amorphous or poorly crystalline states, they can hardly be described as 'typical' insulators. The band structure is like that in crystalline insulators but the band edges are not well defined, being displaced into the forbidden gap. Especially at higher temperatures, this introduces the possibility of electron transfer and the emergence of atypical redox effects.

The compounds which are used most often in catalysts are:

Bases		Amphoteric or acidic oxides	
MgO (MP, 2800°C)	B_2O_3 (MP, 460°C)	SiO_2 (MP, 1700°C)	P_2O_5 (MP, 580°C)
CaO (2580)	Al_2O_3 (2045)	ArO_2 (2715)	
SrO (2430)		ThO_2 (3050)	
BaO (1923)			

Diamond is of considerable interest in the theory of the solid state but has no catalytic properties of value to the chemical industry.

They are all colourless diamagnetic solids and, with the exception of B_2O_3 and P_2O_5, of high melting point. The acidic oxides of high ionic potential are network-formers and essential components of glasses, whereas the bases, of low ionic potential, are among the network modifiers. Thus the acidic oxides lend themselves to the production of amorphous or defective solids of considerable area and stability. All the refractory insulators are widely used, alone or in combination, as supports and stabilizers for more mobile active phases.

The catalytic activities of the basic and acidic oxides are of the same kind as those possessed by the well-studied aqueous bases and acids of chemistry. Protonic (Bronsted) acids ionize relatively easily and donate a proton to some other species which is more basic:

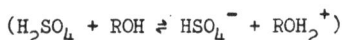

$$AH + B \rightleftharpoons A^- + BH^+$$

$$(H_2SO_4 + ROH \rightleftharpoons HSO_4^- + ROH_2^+)$$

A and B may be in the gas phase or the liquid phase, or one may be a solid and the other a fluid. An aprotic (Lewis) acid is an electron-pair acceptor:

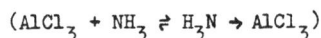

$$X_nM + :Y \rightleftharpoons X_nMY$$

$$(AlCl_3 + NH_3 \rightleftharpoons H_3N \rightarrow AlCl_3)$$

An acid is therefore a proton donor or an electron pair acceptor whereas a base is a proton acceptor or an electron-pair donor but the two aspects are not unrelated, e.g.

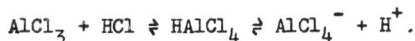

$$AlCl_3 + HCl \rightleftharpoons HAlCl_4 \rightleftharpoons AlCl_4^- + H^+.$$

The reaction of a neutral molecule with a base or an acid leads to the
formation of charged entities which may be more reactive than the reactant.
The mechanisms of many organic reactions catalysed by homogeneous
(dissolved) bases and acids are relatively well known and have been broadly
grouped according to the charge on the reactive intermediate.

Carbanion reactions, in which a negative ion is produced:

$$\equiv C:X \rightleftharpoons (\equiv C:)^- + X^+$$

are catalysed by bases:

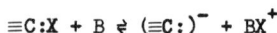

$$\equiv C:X + B \rightleftharpoons (\equiv C:)^- + BX^+$$

Carbonium ion reactions, involving a positive ion, are catalysed
by acids:

$$\equiv C:X + A \rightleftharpoons (\equiv C)^+ + AX^-$$

Alternatively one may have a concerted or 'switch' reaction utilizing
basic and acidic centres:

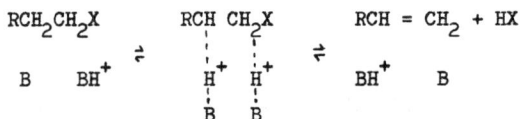

$$
\begin{array}{ccccc}
RCH_2CH_2X & & RCH\ CH_2X & & RCH = CH_2 + HX \\
| & \rightleftharpoons & |\quad\ | & \rightleftharpoons & \\
B\quad\ BH^+ & & H^+\ H^+ & & BH^+\quad B \\
& & |\quad\ | & & \\
& & B\quad B & &
\end{array}
$$

Such heterolytic, carboniogenic processes are typical but homolytic,
radical processes:

$$\equiv C:X \rightleftharpoons \equiv C\cdot + X\cdot$$

are atypical.

8.1. Bases

The pure oxides of magnesium, calcium and strontium as powders
prepared by evacuation at 1000°C show no significant deviations from
stoichiometry; no anion vacancies could be detected by electron-spin
resonance spectroscopy (ESR) although in magnesia, of high area, oxygen
and magnesium vacancies and aggregates of these seem to exist on the
surface [45].

8.1.1. Chemisorption

The alkaline earth oxides chemisorb only very small amounts of
hydrogen, nitrogen and oxygen at moderate temperatures and pressures,
although oxygen leads to bulk peroxides under more extreme conditions.
The small uptake of O_2 by MgO is associated with impurities or with
defects introduced by irradiation. Thus oxygen is chemisorbed by MgO

in proportion to the concentration of Mn^{2+} impurity [46]. Blue MgO, with about 1% of paramagnetic, electron-excess surface centres (similar to F centres), produced by γ radiation or by photolysis in hydrogen, reacts irreversibly with O_2 at 20°C; the blue colour is discharged and an ESR signal associated with O_2^- appears; such O_2^- species are stable at room temperature and do not exchange with the anions of the lattice [47]. The small adsorption of hydrogen is due either to the reduction of ions such as Mn^{3+} or to reaction with the electron-excess centres which have been introduced. Carbon dioxide forms surface carbonate ions on the normal oxides and CO_2^- as well as on the irradiated oxides [48].

Polar molecules, exemplified by water and alcohols, are strongly held.

Water adsorbed on outgassed magnesium oxide hydroxylates the surface [49]:

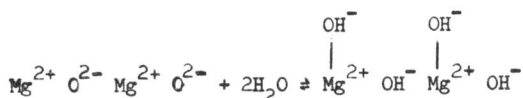

$$Mg^{2+}\ O^{2-}\ Mg^{2+}\ O^{2-}\ +\ 2H_2O \rightleftharpoons Mg^{2+}\ \overset{\displaystyle \underset{|}{OH^-}}{}\ OH^-\ Mg^{2+}\ \overset{\displaystyle \underset{|}{OH^-}}{}\ OH^-$$

A physisorbed layer can be adsorbed on top of a partly filled chemisorbed layer. Desorption proceeds by removal of adjacent hydroxyls and protons leading to residual clusters of (OH) groups and vacant areas. Single crystals of MgO behave similarly although it is claimed that isolated (OH) groups are of less frequent occurrence than on samples of high area [50]; it is suggested that the very strongly held water is as described above (but augmented by hydrogen bonds) and that a more weakly held form is found, again with some hydrogen bonding:

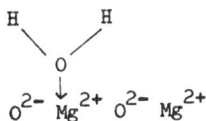

$$O^{2-}\ \overset{\displaystyle \overset{H\ \diagdown\ \ \diagup\ H}{\underset{\downarrow}{O}}}{Mg^{2+}}\ O^{2-}\ Mg^{2+}$$

The chemisorption of alcohols gives analogous adsorbed states; there is a hydrogen bonded complex:

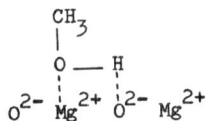

$$O^{2-}\ \overset{\displaystyle \overset{\underset{|}{CH_3}}{\underset{\vdots}{O \text{—} H}}}{Mg^{2+}}\ \overset{\displaystyle \underset{\vdots}{O^{2-}}}{}\ Mg^{2+}$$

sometimes described as strongly physisorbed, and an alkoxide form [51]:

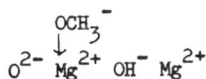

$$O^{2-}\ \overset{\displaystyle \overset{OCH_3^-}{\downarrow}}{Mg^{2+}}\ OH^-\ Mg^{2+}$$

IR methods show that decomposition of the methoxide complexes at higher temperatures forms adsorbed carboxylate and carbonate species. Similar chemisorbed complexes appear on Al_2O_3, SiO_2, TiO_2 and Cr_2O_3 [51].

Chlorine is rapidly adsorbed by MgO at 300 K to give a surface coverage
of about 0.2 chlorine ions per MgO ion pair; oxygen is evolved at 673 K [52]:

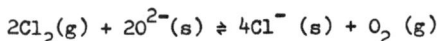

$$2Cl_2(g) + 2O^{2-}(s) \rightleftharpoons 4Cl^-(s) + O_2(g)$$

The first stage can be represented as follows:

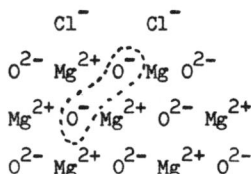

$$
\begin{array}{ccccc}
& Cl^- & & Cl^- & \\
O^{2-} & Mg^{2+} & O^- & Mg & O^{2-} \\
Mg^{2+} & O^- & Mg^{2+} & O^{2-} & Mg^{2+} \\
O^{2-} & Mg^{2+} & O^{2-} & Mg^{2+} & O^{2-}
\end{array}
$$

in which the oxygen complex may be intermediate between $2O^-$ and O_2^{2-}.

The basic oxides chemisorb acids and the strength and the number
of basic centres can be measured by methods quite analogous to those used
in homogeneous chemistry. The strength of the centres are found either
with the use of indicators which have different colours in their acidic and
basic forms or from an investigation of the strength of adsorption of a weak,
stable acid such as phenol [53]. The number of basic centres is obtained
by titration with an acid and an appropriate indicator, e.g. on CaO by
titration with a solution of benzoic acid in benzene using adsorbed bromo-
thymol blue as an indicator. Alternatively, the quantities of adsorbed
acids may be measured directly or calorimetrically from the heat evolution.

8.1.2. Catalysis

The basic insulator-oxides are very poor adsorbers of dihydrogen and
very poor hydrogenation catalysts. The catalysis of H_2/D_2 exchange by
MgO at 78 K [54] appears to occur on active centres which are defects
produced, for instance, by strong heat treatment ($\geqq 500°C$) or intrinsic defects
modified by irradiation. A number of authors associate the activity with the
presence of surface defects like V centres, e.g. an excited charge-
compensated V_1 centre:

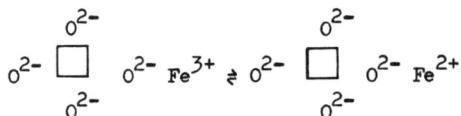

$$
\begin{array}{ccccc}
& O^{2-} & & & O^{2-} \\
O^{2-} \ \square & & O^{2-}\ Fe^{3+} & \rightleftharpoons & O^{2-}\ \square \quad O^{2-}\ Fe^{2+} \\
& O^{2-} & & & O^{2-}
\end{array}
$$

but a careful study of the reaction on the (111) face of MgO crystals gave
results in accord with a V_1 centre of the following type [55]:

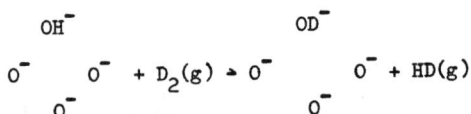

$$
\begin{array}{ccccc}
OH^- & & & OD^- & \\
O^- & O^- + D_2(g) & \rightarrow & O^- & O^- + HD(g) \\
& O^- & & & O^-
\end{array}
$$

The proton appears to be essential and the D_2 molecule is weakly held on the two equatorial O^- ions to give a roughly triangular activated complex and thus a form of Rideal-Eley mechanism:

$$H(ads) + D_2(g) \rightarrow D\ (ads) + HD\ (g)$$

As the V_1 centres are unstable sites, readily poisoned or removed by oxygen and water, this may explain the relative inactivity of MgO under industrial conditions. Nevertheless MgO possesses notable activity in hydrogen-transfer reactions one of which, the hydrogenation of acrolein to allyl alcohol at 400°C:

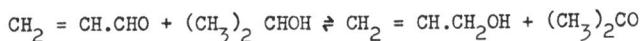

$$CH_2 = CH.CHO + (CH_3)_2\ CHOH \rightleftharpoons CH_2 = CH.CH_2OH + (CH_3)_2CO$$

is an important step in a synthetic route to glycerol [56].

The principal reactions catalysed by the solid bases are the heterogeneous counterparts of the homogeneous base-catalysed reactions, such as the Aldol condensation and its variants:

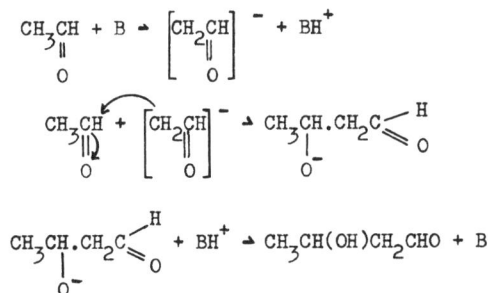

$$CH_3CH\underset{O}{\overset{\|}{}} + B \rightarrow \left[CH_2CH\underset{O}{\overset{\|}{}}\right]^- + BH^+$$

$$CH_3CH\underset{O}{\overset{\|}{}} + \left[CH_2CH\underset{O}{\overset{\|}{}}\right]^- \rightarrow CH_3CH.CH_2C\underset{O^-}{\overset{H}{\diagdown}}_O$$

$$CH_3CH.CH_2C\underset{O^-}{\overset{H}{\diagdown}}_O + BH^+ \rightarrow CH_3CH(OH)CH_2CHO + B$$

For reactions like:

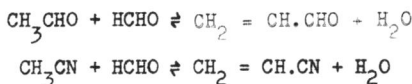

$$CH_3CHO + HCHO \rightleftharpoons CH_2 = CH.CHO + H_2O$$

$$CH_3CN + HCHO \rightleftharpoons CH_2 = CH.CN + H_2O$$

there is a linear relation between the apparent rate constants and the concentration of base in a supported catalyst [53].

There are very few gas phase industrial processes which depend upon the activity of a basic catalyst. In the developing countries the formation of butadiene from ethanol has assumed some importance and is perhaps an example of the activity of basic oxides:

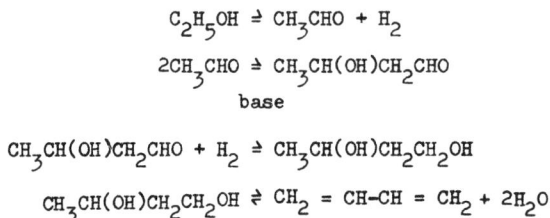

$$C_2H_5OH \rightleftharpoons CH_3CHO + H_2$$

$$2CH_3CHO \underset{base}{\rightleftharpoons} CH_3CH(OH)CH_2CHO$$

$$CH_3CH(OH)CH_2CHO + H_2 \rightleftharpoons CH_3CH(OH)CH_2CH_2OH$$

$$CH_3CH(OH)CH_2CH_2OH \rightleftharpoons CH_2 = CH-CH = CH_2 + 2H_2O$$

8.2. Acids

The acidic catalysts include not only the binary oxides Al_2O_3, SiO_2, P_2O_5, ZrO_2 and ThO_2 but also some ternary (BPO_4, $AlPO_4$, SiO_2-P_2O_5, SiO_2-Al_2O_3, SiO_2-MgO, SiO_2-ZrO_2) and quaternary (e.g. SiO_2-Al_2O_3-ZrO_2, SiO_2-Al_2O_3-MgO) systems, including the zeolitic aluminosilicates. Trivial examples are provided by phosphoric and sulphuric acids supported in porous carbon and by some acid salts (NaH_2PO_4, $CaHPO_4$, $NaHSO_4$).

The catalysts may be amorphous (SiO_2, SiO_2-Al_2O_3), poorly crystalline (the transitional aluminas) or crystalline (the zeolites) and comprise the solids most easily prepared in amorphous forms, very high areas and with very small pores. The transitional aluminas (especially γ alumina) are probably the most used and the cheapest substances employed as supports for less stable phases; they are metastable, containing more or less residual water, produced during the course of the dehydration of $Al(OH)_3$ to α alumina. γ alumina is representative; once thought to be a spinel with cation vacancies, then a hydrogen-aluminium spinel $(H_{0.5}Al_{0.5})Al_2O_4$, it is now known to possess a defective spinel structure with a slight tetragonal distortion.

The strongly acidic ternary system formed by silica with $\sim 10\%$ alumina may acquire its properties through the substitution of Al^{3+} for Si^{4+} but it is not clear precisely how the acidity arises. Many suggestions have been made covering the whole spectrum of possibilities, from electrostatic polarization to ionization, like those previously known in the study of mechanism in physical organic chemistry. The insertion of Al^{3+}, which is most commonly found in octahedral sites, into a tetrahedral Si^{4+} site in a hydrous SiO_2 amorphous structure produces a local negative charge which can be neutralized by the formation of a local OH group or shielded by the polarization of a water molecule, e.g.

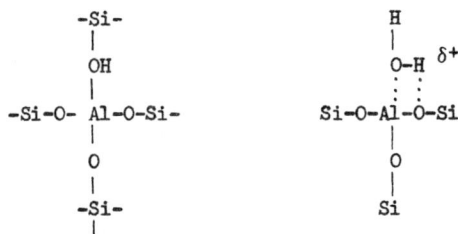

```
        -Si-                            H
         |                              |
         OH                           O-H   δ+
         |                            : :
-Si-O- Al-O-Si-                  Si-O-Al-O-Si
         |                              |
         O                              O
         |                              |
        -Si-                            Si
         |
```

                    Protonic acids

A centre apparently similar to this has been found in quartz and identified by electron-spin resonance spectroscopy [57]. Alternatively the site can be viewed as a Lewis acid reacting with water to give a Bronsted acid:

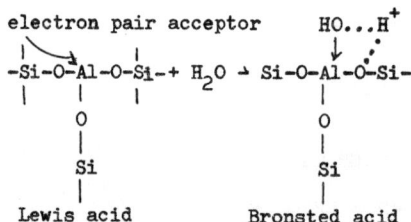

```
  electron pair acceptor        HO...H⁺
      \      |                     ↓ :
   -Si-O-Al-O-Si-+ H₂O  →  Si-O-Al-O-Si-
      |    |    |                   |
      O    O                        O
      |                            |
      Si                           Si
      |                            |
   Lewis acid                Bronsted acid
```

Other models are reviewed in Ref.[53]. The acidic zeolites are prepared from the alkali metal aluminosilicates of the faujasite, chabazite, erionite and mordenite groups with large 8A channels in their crystal structures [58]. On replacement of the monovalent cations by hydrogen, alkaline earth or rare earth cations, the solids become very strong acids, the strongest solid acids known. It can be seen that acidity is again associated with the insertion of ions which perturb the electroneutrality of the framework and the models are of the same kinds as those discussed for amorphous silica-aluminas. The local variations of geometry and Al^{3+} concentration in the silica-alumina catalysts result in active centres having a wide range of acid strengths. In the ion-exchanged crystalline zeolites, despite the regularity of the structure, there also exist sites of different geometry and various acid strengths, both protonic and aprotic.

If polar molecules are removed, some sites will become Lewis acid centres with a strong affinity for electron pairs and possibly able also to accept electrons in redox processes.

Chemisorption

Chemisorption by the acidic oxides is similar to that by the bases, i.e. small or negligible adsorption of the non-polar diatomic gases but strong adsorption of polar molecules. The number and strength of the active centres are found by methods like those used for basic centres, e.g. titration with suitable bases. The strengths of chemisorption of a series of molecules follow roughly the order of increasing basicity. The infra-red spectra of adsorbed molecules such as pyridine, which can chemisorb at aprotic centres and protonic centres to give distinguishable adsorbed species:

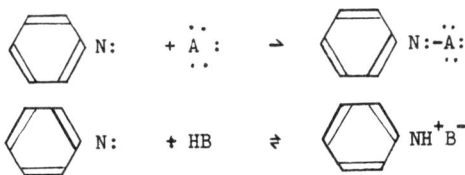

$$\text{Py}-N: \ + \ \ddot{A}: \ \rightarrow \ \text{Py}-N:-\ddot{A}:$$

$$\text{Py}-N: \ + \ HB \ \rightleftarrows \ \text{Py}-NH^+B^-$$

are invaluable in the characterization of surfaces. Substances of low ionization potential (perylene) or high electron affinity (tetracyanoethylene) will chemisorb to form, respectively, radical cations and anions, phenomena which although interesting in themselves and an indication of atypical redox properties, have not yet any industrial application [59].

Catalysis (Reactions 3-6, 9, 10 and 16 in Table III)

The classical reaction type is that intermediated by carbonium ions [60] formed by the removal of a negative group from a saturated reactant or the addition of a proton to an unsaturated molecule. Saturated hydrocarbons form carbonium ions either by loss of a hydride ion, e.g. to a Lewis acid centre:

$$C_nH_{2n+2} + \overset{..}{A}\!: \rightleftharpoons \left[C_nH_{2n+1}\right]^+ + H\!:\!\overset{..}{A}\!:^-$$

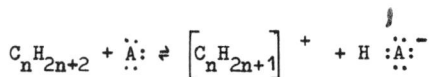

or by transfer to another carbonium ion:

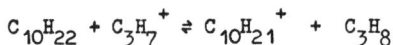

$$C_{10}H_{22} + C_3H_7^+ \rightleftharpoons C_{10}H_{21}^+ + C_3H_8$$

The initiating carbonium ion may arise from protonation of unsaturated impurities present adventitiously or produced by parasitic dehydrogenation, dehydration, etc.,

$$C_3H_6 + H^+ \rightleftharpoons C_3H_7^+$$

$$ROH + H^+ \rightleftharpoons R^+ + H_2O$$

The products of the more simple catalysed reactions of paraffins are affected by the relative stabilities of carbonium ions:

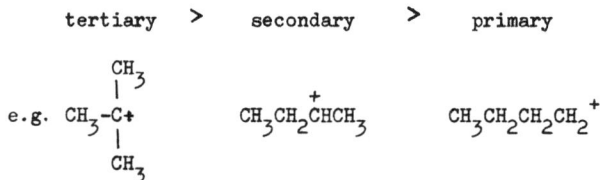

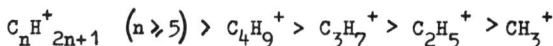

tertiary $>$ secondary $>$ primary

$$\text{e.g.} \quad \underset{\underset{\displaystyle CH_3}{|}}{\overset{\overset{\displaystyle CH_3}{|}}{CH_3\!-\!C\!+}} \qquad CH_3CH_2\overset{+}{C}HCH_3 \qquad CH_3CH_2CH_2CH_2^+$$

$$C_nH_{2n+1}^+ \ (n \geqslant 5) > C_4H_9^+ > C_3H_7^+ > C_2H_5^+ > CH_3^+$$

and by the tendency of the carbonium ions to undergo 'β fission':

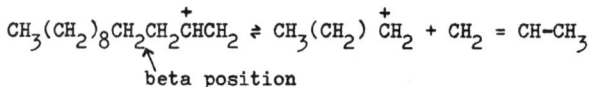

$$CH_3(CH_2)_8CH_2CH_2\overset{+}{C}HCH_2 \rightleftharpoons CH_3(CH_2)\ \overset{+}{C}H_2 + CH_2 = CH-CH_3$$

beta position

Reactions catalysed by acids are found throughout the petroleum refining and the petrochemical industries either as the main reaction (e.g. the cracking of heavy petroleum fractions to lighter hydrocarbons) or as contributors to multifunctional catalysis (e.g. hydrocracking).

The dehydration of an alcohol to an ether or an olefine is a typical acid catalysed reaction in which the acid may be aqueous (phosphoric or sulphuric acids) or solid (γ-alumina, silica-alumina). The reverse reaction, the hydration of an olefine to an alcohol, is more important, as in the production of isopropanol from propylene:

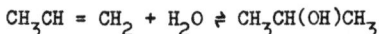

$$CH_3CH = CH_2 + H_2O \rightleftharpoons CH_3CH(OH)CH_3$$

Amines can be made by the intermolecular dehydration of ammonia and an alcohol over alumina:

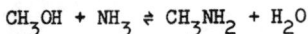

$$CH_3OH + NH_3 \rightleftharpoons CH_3NH_2 + H_2O$$

The reaction of adipic acid with ammonia to give adiponitrile is a principal reaction in the production of nylon:

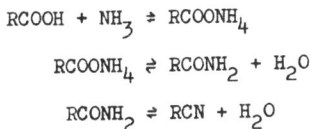

$$RCOOH + NH_3 \rightleftharpoons RCOONH_4$$

$$RCOONH_4 \rightleftharpoons RCONH_2 + H_2O$$

$$RCONH_2 \rightleftharpoons RCN + H_2O$$

The catalyst may be alumina or a solid phosphate.

By far the greatest tonnages of acidic catalyst are consumed in the cracking of petroleum fractions, which utilizes the following reactions in addition to the cracking of paraffins (noted above):

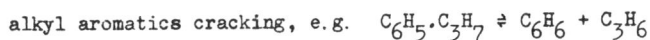

olefine cracking, $C_nH_{2n} \rightleftharpoons$ smaller olefines

cycloparaffin cracking, $C_nH_{2n} \rightleftharpoons$ smaller olefines

alkyl aromatics cracking, e.g. $C_6H_5 \cdot C_3H_7 \rightleftharpoons C_6H_6 + C_3H_6$

isomerization (double bond):

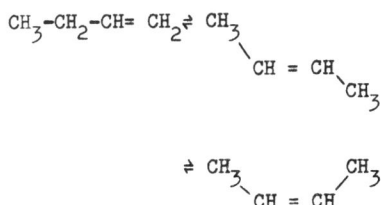

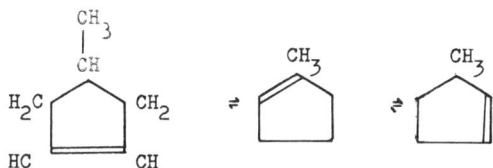

$$CH_3-CH_2-CH= CH_2 \rightleftharpoons CH_3$$
$$CH = CH \diagdown CH_3$$

$$\rightleftharpoons CH_3 \qquad CH_3$$
$$CH = CH$$

isomerization (skeletal):

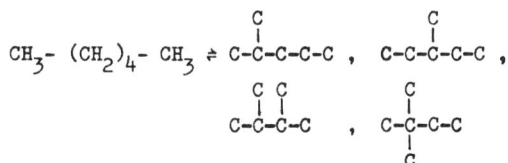

$$CH_3- (CH_2)_4- CH_3 \rightleftharpoons C-\overset{C}{\underset{|}{C}}-C-C-C \; , \quad C-C-\overset{C}{\underset{|}{C}}-C-C \; ,$$

$$C-\overset{C}{\underset{|}{\underset{|}{C}}}-\overset{C}{\underset{|}{C}}-C-C \quad , \quad C-\overset{C}{\underset{|}{\underset{|}{C}}}-C-C$$

The cracking of heavy oils containing paraffins, cycloparaffins and aromatics over silica-alumina and exchanged zeolite catalysts involves all these reactions in the production of fractions boiling in the gasoline range. Hydrogen transfer and transalkylation also occur.

In the production of paraxylene, a precursor of the terephthalic acid used in the manufacture of Terylene, acid catalysts are used to speed the isomerization of xylenes to the equilibrium composition:

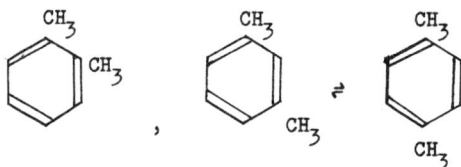

Cycloparaffins also isomerize over acids, e.g. cyclohexane to methyl cyclopentane:

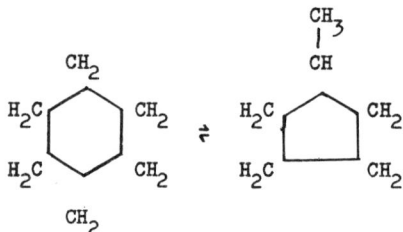

Naturally the reverse reactions are also acid catalysed, such as the alkylation of paraffins and aromatics by olefines or the polymerization of olefines to larger olefines (butenes to octenes).

Acidic oxides in the role of supports must nevertheless exert their typical activity and if the reactions so induced are parasitic it may be necessary to suppress the activity by poisoning with bases. A ubiquitous, deleterious process in almost all organic reactions is the deposition of so-called polymer 'carbon' or 'coke' by slow parasitic reactions leading to encapsulation of the catalyst surface. Reactants or intermediates may condense or polymerize to oligomers which then dehydrogenate and aromatize to form polynuclear aromatic species if the temperature is sufficiently high.

Finally it must be emphasized that basic and acidic properties are possessed by solids, oxides and sulphides which are electronic conductors; oxides such as MnO and ZnO are basic semiconductors whereas CrO_3, MoO_3, WO_3, etc., are acidic and under appropriate conditions will function as basic and acidic catalysts, respectively.

9. ELECTRONIC STRUCTURE AND SURFACE CHEMISTRY

The foregoing review of chemisorption and catalysis emphasizes the correlation with the electronic structure of solids. With the aid of simple ionic models it is possible to see, at least for oxides, the origin of the relationship, and current theories of the band structures of metals provide further insight into the activities of metals.

9.1. Insulators

Little has been done to calculate the proton affinity of solid bases and acids. There are only the broad lines laid down by the early expositors [61] of the ionic theory who pointed to the influence of ionic potential on the strength of acids and bases and explained the properties of magnesium hydroxide and phosphoric acid by the relative bond strengths due to electrostatic polarization by Mg^{2+}, P^{5+}, O^{2-} and H^{+}:

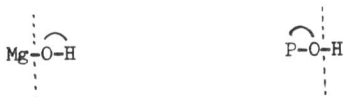

Mg$\frown$O$-$H   P$-$O$\frown$H

Using magnesium oxide (NaCl structure) as a model and the earlier simple type of electrostatic lattice-theory calculation [62], the absence of redox properties is easily understood as well as the association with chemical properties such as irreducibility [41].

9.2. Conductors

The progression from the insulating to the semiconducting lattice can also be followed together with the development of the ability to chemisorb hydrogen. The twin-peaked patterns of adsorption and activity observed for the oxides of the first long period then receive some explanation from the shift of the 3d levels of the cations into the forbidden gap of the insulators (e.g. comparing NiO with MgO) and the ligand-field splitting of the 3d levels [41]. As the lattices of the solids become more covalent the simple models cease to be applicable and more elaborate calculations will be necessary.

The properties of the metals and alloys depend upon the nature of the chemisorbed complexes formed with the aid of the orbitals of the atoms of the surface. In some situations it can be seen that the surface complex is like a surface molecule with molecular orbitals formed following the usual rules and filled by electrons in a manner which depends upon the Fermi level of the metal [63-65]. The processes on metals which are close together in the periodic table seem to depend upon the gradient of this density with energy. More fundamental investigations with simple models [66] indicate that the adsorption of hydrogen atoms will be stronger the narrower the band of electron states in the metal and the nearer the band is to being half full.

The early simple band theories of metals have now been superseded and the new constructs lead perhaps to particularly interesting conclusions as the d-bands of the metals fill between groups 8 and 1B. It will be recalled that at this electron-to-atom ratio the catalytic properties also change markedly. Atoms in alloys of the type Ni-Cu and Pd-Ag are now known not to share a common d-band; instead, particularly in the Ni-Cu alloys, physical properties depend primarily upon clusters (of different sizes) of atoms of one kind formed either randomly or under the influence of a pairwise interaction.

Clustering both in the absence (Ni-Cu) or the presence of d-band filling (Pd-Ag) thus reintroduces the geometric factor because one reaction (e.g. hydrogenolysis) may require a larger cluster of, say, nickel atoms than another (e.g. olefine saturation) [67].

The total effect of electronic structure and geometry upon the surface chemistry of solids therefore remains a most important area for research both theoretical and experimental.

10. MULTIFUNCTIONAL CATALYSTS

Evidently different types of solids have different typical intrinsic activities, usually accompanied by a much smaller activity for atypical reactions. The members of each class can also be ranked in order of increasing activity for the characteristic reactions, typical and atypical. Because each type of exposed surface will induce its own kind of activity almost irrespective of the presence of other phases, a composite catalyst can be 'multifunctional'; indeed the presence of the small atypical activity implies that even a single phase can be multifunctional.

Thus for complex processes which comprise a sequence of reactions it is often possible to compound a multifunctional catalyst containing a number of active phases, at least one phase appropriate to each step in the chain.

In the dehydration of ethanol, γ-alumina is unifunctional:

$$C_2H_5OH \rightleftharpoons C_2H_4 + H_2O$$

but the hydrogenolysis of an alcohol proceeds over a catalyst which may be bifunctional:

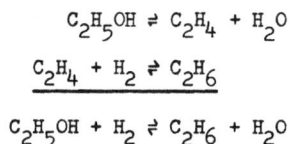

$$C_2H_5OH \rightleftharpoons C_2H_4 + H_2O$$
$$\underline{C_2H_4 + H_2 \rightleftharpoons C_2H_6}$$
$$C_2H_5OH + H_2 \rightleftharpoons C_2H_6 + H_2O$$

The first step requires a dehydrating catalyst (γ-Al$_2$O$_3$), the second step a hydrogenating catalyst (e.g. Ni), so that one suitable catalyst comprises finely divided metallic nickel supported upon high-area alumina. The catalyst will evidently be inhibited by poisons for nickel (e.g. sulphur compounds). Alternatively if alumina is used as a support for nickel in the hydrogenation of a ketone to an alcohol, subsequent unwanted hydrogenolysis can be inhibited by adding alkali to the alumina. The acceleration or the inhibition of acid-catalysed steps in complex reactions by the addition or neutralization of acids is a very common device; polymerization by acidic components leading to catalyst encapsulation by coke is frequently minimized by adding an alkaline component.

The catalytic reforming of naphtha to give gasolines of higher octane number is an excellent example of an important complex industrial process which depends upon a multifunctional catalyst. The reactions occurring in reforming include dehydrogenation, cyclization, aromatization, isomerization and some cracking, all of which occur over catalysts of platinum supported upon γ-alumina containing a small concentration of chlorine. Platinum supplies the hydrogenation-dehydrogenation function, the halogenated

alumina is the acidic component which contributes the property of acid-type
isomerization and cracking. The platinum dehydrogenates the saturated,
non-reactive hydrocarbons to more reactive olefines, e.g. cyclohexane
to cyclohexene, the olefines isomerize on the acidic constituent and are
rehydrogenated on the metal:

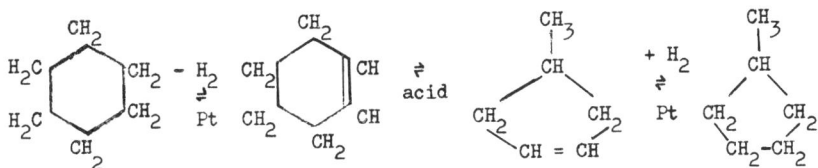

The conversion of ethanol to butadiene over a multifunctional catalyst has
already been mentioned. All hydrocarbon-oxidation catalysts are multifunctional in that C-C or C-H bonds must be broken, the fragments combined
with oxygen and water expelled.

The combination of the concept of the multifunctional catalyst with that
of the specificity of solid phases is of great use, to a first order, in the
interpretation of reaction mechanism and in the design of catalysts for
specific reactions.

11. SUPPORT ACTION

All the refractory solids, conductors and insulators may be used as
supports or stabilizers for active phases which have low melting points
or lattices which are otherwise mobile. Because the supports themselves
possess catalytic activity and solid reactivity the choice depends upon the
nature of the active phases and upon the reaction to be catalysed. Generally,
to engender catalysts of large area and small particle size, the support must
also be of high area and, moreover, easily and cheaply manufactured in
that form.

Oxidation reactions are intrinsically fast and it is not always necessary
to prepare catalysts in large area; for example, many oxidations can be
successfully accomplished with a catalyst (area $< 1\,m^2/g$) made by dipping
lumps of non-porous α-alumina into molten vanadium pentoxide. Hydrogen
cyanide can be manufactured by dehydrogenation of methane and ammonia
(at $> 1000°C$) over platinum supported on the inner surfaces of ceramic
alumina tubes or by oxidative dehydrogenation using platinum on pieces of
natural beryl.

Only the insulator oxides of negligible specific area and without
defective lattices operating at low or moderate temperatures can be
considered to be inert [68]. Such conditions are seldom encountered.
Semiconducting zinc oxide supported upon high area γ-alumina and used
for dehydrogenation can lose activity by forming the inactive, insulator
spinel (ZnO + Al_2O_3 = $ZnAl_2O_4$), a process accelerated by steam.

The parasitic dehydration and hydrogenolysis reactions caused by
alumina supports for active nickel in the hydrogenation of C=O groups have
already been noticed. It is sometimes possible to moderate such effects
by replacing alumina by chromia (Cr_2O_3), which possesses more activity
for hydrogenation and less for dehydration.

Some supports, especially silica, are moderately volatile in steam and cannot be used in steam at high pressures and temperatures. Even where volatilization does not occur, water vapour induces considerable mobility in solid surfaces and the catalyst support must then be carefully chosen to have a high lattice energy.

REFERENCES

[1] PRIGOGINE, I., DEFAY, R., Chemical Thermodynamics (transl. EVERETT, D.H.), Longmans Green, London (1954).
[2] PRIGOGINE, I., OUTER, P., HERBO, C., J. Phys. Colloid Chem. 52 (1948) 321.
[3] CREMER, E., Advances in Catalysis 7, Academic Press, New York (1955) 75.
[4] DEFAY, R., PRIGOGINE, I., BELLEMANS, A., Surface Tension and Adsorption (transl. EVERETT, D.H.), Longmans Green, London (1966).
[5] POLTORAK, O.M., BORONIN, V.C., Zh. Fiz. Khim. 40 (1966) 2671.
[6] GREGG, S.J., SING, K.S.W., Adsorption, Surface Area and Porosity, Academic Press, New York (1967).
[7] COBLE, R.L., BURKE, J.E., Progress in Ceramic Science 3, Pergamon, London (1963) 197.
[8] HENDERSON, B., Defects in Crystalline Solids, Edward Arnold, London (1972).
[9] BURTON, W.K., CABRERA, H., Discuss. Faraday Soc. 5 (1949) 33.
[10] HERRING, C., in Structure and Properties of Solid Surfaces (GOMER, R., SMITH, C.S., Eds), Univ. of Chicago Press, Chicago (1953) 5.
[11] GJOSTEIN, N.A., Metal Surfaces, Am. Soc. for Metals (1963) 99.
[12] VAN HARDEVELD, R., VAN MONFOORT, A., Surf. Sci. 4 (1966) 3996.
[13] BUDNIKOV, P.P., GINSTLING, A.M., Principles of Solid State Chemistry: Reactions in Solids, Maclaren, London (1968).
[14] KUCZYNSKI, G.C., in Ferrites (HOSHINO, Y., IDA, S., SUGIMOTO, M., Eds),Univ. Park Press (1971).
[15] JAKY, K., SOLYMOSI, F., BALTA, I., SZABO, Z.G., Reactivity of Solids (SCHWAB, G.M., Ed.), Elsevier, Amsterdam (1965) 540.
[16] KOOY, C., ibid, 21.
[17] ROGINSKII, S.Z., Dokl. Akad. Nauk SSSR 67 (1949) 97.
[18] HAYWARD, D.O., TRAPNELL, B.M.W., Chemisorption, Butterworth, London (1964).
[19] TARDY, B., TEICHNER, S.J., J. Chem. Phys. 67 (1970) 1968.
[20] SCHAFER, K., Z. Elektrochem. 56 (1952) 398.
[21] BOND, G.C., Catalysis by Metals, Academic Press, New York (1962).
[22] GOLDSCHMIDT, H.S., Interstitial Alloys, Butterworth, London (1967).
[23] EBISUZAKI, Y., O'KEEFE, M., in Progress in Solid State Chemistry 4 (REISS, H., Ed.), Pergamon, London (1967).
[24] SMITH, D.P., Hydrogen in Metals, Univ. Chicago Press, Chicago (1948).
[25] DOWDEN, D.A., J. Chem. Soc. (1950) 242.
[26] JONES, D.W., PESSALL, N., McQUILLAN, A.D., Philos. Mag. 6 (1961) 455; J. Phys. Chem. Solids 23 (1962) 1441.
[27] ELEY, D.D., SHOOTER, D., J. Catal. 2 (1963) 259.
[28] EMMETT, P.H., Ed., Catalysis 4, Reinhold, New York (1956).
[29] BENARD, J., Catal. Rev. 3 (1969) 93.
[30] ELEY, D.D., KNIGHTS, C.F., Proc. R. Soc. (London) Ser. A. 294 (1966) 1.
[31] MOSS, R.L., WHALLEY, L., Advances in Catalysis 22, Academic Press, New York (1972) 115.
[32] HANNAY, N.B., Semi-conductors, Reinhold, New York (1960).
[33] GERMAIN, J.E., Intra-science Chem. Reports 6 (1972) 101.
[34] BEEBE, R.A., DOWDEN, D.A., J. Am. Chem. Soc. 60 (1938) 2912.
[35] MORGAN, A.E., Surf. Sci. 43 (1974) 150.
[36] KRYLOV, O.V., Catalysis by Non-metals, Academic Press, New York (1970).
[37] LEE, E.H., Catalysis Reviews 8 (HEINEMANN, H., Ed.), Marcel Dekker, New York (1974) 285.
[38] ALSOP, B., DOWDEN, D.A., J. Chem. Phys. 51 (1954) 678.
[39] SHANNON, I.R., KEMBALL, C., LEACH, H.F., Chemisorption and Catalysis (HEPPLE, P., Ed.), Inst. Petroleum, London (1971) 46.
[40] GERMAIN, J.E., Catalytic Conversion of Hydrocarbons, Academic Press, New York (1969).

[41] DOWDEN, D.A., Catalysis Reviews 5 (HEINEMANN, H., Ed.), Marcel Dekker, New York (1971) 1.
[42] GRAVELLE, P.C., TEICHNER, S.J., Advances in Catalysis 20, Academic Press, New York (1969) 168.
[43] KAZANSKY, V.B., et al., Catalysis 2 (HIGHTOWER, J.W., Ed.), North-Holland, Amsterdam (1973) 1423.
[44] COLPAERT, M.N., et al., Surf. Sci. 36 (1973) 513.
[45] LUNSFORD, J.H., JAYNE, J.P., J. Phys. Chem. 70 (1966) 3464.
[46] TENCH, A.J., et al., Trans. Faraday Soc. 65 (1969) 2740.
[47] TENCH, A.J., HOLROYD, P., J. Chem. Soc. D (1968) 471.
[48] LUNSFORD, J.H., JAYNE, J.P., J. Phys. Chem. 69 (1965) 2182.
[49] WEBSTER, R.K., JONES, T.L., ANDERSON, P.J., Proc. Br. Ceram. Soc. (1965) No.5 153.
[50] RICE, R.W., HALLER, G.L., Catalysis 1 (HIGHTOWER, J.W., Ed.), North-Holland, Amsterdam (1973) 317.
[51] KAGEL, R.O., GREENLER, R.G., J. Chem. Phys. 49 (1968) 1638.
[52] TENCH, A.J., KIBBLEWHITE, J.F., Chem. Phys. Lett. (1972) 14 (2) 220.
[53] TANAKA, K., Solid Acids and Bases, Academic Press, New York (1970).
[54] LUNSFORD, J.H., J. Phys. Chem. 68 (1964) 2312.
[55] BOUDARD, M., et al., J. Am. Chem. Soc. 94 (1972) 6622.
[56] BALLARD, S.A., FINCH, H.D., WINKLER, D.F., Advances in Catalysis 9, Academic Press, New York (1957) 754.
[57] O'BRIEN, M.C.M., Proc. R. Soc. (London) Ser. A. 231 (1955) 404.
[58] BRECK, D.W., Zeolite Molecular Sieves, Wiley, New York (1974).
[59] FLOCKHART, B.D., Surface Defect Properties of Solids 2, Chem. Soc. London (1973) 69.
[60] GOLD, B., Carbonium Ions, Academic Press, New York (1967).
[61] VAN ARKEL, A.F., DE BOER, J.H., Die chemische Bindung als elektrostatische Erscheinung, Herzel, Leipzig (1931).
[62] MOTT, N.F., GURNEY, R.W., Electronic Processes in Ionic Crystals, Oxford Univ. Press (1940).
[63] DOWDEN, D.A., Chemisorption and Catalysis (HEPPLE, P., Ed.), Inst. Petroleum, London (1971) 1.
[64] DOWDEN, D.A., Catalysis 1 (HIGHTOWER, J.W., Ed.), North-Holland, Amsterdam (1973) 621.
[65] GRIMLEY, T.B., J. Vac. Sci. Technol. 8 (1971) 31.
[66] SCHRIEFFER, J.R., in Proc. Enrico Fermi Summer School in Physics, Varenna, Course No. LVII, 1973.
[67] SINFELT, J.H., Advances in Catalysis 23, Academic Press, New York (1973) 91.
[68] DOWDEN, D.A., "Factors affecting the choice of catalyst supports", Engineering of Gas-Solid Reactions, Proc. Inst. Chem. Eng. Symp. Series No.27 (1968) 18.

STATIC ELECTRIFICATION

B. MAKIN*
Department of Electrical Engineering,
University of Southampton,
United Kingdom

Abstract

STATIC ELECTRIFICATION.
1. Electrification in liquids: Introduction; Electrification of liquids flowing through pipes; Mechanisms of charge generation in moving liquids; Spray electrification; Electrohydrodynamic coupling; Electrostatic colloid spraying and propellant study. 2. Electrostatic discharges: Discharge processes; Charge generation mechanisms; Examples of electrostatic hazards; Supertanker explosions; Static detectors and eliminators. 3. Electrification at solid/solid and gas/solid interfaces: Contact charging; Triboelectrification; Corona charging and ozone formation; Adhesion of charged particles; Electrostatic imaging.

1. ELECTRIFICATION IN LIQUIDS

1.1. Introduction

The electrification of liquids is a most complex problem requiring detailed understanding of the chemistry and the dynamics of the system. The subject is particularly important in the petrochemical industry, where in transporting liquids it has been discovered that a charge transfer can take place at a liquid/solid interface and that charge carriers of predominantly one sign can be transported in the fluid. Depending on the properties of the liquid, this phenomenon can lead to eventual electrical breakdown with the possible ignitable hazard. Examples of this are found in the aviation industry, where flow rates of insulating fluids like kerosene are high and where there have been many disasters [1].

This major problem has received attention from several disciplines, and this has enabled scientists and engineers to obtain a clearer understanding of the mechanism of charge injection into liquids, the nature of the distribution of ionic charges at an interface, and the dynamic properties of transferring charge in both conducting and insulating materials. An engineering solution to this problem has been obtained by injecting chemical additives into the liquid but the basic electrochemistry of the charge transfer process remains uncertain.

A similar electrification problem is encountered in the atomization of liquids at a metal/liquid/air interface. The atomization produces a large quantity of charged liquid droplets. There are applications for which it is important to transfer the maximum amount of charge onto the droplets and there are other situations where it is necessary to transfer zero charge. The progress in these objectives is reviewed here together with details of their applications.

---

* Present address: Department of Electrical Engineering and Electronics, The University, Dundee, United Kingdom.

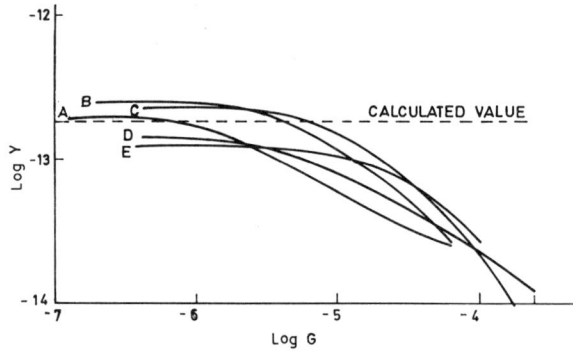

FIG.1.   Log Y-log G relationships for toluene flowing in large-diameter metal pipes. Pipe diameters: A, 1.62 cm; B, 2.88 cm; C, 5.39 cm; D, 8.35 cm; E, 10.90 cm. (From Gibson and Lloyd [6].)

1.2. Electrification of liquids flowing through pipes

The electrification of liquids flowing through metal pipes has been well studied [2, 3]. The standard experimental technique is to insulate the receiving collector and measure the streaming current. Many attempts have been made to analyse the results using non-dimensional analysis [4,5]. It has been shown that groups of parameters can be formed such that for liquids of all resistivities

$$Y = A_1 f(G)$$

where

$$Y = \frac{i_\infty}{V^{1.88} d^{0.88}} \quad , \quad G = \frac{d^2}{R^{1.75} \tau \nu}$$

and

$$A_1 = \frac{0.35 \, R_1 \, T\epsilon\epsilon_0 \, S^{0.25}}{nF\nu^{0.88}} \left(1 - \frac{C_s}{C_0}\right)$$

i_∞ : streaming current for infinite pipe (A)
V : velocity (cm/s)
d : pipe diameter (cm)
R : Reynolds number
τ : relaxation time (= $\epsilon\epsilon_0\rho$) seconds
ρ : resistivity ($\Omega\cdot$cm)
ϵ_0 : permittivity of free space (= 8.85×10^{-14} F/cm)
ν : kinematic viscosity (cm$^2\cdot$s^{-1})

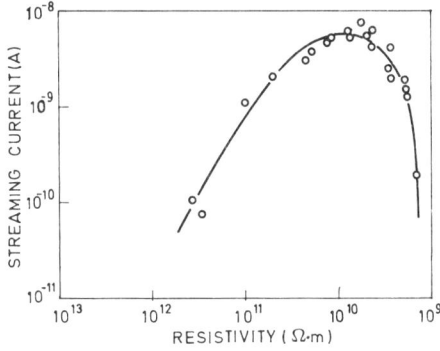

FIG.2. Effect of resistivity on charging tendency of JP-4 fuel (from Leonard and Carhart [7]).

R_1: gas constant (= 8.31 J·mol^{-1}·K^{-1})
T : temperature (°K)
n : transference number
F : Faraday (= 96 500 C·mol^{-1})
S : Schmidt number (ν/D)
D : diffusivity (cm^2/s)
C_0: concentration of discharging ions in bulk fluid (mol/cm^{-3})
C_s: concentration of discharging ions at tube wall (mol/cm^{-3})

for liquids of high resistivities in turbulent flow Y = A_1. The experimental results of Gibson and Lloyd [6] confirm these expressions for small-diameter metal pipes with toluene (Fig.1).

There is experimental evidence which suggests that the velocity dependence is larger for larger-diameter pipes; for diameters above 1 cm it has been shown that the dependence of streaming current on velocity is $i_\infty \propto V^m d^n$ where 1.4 < n < 2.0 and 0.88 < m < 2.4. In the curves for toluene shown in Fig.1 it was found that $i_\infty \propto V^{2.4} d^{1.6}$. Examination of the expression for streaming current shows only a weak dependence on Schmidt number. The more significant parameters are electrical resistivity and liquid viscosity. The presence of ionic additives can change the electrical resistivity by several orders. The effect of modifying the resistivity can produce changes in the streaming current as shown by Leonard and Carhart [7] for the case of adding antistatic additives to aviation fuel JP-4 (Fig.2).

Workers have found that the type of additive is important in determining the polarity of the streaming current. For the flow of hydrocarbons through steel pipes, additions of alcohols, acids, nitrobenzene and Shell additive ASA-3 produced positive tube currents, while solutions of ketones, amines and esters produced negative tube currents. In general it has been demonstrated that within an order of magnitude, the streaming current is modified by the electrical resistivity of the solution. This form of qualitative relationship is currently used for determining the safe filling rates for oil tankers.

Another area of investigation that is becoming increasingly important is in the flow of liquids through insulating pipes. High-resistivity materials like PTFE,[1] polyethylene and polypropylene introduce long charge relaxation

---

[1] Polytetrafluoroethylene (Teflon).

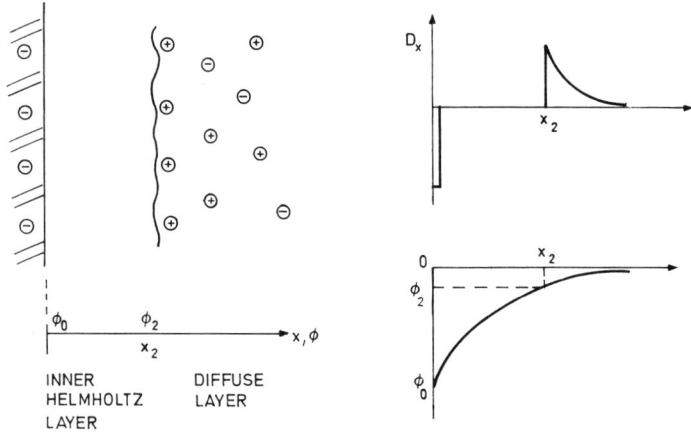

FIG.3.   Schematic of the double layer showing distribution of charge density (D_x) and potential (ϕ).

times and there is a more complex solid/liquid interface. A general observation is that for liquids with resistivities lower than 10^8 $\Omega \cdot$m the streaming
currents are no different from the case with metal pipes. However, for high-
resistive liquids $> 10^{10}$ $\Omega \cdot$m, the streaming current decreases with time.
A significant effect arising from insulating pipes is the presence of a high
charge density on both the inner and outer surfaces of the insulator. It has
been demonstrated that incendive discharges are possible from insulating
surfaces and therefore with highly inflammable liquids this aspect is important.

1.3.   Mechanisms of charge generation in moving liquids

The classical solution associated with the generation of charge during
the motion of liquids through pipes is based on shearing the charge distribution at a metal/liquid interface. The distribution of charges is given by the
model of Gouy [8] and Chapman [9] who postulated a diffuse double layer.
They assumed a uniform layer of unipolar charges secured to the surface
with a diffuse distribution in the remainder of the double layer (Fig.3).
The ions are assumed to be point charges with a Boltzmann distribution.
The inner layer may be regarded as a parallel plate capacitor with a capacitance per unit plate area given by $C_i = \epsilon \epsilon_0 / x_2$. For the diffuse layer it is
necessary to integrate over the charge distribution and for a 1:1 electrolyte

$$C_d = \left(\frac{\epsilon \epsilon_0}{\delta} \right) \cosh \left(\frac{e\phi_2}{2kT} \right)$$

The double layer thickness δ is given by the Debye length where n is the
number of ions/volume and

$$\delta = \left[\frac{\epsilon \epsilon_0 kT}{2e^2 n} \right]^{\frac{1}{2}}$$

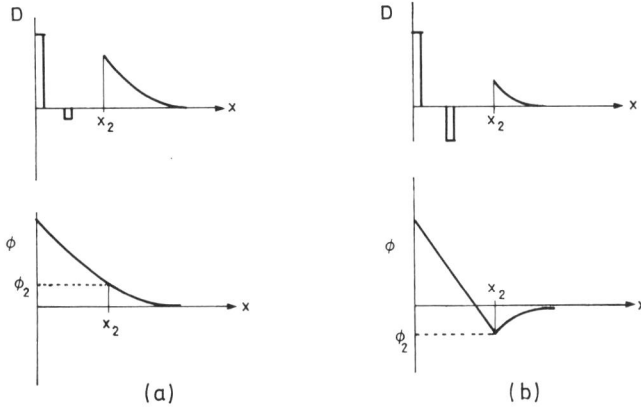

FIG.4.   Distribution of charge density and potential in the double layer for different concentration of adsorbed ions on the electrode.

Also, the distribution of charge and potential in the diffuse layer is given by

$$D_x = (4\epsilon_0 \epsilon_r kTn)^{\frac{1}{2}} \sinh \left(\frac{e\phi_x}{2kT} \right)$$

and

$$\phi = \phi_0 \exp \left(- \frac{x}{\delta} \right)$$

The inner Helmholtz layer is of the order of molecular dimensions and there is negligible error associated with the change of origin. For hydrocarbons with with a relative permittivity of 2 the inner layer capacitance is $\simeq 4\mu F/cm^2$. The diffuse capacitance is determined by the ionic concentration and can have a comparable value at low concentrations but will increase with concentration. The double layer thickness can also be shown to be given by

$$\delta = \sqrt{\Delta_m \cdot \tau}$$

where Δ_m is the molecular diffusivity m^2/s and $\tau = \epsilon\epsilon_0/\sigma$ the relaxation time.
    For an aqueous solution (10^{-3} molar): $\sigma = 0.013 \ \Omega^{-1}\cdot m^{-1}$; $\epsilon = 80$; $\tau = 5\times10^{-8}$s; $\Delta_m = 1.9\times10^{-9} \ m^2/s$, giving $\delta = 10^{-8} m$.
    For a typical hydrocarbon: $\sigma = 10^{-12}\Omega^{-1}\cdot m^{-1}$; $\tau = 18$ s; $\delta = 2\times10^{-4}$ m.
A modification of the theory was proposed by Stern to accommodate the possibility of ions being adsorbed to the surface. The number of adsorbed ions can modify the charge and potential distribution. In Fig.4 the number of counter-adsorbed ions in (a) is small in comparison with (b) where the effect is to reverse the potential in the inner layer, producing intense fields.
    The electrokinetic effects in pipe flow is to transport the outer part of the diffuse layer and it is necessary to define a plane of slip where the

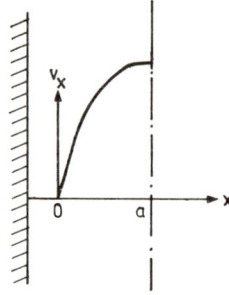

FIG.5.   Velocity distribution of a liquid flowing through a pipe.

double layer is sheared. The potential at the plane of slip is referred to as the ζ potential. The streaming current for the flow of liquid through a pipe radius a, length ℓ can be found assuming Poiseuille laminar flow such that (see Fig.5)

$$v_x = p \left(\frac{2ax - x^2}{4\eta\ell} \right)$$

$$I_s = \int_0^a \rho \, d \left| \frac{d\,vol}{dt} \right| = \frac{\pi\epsilon pa^2}{\eta\ell} \int_0^a x \frac{d^2\phi}{dx^2} \, dx$$

$$I_s = \frac{\epsilon pA\zeta}{\eta\ell}$$

Providing the double layer is confined to within the laminar sublayer, the streaming current is proportional to the pressure drop in the pipe. With hydrocarbons this is not always the case as the double layer can become larger than the sublayer.

Another possible mechanism for contribution to the streaming current can arise through electrode reactions. Parsons [10] has estimated that in the case of kerosene flowing through a steel pipe, the measured streaming current of 10^{-9} A could be accounted for by discharging all the ions in solution. This extra possible mechanism makes for a most complicated interface problem. Only by considering these processes can one hope to interpret the phenomena encountered with streaming and spray electrification.

1.4. Spray electrification

It has long been known that when water disintegrates into drops there is a charge transfer and a net charge can be obtained in the water mist. Early experiments with waterfalls showed that small droplets were negatively charged and the large drops positively charged. More recent studies by Jonas, Mason, Iribarne and Voss have been undertaken to investigate the significance of chemical additives when a water surface is disrupted by either mechanical impact or bursting air bubbles. This work is closely related to the supertanker accidents when three ships exploded during the washing process. It has been found that an impacting water jet produces charged particles which can generate large electric fields. The chemistry of the water and the surface influence the rate of charge generation and the polarity. Originally it was anticipated that by careful control of the wash water it might be possible to eliminate the charged mist.

An intensive investigation conducted by Mason and Iribarne [11] has produced a successful model for the charged fragments produced by either the disruption of a liquid surface using air bubbles or by the impact of a liquid surface with a liquid droplet. Their experiments were carried out in the absence of electric fields.

It was generally found that for concentrations less than 10^{-4} M, the charge produced by bubbles bursting in sodium chloride solution was negative. For larger concentrations the charge was slightly positive. A maximum negative charge was obtained at a radius of 150 μm. A theoretical inter-pretation of their results was obtained by assuming a shearing of the double layer at the air/liquid interface. In the model, it is necessary to determine the charge density in a Gouy diffuse double layer and integrate it over the depth of liquid which is removed. The thickness of the double layer δ is obtained from the conductivity κ:

$$\kappa = \frac{3}{10^{8} c^{\frac{1}{2}}}$$

where c is the concentration and where $\delta \propto \kappa^{-1}$.

For pure water (c = 10^{-7} N) at 25°C, $\delta \approx 1\ \mu$m. The surface charge density (σ) in the double layer:

$$\sigma \propto \sqrt{c} \sinh \left(\frac{e\phi}{2kT}\right)$$

where ϕ is the potential. An exponential distribution of potential with depth in the liquid together with the limiting solution of $\tanh x = \sinh x = x$ gives the expression for the charge carried by a droplet of radius R:

$$Q = \frac{4\pi R^{2}\sigma}{10} \propto R^{2} \sqrt{c} \exp\left[-\tfrac{1}{2} - 10^{6} \sqrt{c}\ \frac{R}{9}\right]$$

The difference in polarity of ejected charge was used to postulate a model of the formative processes. For small concentrations it was assumed that the layer of negative charge rapidly moved down the cavity surface, rising in the centre to form negative ejected particles (Fig.6).

From a consideration of the relaxation time for the double layer (θ),

$$\theta = \frac{7 \times 10^{-9}}{(\mu_{+} + \mu_{-})\ c}$$

for pure water $\theta \approx 10^{-4}$ seconds.

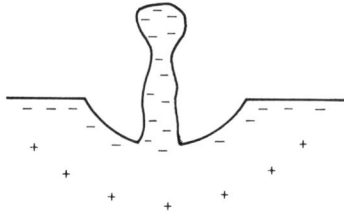

FIG.6.   Schematic of atomization process showing the formation of a negative charged particle at low ion concentrations.

For a bubble size $100\ \mu$m with a velocity of 2 m/s the formation time τ must be of the order 10^{-5} s. For large concentrations ($c > 10^{-4}$ M) the double layer relaxation time is of the order 1 μs and hence an equilibrium condition will be rapidly established. It is thought that the outer layer of negative charge is relatively immobile and in the region of the neck an inner core of rapidly moving positive charge produces a net positive charge on disruption. A theoretical model using pipe flow has been developed which agrees quantitatively with this hypothesis.

An experimental programme by Voss (Shell) [12] which is more realistic for tank cleaning has been directed to the situation. A jet of water collides with a grounded sphere which is enclosed by an insulated container. His results showed that a fine mist was negatively charged while the mist for all the spray was positively charged. By inducing a voltage onto the mist it was possible to neutralize the collected current for any particular velocity. Tests using surface-active agents showed they had a strong effect. An anionic agent |sodium di (2-ethylhexyl) sulphosuccinate| and cationic agent |cetylpyridinium bromide| showed exactly opposite characteristics.

An experimental observation found that the generated current was velocity dependent and proportional to the throughput. It was also thought that the charge would be dependent on shear stress in the double layer which is similarly proportional to velocity. They showed $i = kv^2$, and if one assumes that k is related to the thickness of the double layer then $k \propto \kappa^{-\frac{1}{2}}$. Experimental results gave a power dependency of -0.4. They found that the material of the sphere did not influence the results provided that it was a conductor. Metals tested were stainless steel, gold, copper, brass, aluminium and platinum. An insulator did have an enhanced effect, which was thought to be due to an induced electric field.

In the full-scale experiments on tankers a submicron mist is formed with a particle density of 10^{-3} kg/m^3, and a number density of 4×10^{13}/m^3. Approximately 80% of the particles are less than 0.8 μm. During the cleaning process the polarity of the mist reverses from a negative field while the jet is contacting an oil surface to a positive field on a clean metal surface. In all the experiments at full scale, sea-water is used. The standard procedure is to add a small concentration of surfactants, which generally produce a negative field.

1.5. Electrohydrodynamic coupling

The phenomenon of charge injection into liquids developed following intensive studies by Felici [13] and others to purify polar liquids. In particular, much effort has been directed to purifying nitrobenzene for use as the insulant in electrostatic generators. Ion-exchange membranes were used with Kerr cells which allowed for a detailed examination of the purification sequence. More recent developments have shown that charges can be injected into insulating liquids producing bulk transport [14]. In general, it has been shown for the case of low-conductivity liquids when there are no thermal gradients, conditions for the hydrodynamic stability can then be found for either (i) dissociation process in the bulk, i.e. space charge limited, or (ii) injection of charge carriers at the electrodes. The stability criteria are:

$$M^2 C^2 R = 22 \quad \text{(injection limited current)}$$

$$M^2 R = 161 \quad \text{(space charge limited)}$$

$$M = \frac{1}{\mu}\left(\frac{\epsilon}{\rho}\right)^{\frac{1}{2}} \; ; \quad R = \frac{\mu\phi\rho}{\eta} \; ; \quad C = \frac{q_0 d^2}{\epsilon\phi}$$

(μ = carrier mobility, ρ = liquid density, η = liquid viscosity, q_0 = charge density of electrode and ϕ = voltage).

The effect of the instability in liquid is to produce enhanced mobilities. This may arise from clusters of solvent molecules or from particles.

In flowing systems which have charge-injecting electrodes it is possible to analyse the flow pattern and to investigate the coupling constants. For the flow past a two-dimensional cylinder (radius a) mounted centrally across a duct of semi-width ℓ, the coupling constant in the bulk flow is

$$K_1 = \frac{\ell V_0}{2\mu\phi_0} \ln\left(\frac{2\ell}{\pi a}\right)$$

where V_0 is the free stream velocity and ϕ_0 the potential difference between the cylinder and the duct walls. With insulating liquids for $K_1 > 10$ most of the charges are transported in the bulk flow.

1.6. Electrostatic colloid spraying and propellant study

The production of charged liquid droplets can be obtained by several atomization schemes. There are many uses for such a medium, e.g. industrial painting, printing and electrostatic rockets. In all these applications it is essential to charge the particle to its maximum value. Rayleigh showed that there is a limiting charge beyond which a particle would be unstable:

$$q_{max} = 8\pi(\epsilon\gamma)^{\frac{1}{2}} a^{3/2}$$

where γ is the surface tension. Charge can be acquired by contact, induction, collision with electrons or ions and electrochemical (electrode) processes.

The electrostatic rocket engine is a device where atomization of a liquid occurs under the influence of an electric field. The liquid film is ruptured and a net charge is transferred from a metal electrode placed in the liquid on the drop. This process is carried out in vacuo, and spectrometric techniques can be used to analyse the particles. Much effort has been made to improve the charging by developing smooth electrodes with tip radius $\simeq 1\ \mu m$, and by doping the liquid to produce highly charged stable systems. Current practice is to use solutions of glycerol/NaI, which has a low vapour pressure, producing a charge/mass ratio of up to $10\ 000$ C/kg. Because of the large electric fields at the metal/liquid interface (10^9 V/m), charge exchange reactions can occur producing a deterioration of the precisely machined substrate. Attempts have been made to find a replacement propellant which will accept holes from the electrode and atomize producing singly charged positive droplets.

2. ELECTROSTATIC DISCHARGES

2.1. Discharge processes

Electrostatic discharges arise from a charge separation process and they can be classified into the categories of nuisance, destructive or incendive. The nominal breakdown strength of air is 3×10^6 V/m and for fields in excess of this value there is generally an arc or streamer discharge, and any associated charged capacitance or inductance in the system will be dissipated into the gap in times less than $\simeq 1$ ms. The maximum surface charge density which can withstand the breakdown field is generally given by $\sigma = \epsilon E = 3 \times 10^{-5}$ C/m^2. If the surface has any asperities it will produce enhanced fields resulting in local ionization of the air. At lower field strengths there is a corona dissipative discharge which consists of pulses of ionization (Trichel pulses at frequencies of $\simeq 20$ kHz) which continually lower the field, forming a space charge. The onset of corona is determined by the geometry of the electrode and by the voltage being greater than a threshold voltage. Corona more readily occurs from sharp electrodes. In an incendive atmosphere it is generally found to be nearly impossible to ignite a hydrocarbon mixture with only a corona discharge.

2.2. Charge generation mechanisms

There are different ways of transferring charge to a body. The most direct method is by contacting another charged body or high-voltage source. The charge transfer will be determined by the relaxation time $\tau = \epsilon_0 \epsilon_r \rho$. Another mechanism is rubbing or triboelectrification, where electron transfer can take place between two dissimilar materials.

When a material transfer takes place in the presence of an electric field, net charge is acquired by the material and this process is called induction. This occurs in the atomization of liquids or when solid particles leave a pipe. Moreover, when any object, charged or neutral, is resident in an electric field, polarization will tend to produce a modified distribution of charge which can be very important when considering breakdowns from insulated objects.

2.3. Examples of electrostatic hazards

Those discharges which can be described as a nuisance relate to the human body which has a capacitance of $\simeq 300$ pF. By typically walking over good insulating material the human body potential can rise to 10 kV, producing a discharge of 15 mJ. This level of energy is not lethal, but is sufficient to ignite say a propane/air mixture where the minimum ignition energy $\simeq 0.3$ mJ.

An example of a destructive discharge is in the semiconductor fabrication technology, particularly with MOS devices. The thin oxide layer can be readily destroyed by electrostatic discharges from personnel. It is common practice to build in resistive and Zener circuits, but the high wastage rates suggest this is still a problem area. Another occurrence in this category is in the handling of light-sensitive photographic material. Electrical discharges will fog the film.

In dust-laden atmospheres large space-charge densities are commonly generated. With materials like sugar, milk powders and flour, the maximum observed charge density $\simeq 3 \times 10^{-2}$ C/kg [15]. The general ignition energy for powders is $\simeq 10$ mJ. Combustible powders can be the most hazardous material, as in the case of coal, when it is possible to extract the heat of combustion. In confined environments the discharge can lead to a detonation, with disastrous effects.

In the transport of insulated sheet material over rollers, charges can be stripped producing high voltages on the material. These will break down to the same grounded roller or a subsequent roller if the relaxation time is long enough. Measured surface charge densities of 2×10^{-5} C/m^2 with paper and plastic film indicate that the process is close to the maximum. These high charge densities are responsible for many fires in a conventional gravure printing press which uses organic solvent baths.

When pumping hydrocarbon fuels through pipes there is a maximum bulk space-charge density of 5×10^{-6} C/m^3 providing the characteristic relaxation length $\lambda = v \epsilon \rho < $ tube length. If a filter is introduced into the pipe, the charge density can be increased up to 10^{-2} C/m^3. Similarly, by spraying diesel oil/water mixtures through nozzles, typical observed charge densities $\simeq 10^{-5}$ C/kg. The hazard associated with fuelling is in the collection receiver where the normal fields build up above the surface of the liquid. As the liquid rises up to a grounded object or superstructure there is a possibility of a discharge. For filling tankers, guidance is given for the flow rates to prevent the build-up of dangerous fields. The analytical solution of the problem for a rectangular box is complex [16].

An example of the combination of tribocharging and induction charging has been discovered with helicopters. The efflux of carbon particles from the engine can produce a net charge on the helicopter generating potentials of 10 kV and a discharge current of 3 μA. If a thundercloud should pass over a helicopter it can induce potentials up to 100 kV, and with precipitation the charging currents are typically 50 - 100 μA [17].

2.4. Supertanker explosions

The explosions which occurred in three VLCC^2 supertankers in 1969 promoted an intense investigation into the generation mechanisms of static during the washing cycle [18-21]. The incidents took place when the largest tanks (24 000 m^3) were being cleaned, using high-intensity water jets at a flow rate of 180 m^3/h and with a velocity of 40 m/s. It was discovered that the water disintegrated on impact with the wall to form a charged water mist with a space-charge density of 10^{-8} C/m^3 and a maximum potential at the centre of the tank of 40 kV. Approximating the tank to an empty sphere, the

---

^{2} Very large crude carriers.

maximum potential $V_{max} = \rho a^2/6\epsilon_0$ and the maximum field at the surface
$E_0 = \rho a/3\epsilon_0 \simeq 13$ kV/m. If a grounded structure is directed to the centre of the
tank, large electric fields are produced at the surface and measured values
of 700 kV/m have been observed.

Many suggestions were put forward to explain where an electrical
discharge could occur with an energy > 1 mJ. All possibilities have been
considered including discharges from the space-charge cloud to the walls
similar to discharging a thundercloud; corona discharges, sparking produced
by falling metal objects. The most feasible argument, which is being
generally accepted, incorporates the enhanced electric field produced by a
protruding structure. It has been demonstrated that if a falling object passes
close to a protrusion, a spark can be obtained with sufficient energy to ignite
a stoichiometric hydrocarbon gas [22]. Furthermore, it has been shown that
the falling objects could be slugs of water which are produced by the water
jet. Typical figures show that for long cylindrical objects the capacitance is
10 pF/m length. When the object falls through a space-charge cloud it can
be charged by induction at the exterior surface and by collection of charged
water droplets during its trajectory. As the slug approaches a grounded
object through an increasing electric field, the charge distribution becomes
polarized. During the discharge process approximately 10^{-7} C are transferred between the water slug and the probe. Estimates made of the charge
accumulation show that approximately 80% of the charge transfer arises from
polarization. Measurements of the energy in the discharge are restricted
because it is a single electrode reaction. From combustion studies it is
evident that the energy is in excess of the minimum ignition energy and
qualified guesses estimate it to be 4×10^{-4} J.

2.5. Static detectors and eliminators

Static is detected by measuring the voltage with a conventional contacting
electrostatic voltmeter or by measuring the field with a non-contacting
field mill. The mill is a useful qualitative device but it is sensitive to
separation, and interpretation of the output is difficult. In supertankers,
field mills are used for measuring charged mist but because the instrument
has a grounded body the field is enhanced and care is required in interpreting the result. One device for measuring static is to neutralize the
source with a supply of oppositely charged ions. Another device consists of an
isotope source (β-emitter or tritium source) mounted in an insulated
container. The current flow of the ions gives a measure of the field but it is
also sensitive to distance and needs calibration. Discharges can be detected
by capacitive coupling with an antenna [23]. The frequencies used are
$\simeq 4$ MHz and it is possible to determine the difference between brush discharges and sparks and the energy of the discharge.

When static has been diagnosed it is usually necessary to eliminate the
problem. Several schemes are available which use the principle of either
suppressing the charge separation at the source or subsequently neutralizing
the charge. The textile industry has experienced difficulties for many years,
and a standard solution is to use an α source from ^{210}Po which is safe even
in explosive atmospheres. More active devices used in the printing and
plastic industries use either a.c. or d.c. corona wires. With a.c., the voltage
is typically 5 kV and no control system is required. When d.c. is used, a
feedback control system is required to switch off the supply and prevent the

object from becoming oppositely charged. The d.c. system is used on helicopters where two EHT ± 100 kV generators are required to overcome the
polarity charges produced by thunderclouds.

With the supertanker problem, efforts have been made to control the
concentration of the wash water to prevent particle charging. However, only
a small quantity of charge per droplet will ultimately generate the same
space potentials because of the long charge relaxation time. The most
accepted solution is to subdivide the volume electrically to reduce the space
potentials and fields [24]. It is known that small volumes are quite safe and
any discharges associated with water slugs would no longer be incendive.
For large supertankers this represents an expensive modification and the
more easily adaptable solution is to inert the atmospheres and ignore the
static effects.

3. ELECTRIFICATION AT SOLID/SOLID AND GAS/SOLID INTERFACES

3.1. Contact charging

When two dissimilar materials are brought together there is a contact
potential, and from solid-state principles we interpret the equilibrium
phenomena as equating the Fermi levels of each material. There is a transfer
of electrons and the materials become charged. On separating the materials,
there is a back flow of charge which may be an order of magnitude larger
than the final current. Recent trends in the electrification of solids have
been directed to investigating the charging of polymers. Based on the concepts of solid-state theory, a model has been developed describing the
electrical properties of a polymer with a true work function and the existence
of surface states.

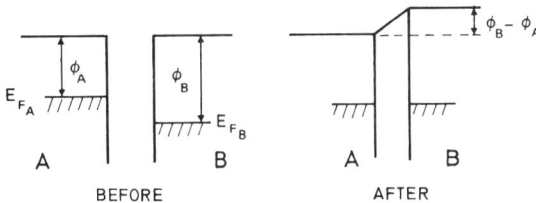

FIG. 7.   Distribution of energy levels on contact of two different materials.

The simple case of contact between two different metals A and B where
$\phi_A < \phi_B$ is shown in Fig.7. During contact, electrons are transferred from
A to B until the Fermi levels are balanced and B becomes negatively charged.

For the case of a metal/semiconductor contact, Cowley and Sze [25]
found the surface charge density on the semiconductor (σ) to be

$$\sigma = \frac{eD_s (\phi_M - \phi_{sc})}{\left(1 + \dfrac{e^2 z D_s}{\epsilon_0}\right)}$$

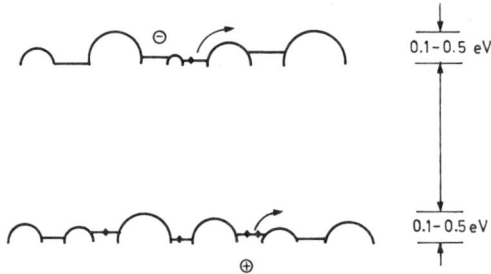

FIG. 8.   Model of valence and conduction bands of a polymer.

where D_s is the density of surface levels within the gap band and z is the
gap between the faces ($\simeq 4$ Å). For ionic semiconductors (ZnS, Al$_2$O$_3$)
$D_s < 10^{12}$ cm^{-2}·eV^{-1} and therefore $\sigma \propto D_s (\phi_M - \phi_{sc})$. For clean covalent semiconductors (Ge, Si, GaAs) the surface state density can be high and
$\sigma = \epsilon_0(\phi_M - \phi_{sc})/ez$. Typical values for $\phi_M - \phi_{sc}$ are up to 1.0 eV, generating
surface charge densities up to $\pm 10^{-6}$ C/cm^2.

In the contact between a polymer and a metal there have been two rival
theories: that the charges transferred are either ions [26] or electrons [27].
From the results of Davies and following Bauser [28], it seems evident that
the results can be explained using the concept of electronic transfer.
Davies [29] has tested several polymers with a range of metal surfaces of
varying work function. His results show the charge to be proportional to the
work function difference. The experiments were carried out in vacuo and
the reproducible nature of the results invalidated the possibility of ionic
transfer from impurities and surface states. In a review paper by Krupp [30],
the model of a polymer is put forward consisting of localized levels distributed
over a narrow range ($\simeq 0.1$ eV) replacing the normal conduction and valence
bands (see Fig.8). Each level represents a molecular unit (e.g. a benzene
ring or a hydrocarbon chain). Conduction is via electrons and hole-hopping.
When this structure is put in contact with a metal, extended contact times
are required to generate the measured surface densities (10^{13} days). The
phenomenon can be speeded up by assuming a distribution of surface states
and the introduction of bulk defect states within the energy gap. Observed
linear relationships between surface charge and metal work function indicates
a uniform energy distribution of surface states [30]. For pure polymers
$\sigma \simeq 10^{-9}$ C/cm^2 gives a surface state density $D_s = 10^{10}$ cm^{-2}·eV^{-1}. Exposure of
the sample to air or flame cleaning increases σ and D_s by two orders of
magnitude.

The back flow of charge on separating the surfaces depends on many
factors: surface preparation, geometry of contact, speed of separation,
temperature, gas atmosphere. When experiments are carried out in vacuo
the major transfer will be via tunnelling [26].

3.2. Triboelectrification

The topic is generally related to contact charging in that charge is
transferred during a frictional contact. The subject is very complicated and

little has been quantified to date. There is a resurgent interest in the subject because of the rapid development and application of powders and man-made fibres. In the transport of a powder, frictional charging is dominant. During contact and sliding, particles will be abraded, resulting in material transfer. The irregularity of the surface will produce large pressures with local temperature differences. Harper has reported examples where the asymmetry of the rubbing between two materials and the contact pressure will affect the charge transfer.

3.3. Corona charging and ozone formation

Particle charging is most important in the applications of electrostatics. The maximum charge on a solid particle is restricted by surface breakdown. When a conducting particle of radius a is resting on a metal plane, the charge induced in the particle is given by

$$Q' = 1.65 \times 4\pi\epsilon \, a^2 \, E$$

where E is the field at the plate without the particle [31]. For a sphere we can take an average value of 1.65 E and assume the maximum field is 30 kV/cm. Then for a 1-μm particle the maximum charge is 3.3×10^{-16} C. If the same experiment is repeated in vacuo the maximum local fields $\simeq 10^9$ V/m, producing a maximum charge $\simeq 10^{-13}$ C.

In charging insulating particles the charge is related to the relaxation time $\tau_e = \epsilon/\sigma$ and the contact time by

$$Q = Q' \left[1 - \exp\left(- \frac{t}{\tau_e} \right) \right]$$

The maximum charge that can be stored on a particle of relative permittivity ϵ, in an electric field E_0, is given by Pauthenier as

$$Q_0 = 12\pi\epsilon_0 \, \frac{\epsilon_1}{2 + \epsilon_1} \, E_0 \, a^2$$

Hence for a 1-μm particle at breakdown fields (30 kV/cm) $Q_0 \simeq 10^{-15}$ C.

The most widely used charging mechanisms are a combination of contact and ion bombardment. At atmospheric pressures the corona discharge is a readily available source of ions. It has been shown [32] that for a particle in a uniform ion current density $J(A/m^2)$ the charging rate is given by

$$Q = Q_0 \frac{t}{t + \tau} = Q_0 \frac{1}{1 + \dfrac{4\epsilon_0}{k \, net}}$$

where k is ion mobility $(m^2/V \cdot s)$ and n is ion number density (m^{-3}), and

$$Q = Q_0 \frac{1}{\left(1 + \dfrac{4\epsilon_0 E_0}{Jt} \right)}$$

Typical experimental values are 100 kV at 600 μA for 100 cm^2 with 6 cm separation producing $E_0 = 1.6 \times 10^6$ V/m, $n = 10^{15}$m^{-3}, and after an exposure time of 10 ms, 90% of the maximum charge will be collected by ion bombardment. This example represents the charging of a single particle but in practice we have a cloud of particles and a space-charge effect which tends to reduce the efficiency. In general, an increase in current density will create more corona ions and hence speed up the process. The application of powder coating and electrostatic precipitation requires fast, efficient charging processes for the particles to move in the lines of electric field.

In any corona discharge, the collision between the ions and oxygen molecules produce ozone. This is a toxic material and legislation limits the maximum safe concentration. It has been shown [33] that the ozone production is proportional to the corona current or the power dissipated in the corona sheath;

$$[O_3] = \frac{kE_{av}aI_c}{Q}$$

where E_{av} is the average electric field in the corona sheath, Q is the air flow (m^3/s) and [O$_3$] is the ozone concentration on a volume basis. An experimental study was carried out to increase the temperature of the corona source and it was found that the corona current increased at the same field but the ozone concentration was greatly reduced. The analysis showed $E_{av} = f(T^{-0.8})$ and a decomposition function varied with temperature like T^{-2}.

3.4. Adhesion of charged particles

An important area of surface physics is related to the deposition of particles on a surface. This aspect is most important in the applications of powder coating and xerography. The force field retaining particles to a surface is a combination of adhesive (van der Waals) forces and electrostatic forces. The magnitude of the force is critical in determining the overall efficiency of the process. Adhesion is complicated and little quantitative work has been attempted. Beischer has experimented with quartz spheres and non-magnetic iron oxide particles 0.25 μm in size and has obtained the expression for the adhesive force:

$$F = 1.6\ a \times 10^{-3}\ (N)$$

where a is particle radius (m). This is equivalent to saying that particles larger than 0.01 μm will not be deteared in a field of 10^8 V/m.

More recent interest has been directed to the structure in a layer of insulating charged particles which are deposited in an electric field together with a flow of gas ions. The packing structure and the interparticle contacts are influenced by the electrification, and layers of different strength and composition are obtained. In a continuous coating with insulators, the charge does not leak away and it was traditionally thought that a repulsive space-charge layer provided the mechanism for a limiting asymptotic thickness. At high deposition fields the powder can produce back ionization with the emission of counter-ions which are believed to neutralize the incident powder. Other thickness-limiting mechanisms are now believed to be associated with continuous interparticle explosions where agglomeration of particles break down under their repulsive forces. This idea is analogous to the Rayleigh limit for liquid droplets where the adhesive interparticle force is equivalent to surface tension.

3.5. Electrostatic imaging

The subject of electrophotography embraces many different electrostatic processes. The basic stages are:

(a) Sensitization;
(b) Latent image production;
(c) Development; and
(d) Fixing

Modern Xerox machines incorporate amorphous selenium photoconductors by which stages (a) and (b) can be efficiently accomplished. The sensitization is achieved by a corona discharge where a positive surface layer of charge is uniformly distributed over the selenium. By exposing the surface to light, the photons generate electron/hole pairs which, in the presence of an electric field, produce a current. Electrons can neutralize the positive surface charges forming the latent image. A discrete line of charge on the selenium surface has a high fringing field. The development process cascades carrier and toner particles at the charged image. Toner materials are selected to have a preferential negative charge so that they are attracted to the charged areas. Also because of polarization, the force on a dipole $\propto$ dipole moment $\cdot$ $\cdot dE/dx \propto \nabla E^2$. This force allows for edge contour effects to be efficiently covered with toner particles. For a typical 10-μm toner particle in a field of 10^6 V/m charged to 5×10^{-15}C, the Coulomb attractive force $\simeq 5\times10^{-9}$N; gravitational force $\simeq 5\times10^{-12}$N; and the force on a dipole $\simeq 3\times10^{-12}$N.

REFERENCES

[1] KLINKENBERG, A., VANDERMINNE, J.L., Electrostatic in the Petroleum Industry, Elsevier, Amsterdam (1958).
[2] KLINKENBERG, A., in Proc.2nd Conf.Static Electrification, 1967, Inst.Phys., London (1967) 63.
[3] GIBSON, N., in Proc.3rd Conf.Static Electrification, 1971, Inst.Phys., London (1971) 71.
[4] KOSZMAN, I., GAVIS, J., Chem.Eng. Sci. 17 (1962) 1013, 1023.
[5] KLINKENBERG, A., Chim.Ind. 82 (1959) 149.
[6] GIBSON, N., LLOYD, F.C., Chem.Eng. Sci. 25 (1970) 87.
[7] LEONARD, J.T., CARHART, H.W., US Naval Res.Lab.Rep. No.6952 (1969).
[8] GOUY, G., J.Phys. (Paris) 9 (1910) 457.
[9] CHAPMAN, D., Philos.Mag. 25 (1913) 475.
[10] PARSONS, R., in Proc. 3rd Conf.Static Electrification, 1971, Inst.Phys., London (1971) 124.
[11] MASON, B.J., IRIBARNE, J.V., Trans.Faraday Soc. 63 (1967) 2234.
[12] VOSS, B., in Proc.3rd Conf.Static Electrification, Inst.Phys.London (1971) 184.
[13] FELICI, N.J., Direct Current 4 (1959) 192.
[14] ATTEN, P., in Conduction and Breakdown Conf., Univ.Dublin (1972) 85.
[15] SCHON, G., "Static electricity and its ignition hazards", ISA, Electrical Safety Conf., Wilmington, 1971.
[16] LYLE, A.R., STRAWSON, H., "Estimation of electrostatic hazards in tank filling operations", Proc.3rd Conf.Static Electrification, 1971, Inst.Phys., London (1971) 234.

[17] ODAM, G.A.M., "Electrostatic charging of aircraft in flight", Advances in Static Electricity, 1st Int. Conf.Vienna (1970) 248.

[18] Van der MEER, D., "Electrostatic charge generation during washing of tanks with water sprays - I", Proc. 3rd Conf.Static Electrification 1971, Inst.Phys., London (1971) 153.

[19] Van de WEERD, J.M., "Electrostatic charge generation during washing of tanks with water sprays - II", ibid.

[20] SMIT, W., "Electrostatic charge generation during washing of tanks with water sprays - III", ibid.

[21] VOSS, B., "Electrostatic charge generation during washing of tanks with water sprays - IV", ibid.

[22] HUGHES, J.F., BRIGHT, A.W., MAKIN B., PARKER, I.F., A study of electrical discharges in a charged water aerosol, J.Phys.D. 6 (1973) 966.

[23] CHUB, J., in Proc.Int.Conf.Static Electrification, Frankfurt, 1973.

[24] LINDBAUER, R., ibid.

[25] COWLEY, A.M., SZE, S.M., J.Appl.Phys. 36 (1965) 3212.

[26] HARPER, W.R., Contact and Frictional Electrification, Clarendon Press, Oxford (1967).

[27] DAVIES, D.K., Static Electrification, Inst.Phys.Series No.4, London (1967) 29.

[28] BAUSER H., KLOPFFER, W., RABENHORST, H., Advances in Static Electricity, 1st Int.Conf. Vienna (1970) 2.

[29] DAVIES, D.K., ibid., p.10.

[30] KRUPP, H., Static Electrification, Inst.Phys.Series No.11, London (1971) 1.

[31] CHO, A.Y.H., J.Appl.Phys. 35 (1964) 2561.

[32] CORBETT, R.P., Ph.D. Thesis, University of Southampton (1971).

[33] MAKIN, B., IEEE—IAS Meeting, Milwaukee, 1973 (to be published by Industry and Applications Soc.).

FACULTY AND PARTICIPANTS

An asterisk indicates that a lecturer's contribution is not published in these Proceedings

DIRECTORS

V. Celli	International Centre for Theoretical Physics, Trieste, Italy (Present address: Department of Physics, University of Virginia, Charlottesville, Va. 22901, United States of America)
G. Chiarotti	Istituto di Fisica, Università degli Studi di Roma, Piazzale delle Scienze 5, I-00100 Rome, Italy
F. García-Moliner	Departamento de Física, Facultad de Ciencias, C-XII, 6, Universidad Autónoma de Madrid, Canto Blanco, Madrid 34, Spain
S. Lundqvist	Institute of Theoretical Physics, Chalmers University of Technology, Fack, S-402 20 Göteborg, Sweden
N. H. March*	Department of Physics, Imperial College of Science & Technology, Prince Consort Road, London SW7 2AZ, United Kingdom
J. M. Ziman*	H. H. Wills Physics Laboratory, University of Bristol, Royal Fort, Tyndall Avenue, Bristol BS8 1TL, United Kingdom

LECTURERS

S. Andersson	Department of Physics, Chalmers University of Technology, Fack, S-402 20 Göteborg 5, Sweden
M. V. Berry	H. H. Wills Physics Laboratory, University of Bristol, Royal Fort, Tyndall Avenue, Bristol BS81TL, United Kingdom
W. Brenig	Physik-Abteilung der Technischen Universität München, James-Franck-Strasse, 8040 Garching b. München, Federal Republic of Germany

F. Clementi*   Department of Pharmacology,
                                      University of Milan,
                                      Via Festa del Perdono 7,
                                      Milan, Italy

G. Dearnaley   Nuclear Physics Division, H. 8,
                                      AERE Harwell,
                                      Didcot, Oxfordshire,
                                      United Kingdom

D. Dowden   Imperial Chemical Industries Ltd.,
                                      Agricultural Division,
                                      PO Box 6,
                                      Billingham,
                                      Cleveland TS23 1LE,
                                      United Kingdom

K. Dransfeld*   Max-Planck Institut,
                                      Grenoble, France

A.J. Forty   Department of Physics,
                                      University of Warwick,
                                      Coventry, War., United Kingdom

R. Gomer   James Franck Institute,
                                      University of Chicago,
                                      5640 Ellis Avenue,
                                      Chicago, Ill. 60637,
                                      United States of America

R. Jones   Institut für Festkörperforschung,
                                      Postfach 365,
                                      D 517 Jülich 1,
                                      Federal Republic of Germany

B. Makin   Department of Electrical Engineering,
                                      University of Southampton,
                                      United Kingdom
                                      (Present address: Department of Electrical Engineering
                                                           and Electronics,
                                                        University of Dundee,
                                                        Dundee DD1 4HN, United Kingdom)

A. Many   Racah Institute of Physics,
                                      Hebrew University of Jerusalem,
                                      Jerusalem 91000, Israel

H. Nahr   Physics Institute IV,
                                      University of Erlangen,
                                      Erwin-Rommel-Strasse 1,
                                      Erlangen,
                                      Federal Republic of Germany

C.A. Neugebauer*   General Electric Company,
                                      Research and Development Center,
                                      PO Box 8,
                                      Schenectady, N.Y. 12301,
                                      United States of America

R. Parsons

Department of Physical Chemistry,
University of Bristol,
Cantock's Close,
Bristol BS8 1TS, United Kingdom

J. Schnakenberg

Institut für Theoretische Physik
  der Technischen Universität,
Templer Graben 64,
Aachen,
Federal Republic of Germany

J.R. Schrieffer[*,1]

Department of Physics,
University of Pennsylvania,
Philadelphia, Pa. 19174,
United States of America

G. Scoles[*]

Chemistry Department,
University of Waterloo,
Waterloo, Ont. N2L 3G1,
Canada

G.A. Somorjai

Department of Chemistry,
University of California,
Berkeley, Calif. 94720,
United States of America

D. Tabor

Cavendish Laboratory,
Cambridge University,
Madingley Road,
Cambridge CB3 0HE, United Kingdom

M. Tomášek

J. Heyrovský Institute of Physical Chemistry
  and Electrochemistry,
Máchova 7,
121 38 Prague 2, Czechoslovakia

E. Tosatti[*]

Institut für Theoretische Physik,
Universität Stuttgart,
Azenbergstrasse 12,
7 Stuttgart 1,
Federal Republic of Germany

EDITOR

Miriam Lewis

Division of Publications, IAEA,
Vienna, Austria

---

[1] Professor Schrieffer's lectures on Electron Theory of Chemisorption and Catalysis are published in the Proceedings of Course LVIII of the Enrico Fermi Summer School in Physics, Varenna, 1973.

PARTICIPANTS

G. Abbate	Istituto di Fisica Sperimentale, Università di Napoli, Via A. Tari, Napoli, Italy	Italy
V.K. Agarwal	Dept of Physics, University of Roorkee, Roorkee, India	India
K.G. Aggarwal	Dept of Physics, P.U. Regional Centre, Rohtak, Haryana, India	India
M. A. Alam	Dept of Physics, Punjab University, New Campus, Lahore, Pakistan	Pakistan
M.A. Alario Franco	Instituto de Química Inorgánica, Facultad de Ciencias, CSIC, Madrid-3, Spain	Spain
T. Andersson	Dept of Physics, Chalmers University of Technology, Fack, S-402 20 Göteborg, Sweden	Sweden
N. Angelescu	Institute for Atomic Physics, PO Box 35, Bucharest, Romania	Romania
M.E. Arellano	Instituto de Investigaciones Físicas, Universidad Mayor de San Andrés, La Paz, Bolivia	Bolivia
B.R. Balaguer	Instituto Mexicano del Petróleo, Av. de los Cien Metros No. 152, Mexico 14, DF, Mexico	Mexico
I. Bartos	Institute of Solid State Physics, Czechoslovak Academy of Sciences, Cukrovarnicka 10, Prague 6, Czechoslovakia	Czechoslovakia
C.M. Bertoni	Istituto di Fisica, Università di Modena, Via Vivaldi 70, 41100 Modena, Italy	Italy
O. Bisi	Istituto di Fisica, Università di Modena, Via Vivaldi 70, 41100 Modena, Italy	Italy
A. Bobbio	Istituto Elettrotecnico Nazionale "G. Ferraris", Torino, Italy	Italy

A. ten Bosch	Freie Universität Berlin, Ihnestrasse 24, I Berlin 33, Federal Republic of Germany	Federal Republic of Germany
R. Brako	Institut "Ruder Bošković", PO Box 1016, 41001 Zagreb, Yugoslavia	Yugoslavia
G.P. Brivio	Istituto di Fisica, Università di Milano, Via Celoria 16, 20133 Milano, Italy	Italy
C. Calandra	Istituto di Fisica, Università di Modena, Via Vivaldi 70, 41100 Modena, Italy	Italy
J.C. Campuzano	Dept of Physics, University of Wisconsin-Milwaukee, Milwaukee, Wis.53201, USA	United States of America/ Paraguay
D. Castiel	Université Paris-Sud, Laboratoire de Physique des Solides, Bâtiment 510, 91405 Orsay, France	France/Canada
S.U. Cheema	PINSTECH, PO Nilore, Rawalpindi, Pakistan	Pakistan
P. Chiaradia	Istituto di Fisica "G. Marconi", Università di Roma, Piazzale delle Scienze 5, Rome, Italy	Italy
L.S. Cota Araiza	Instituto de Física, Apartado Postal 20-364, Mexico 20, DF, Mexico	Mexico
M. Cristu	Institute for Atomic Physics, PO Box 35, Bucharest, Romania	Romania
C.B. Cuden	Universidade Federal da Paraiba, Instituto Central de Fisica, João Pessoa, Paraiba, Brazil	Brazil/Canada
M.P. Das	Dept of Physics, University of Roorkee, Roorkee, India	India
F.J.L. Delanaye	Institut de Physique, Université de Liège, B-4000 Sart/Tilman, Liège 1, Belgium	Belgium

J. Derrien	UER Luminy, 13288 Marseille, France	France
U. Desnica	Institut "Ruder Bošković", PO Box 1016, 41001 Zagreb, Yugoslavia	Yugoslavia
I.-A. Dorobantu	Institute of Atomic Physics, PO Box 35, Bucharest, Romania	Romania
M. Elices	Laboratorio de Ciencia y Materiales, OP, 3 Alfonso XII, Madrid-7, Spain	Spain
Y.A. El-Tantawy	Dept of Chemistry, Faculty of Science, Alexandria University, Alexandria, Egypt	Egypt
H. Espaillat	Escuela de Química, Facultad de Ciencias, Universidad Central de Venezuela, Caracas, Venezuela	Venezuela
V.E. Godwin	Fourah Bay College, Mount Aureol, Freetown, Sierra Leone, West Africa	Sierra Leone
A. Griffin	Institut für Festkörperforschung, 517 Jülich, Federal Republic of Germany, and Dept of Physics, University of Toronto, Toronto, Canada	Federal Republic of Germany/Canada
C. Guillot	Commissariat à l'énergie atomique, Centre de Saclay, BP N° 2, F-91190 Gif-sur-Yvette, France	France
O.R.L. Gunnarsson	Institute of Theoretical Physics, Chalmers University of Technology, Fack, S-402 20 Göteborg 5, Sweden	Sweden
A.J. Hamdani	Institute of Physics, University of Islamabad, PO Box 1090, Islamabad, Pakistan	Pakistan
A. Hamnett	Queen's College, University of Oxford, Oxford, England	United Kingdom
V. Hari Babu	Dept of Physics, University College of Science, Osmania University, Hyderabad-500007, India	India

P. Hertel	Universität Hamburg, Fachbereich Physik, Jungiusstrasse 9, 2 Hamburg 36, Federal Republic of Germany	Federal Republic of Germany
H.M. Hjelmberg	Institute of Theoretical Physics, Chalmers University of Technology, Fack, S-402 20 Göteborg, Sweden	Sweden
U.T. Hochli	IBM Research Laboratory Zürich, Säumerstrasse 4, CH-8803 Rüschlikon, Switzerland	Switzerland
M.O. Hunger	Escuela de Física, Matematicas y Computación, Universidad Central, Facultad de Ciencias, Caracas, Venezuela	Venezuela
G. Iernetti	Istituto di Fisica, Università di Trieste, Via Valerio 2, Trieste, Italy	Italy
M. Jaramillo	Departamento de Física, Universidad de Antioquía, Apdo. Aéreo 1226, Medellín, Colombia	Colombia
H.O.K. Kirchner	II. Physikalisches Institut der Universität Wien, Strudlhofgasse 4, A-1090 Vienna, Austria	Austria
I.Z. Kostadinov	Dept of Physics, University of Sofia, Bul. Anton Ivanov 5, Sofia 26, Bulgaria	Bulgaria
D. Kostoski	Laboratory of Solid-State Physics, Institut "Boris Kidrić", PO Box 522, 11001 Belgrade, Yugoslavia	Yugoslavia
G. Knezevic	Prirodno-Matematicki Fakultet, (Faculty of Natural Sciences), Putnika 43, Sarajevo, Yugoslavia	Yugoslavia
G. Lakshmi (Miss)	Dept of Physics, Indian Institute of Technology, Madras-600036, India	India
C. Lamy	Laboratoire d'électrolyse du CNRS, 1 place A. Briand, F-92190 Bellevue, France	France

Z. Lenac	Institut "Ruder Bošković", PO Box 1016, 41001 Zagreb, Yugoslavia	Yugoslavia
J. Llabres	Departamento de Física C-IV, Universidad Autonoma, Canto Blanco, Madrid, Spain	Spain
E. Louis	Departamento de Física XII, Universidad Autonoma, Canto Blanco, Madrid, Spain	Spain
M.I. Mansour	Home: 8 El-Attar Street, Shoubrah, Cairo, Egypt	Egypt
Kh. Mannan	Dept of Physics, University of Dacca, Dacca-2, Bangladesh	Bangladesh
V. Marigliano Ramaglia	Istituto di Fisica Teorica, Università di Napoli, Mostra d'Oltremare, Pad. 19, Napoli, Italy	Italy
N. Martensson	Institute of Physics, University of Uppsala, PO Box 530, S-751 21 Uppsala 1, Sweden	Sweden
E.N. Martinez	Max-Planck-Institut für Physik und Astrophysik, Föhringer Ring 6, 8 München 40, Federal Republic of Germany	Federal Republic of Germany/Argentina
O.P. Mehta	School of Basic Sciences and Humanities, University of Udaipur, Udaipur, Rajasthan, India	India
M. Menyhard	Research Institute for Technical Physics of the Hungarian Academy of Sciences, PO Box 76, 1325 Budapest, Hungary	Hungary
A. Morawski	Institute of Physics, Polish Academy of Sciences, Lotnikow 32, Warsaw, Poland	Poland
A. Mufti	PINSTECH, PO Nilore, Rawalpindi, Pakistan	Pakistan
M. Musa (Mrs.)	Institute of Physics of the CSEN, B-dul Pacii 222, Bucharest 7, Romania	Romania

N. Nafari	Dept of Physics, Arya-Mehr University of Technology, PO Box 3406, Tehran, Iran	Iran
S.S. Nandwani	Dept of Physics, University of Roorkee, Roorkee, India	India
B. Navinsek	Institut "Jozef Stefan", Jamova 39, 61001 Ljubljana, Yugoslavia	Yugoslavia
F. Nizzoli	Istituto di Fisica, Università di Modena, Via Vivaldi 70, 41100 Modena, Italy	Italy
K. Nuroh	Dept of Mathematics, University of Science and Technology, Kumasi, Ghana	Ghana
R. Nyholm	Institute of Physics, University of Uppsala, PO Box 530, S-751 21 Uppsala 1, Sweden	Sweden
A.M. Ozorio de Almeida	Instituto de Física "Gleb Wataghin" Universidade Estadual de Campinas, Campinas, SP, Brazil	Brazil
S. Parangtopo	Dept of Physics, Faculty of Natural Sciences, University of Indonesia, Jl. Salemba 4, Jakarta, Indonesia	Indonesia
T.S. Park	Dept of Physics, Kyongpook National University, 1370 Sanhyok Dong, Taegu, Republic of Korea	Republic of Korea
J.A.A.J. Perenboom	Fysisch Laboratorium, Katholieke Universiteit, Toernooiveld, Nijmegen, Holland	The Netherlands
M. Pfuff	Universität Hamburg, Fachbereich Physik, Jungiusstrasse 9, 2 Hamburg 36, Federal Republic of Germany	Federal Republic of Germany
P. Picco	Istituto di Fisica, Università di Milano, Via Celoria 16, 20133 Milano, Italy	Italy

F. Pratesi   Istituto di Chimica Fisica,   Italy
                             Università di Firenze,
                             Largo Enrico Fermi 2,
                             50125 Firenze, Italy

P. Prelovšek   Institut "Jozef Stefan",   Yugoslavia
                             Jamova 39,
                             61001 Ljubljana, Yugoslavia

A.M. Prodan   Institut "Jozef Stefan",   Yugoslavia
                             Jamova 39,
                             61001 Ljubljana, Yugoslavia

S.K. Rangarajan   Materials Science Division,   India
                             National Aeronautical Laboratory,
                             PO Box 1779,
                             Bangalore-17, India

P. Rujan   Institute for Theoretical Physics,   Hungary
                             Roland Eötvös University,
                             Puskin u. 5-7,
                             1088 Budapest, Hungary

A. Sattar Syed   Regional Laboratories,   Bangladesh
                             Bangladesh Council of Scientific
                                and Industrial Research,
                             Dhanmondi,
                             Dacca, Bangladesh

G. Sauer   Institute for Theoretical Physics,   Austria
                             University of Graz,
                             A-8010 Graz, Austria

M.Y. Sawan   Dept of Chemistry,   Egypt
                             Faculty of Science,
                             Alexandria University,
                             Alexandria, Egypt

M.M. Schmeits   Institut de Physique,   Belgium
                             Université de Liège,
                             B-4000 Sart Tilman,
                             Liège 1, Belgium

J.N. Schmit   Institut de Physique,   Belgium/Luxembourg
                             Université de Liège,
                             B-4000 Sart Tilman,
                             Liège 1, Belgium

A. Selloni (Miss)   Istituto di Fisica,   Italy
                             Università di Roma,
                             Piazzale delle Scienze 5,
                             Roma, Italy

G. Seriani   Istituto di Fisica Teorica,   Italy
                             Università di Trieste,
                             c/o ICTP,
                             Miramare, PO Box 586,
                             Trieste, Italy

K.A. Shoaib	PINSTECH, PO Nilore, Rawalpindi, Pakistan	Pakistan
G. Sloccari	Istituto di Fisica, Università di Trieste, Via Valerio 2, Trieste, Italy	Italy
D. Šokčević	Institut "Ruder Bošković", PO Box 1016, 41001 Zagreb, Yugoslavia	Yugoslavia
A. Stepanescu	Istituto Elettrotecnico Nazionale "G. Ferraris", Torino, Italy	Italy
M. Steslicka (Mrs)	Institute of Experimental Physics, University of Wrocław, ul. Cybulskiego 36, 50-205 Wrocław, Poland	Poland
M.A. Subhan	Atomic Energy Centre, PO Box 164, Ramna, Dacca, Bangladesh	Bangladesh
A.H. Sukiennicki	Institute of Physics, Technical University, Koszykowa 75, 00-662 Warsaw, Poland	Poland
G.D. Surender	Dept of Chemical Engineering, Indian Institute of Science, Bangalore-560012, India	India
A. Tagliacozzo	Istituto di Fisica Teorica, Università di Napoli, Mostra d'Oltremare, Pad. 19, Napoli, Italy	Italy

G.H. Talat	Dept of Physics,
	National Research Centre,
	Dokki,
	Cairo, Egypt
	
	1974-1975:
	    Quantum Radio Physics Laboratory,
	    Lebedev Physics Institute,
	    Leninskij Prospect 53,
	    Moscow, USSR
		Egypt

G. Tay	University of Science and Technology, Kumasi, Ghana, West Africa	Ghana

C. Tejedor   Departamento de Física XII,   Spain
                                Universidad Autónoma,
                                Canto Blanco,
                                Madrid, Spain

M. Tomak   Dept of Physics,   Turkey
                                Middle East Technical University,
                                Ankara, Turkey

M.S. Tomas   Institut "Ruder Bošković",   Yugoslavia
                                PO Box 1016,
                                41001 Zagreb, Yugoslavia

L. Ungier   Research Laboratories of Catalysis   Poland
                                    and Surface Chemistry,
                                Polish Academy of Sciences,
                                ul. Krupnicza 41,
                                30060 Krakow, Poland

J.A. Valles-Abarca   Departamento de Electricidad y Electrónica, Spain
                                Universidad Complutense,
                                Ciudad Universitaria,
                                Madrid-3, Spain

R. Van Santen   Koninklijke/Shell-Laboratorium,   The Netherlands
                                Badhuisweg 3,
                                Amsterdam, The Netherlands

J.V. Vukanic   Ion Physics Laboratory,   Yugoslavia
                                Institut "Boris Kidrič",
                                PO Box 522,
                                Belgrade, Yugoslavia

J. Walraven   Natuurkundig Laboratorium der   The Netherlands
                                    Universiteit van Amsterdam,
                                Valckenierstraat 65,
                                Amsterdam-C, The Netherlands

E. Wikborg   Institute of Theoretical Physics,   Sweden
                                Fack,
                                S-402 20 Göteborg 5, Sweden

and physicists present at the Centre.

The following conversion table is provided for the convenience of readers and to encourage the use of SI units.

FACTORS FOR CONVERTING UNITS TO SI SYSTEM EQUIVALENTS *

SI base units are the metre (m), kilogram (kg), second (s), ampere (A), kelvin (K), candela (cd) and mole (mol).
[For further information, see International Standards ISO 1000 (1973), and ISO 31/0 (1974) and its several parts]

Multiply	by		to obtain
Mass			
pound mass (avoirdupois)	1 lbm	= 4.536×10^{-1}	kg
ounce mass (avoirdupois)	1 ozm	= 2.835×10^{1}	g
ton (long) (= 2240 lbm)	1 ton	= 1.016×10^{3}	kg
ton (short) (= 2000 lbm)	1 short ton	= 9.072×10^{2}	kg
tonne (= metric ton)	1 t	= 1.00×10^{3}	kg
Length			
statute mile	1 mile	= 1.609×10^{0}	km
yard	1 yd	= 9.144×10^{-1}	m
foot	1 ft	= 3.048×10^{-1}	m
inch	1 in	= 2.54×10^{-2}	m
mil (= 10^{-3} in)	1 mil	= 2.54×10^{-2}	mm
Area			
hectare	1 ha	= 1.00×10^{4}	m^{2}
(statute mile)^{2}	1 mile^{2}	= 2.590×10^{0}	km^{2}
acre	1 acre	= 4.047×10^{3}	m^{2}
yard^{2}	1 yd^{2}	= 8.361×10^{-1}	m^{2}
foot^{2}	1 ft^{2}	= 9.290×10^{-2}	m^{2}
inch^{2}	1 in^{2}	= 6.452×10^{2}	mm^{2}
Volume			
yard^{3}	1 yd^{3}	= 7.646×10^{-1}	m^{3}
foot^{3}	1 ft^{3}	= 2.832×10^{-2}	m^{3}
inch^{3}	1 in^{3}	= 1.639×10^{4}	mm^{3}
gallon (Brit. or Imp.)	1 gal (Brit)	= 4.546×10^{-3}	m^{3}
gallon (US liquid)	1 gal (US)	= 3.785×10^{-3}	m^{3}
litre	1 l	= 1.00×10^{-3}	m^{3}
Force			
dyne	1 dyn	= 1.00×10^{-5}	N
kilogram force	1 kgf	= 9.807×10^{0}	N
poundal	1 pdl	= 1.383×10^{-1}	N
pound force (avoirdupois)	1 lbf	= 4.448×10^{0}	N
ounce force (avoirdupois)	1 ozf	= 2.780×10^{-1}	N
Power			
British thermal unit/second	1 Btu/s	= 1.054×10^{3}	W
calorie/second	1 cal/s	= 4.184×10^{0}	W
foot-pound force/second	1 ft·lbf/s	= 1.356×10^{0}	W
horsepower (electric)	1 hp	= 7.46×10^{2}	W
horsepower (metric) (= ps)	1 ps	= 7.355×10^{2}	W
horsepower (550 ft·lbf/s)	1 hp	= 7.457×10^{2}	W

* Factors are given exactly or to a maximum of 4 significant figures

Multiply		by	to obtain
Density			
pound mass/inch^3	1 lbm/in^3	= 2.768 × 10^4	kg/m^3
pound mass/foot^3	1 lbm/ft^3	= 1.602 × 10^1	kg/m^3
Energy			
British thermal unit	1 Btu	= 1.054 × 10^3	J
calorie	1 cal	= 4.184 × 10^0	J
electron-volt	1 eV	≈ 1.602 × 10^{-19}	J
erg	1 erg	= 1.00 × 10^{-7}	J
foot-pound force	1 ft·lbf	= 1.356 × 10^0	J
kilowatt-hour	1 kW·h	= 3.60 × 10^6	J
Pressure			
newtons/metre^2	1 N/m^2	= 1.00	Pa
atmosphere^a	1 atm	= 1.013 × 10^5	Pa
bar	1 bar	= 1.00 × 10^5	Pa
centimetres of mercury (0°C)	1 cmHg	= 1.333 × 10^3	Pa
dyne/centimetre^2	1 dyn/cm^2	= 1.00 × 10^{-1}	Pa
feet of water (4°C)	1 ftH$_2$O	= 2 989 × 10^3	Pa
inches of mercury (0°C)	1 inHg	= 3.386 × 10^3	Pa
inches of water (4°C)	1 inH$_2$O	= 2.491 × 10^2	Pa
kilogram force/centimetre^2	1 kgf/cm^2	= 9.807 × 10^4	Pa
pound force/foot^2	1 lbf/ft^2	= 4.788 × 10^1	Pa
pound force/inch^2 (= psi)^b	1 lbf/in^2	= 6.895 × 10^3	Pa
torr (0°C) (= mmHg)	1 torr	= 1.333 × 10^2	Pa
Velocity, acceleration			
inch/second	1 in/s	= 2.54 × 10^1	mm/s
foot/second (= fps)	1 ft/s	= 3.048 × 10^{-1}	m/s
foot/minute	1 ft/min	= 5.08 × 10^{-3}	m/s
mile/hour (= mph)	1 mile/h	= $\begin{cases} 4.470 × 10^{-1} \\ 1.609 × 10^0 \end{cases}$	m/s km/h
knot	1 knot	= 1.852 × 10^0	km/h
free fall, standard (= g)		= 9.807 × 10^0	m/s^2
foot/second^2	1 ft/s^2	= 3.048 × 10^{-1}	m/s^2
Temperature, thermal conductivity, energy/area·time			
Fahrenheit, degrees −32	°F −32	$\dfrac{5}{9}$	°C
Rankine	°R		K
1 Btu·in/ft^2·s·°F		= 5.189 × 10^2	W/m·K
1 Btu/ft·s·°F		= 6.226 × 10^1	W/m·K
1 cal/cm·s·°C		= 4.184 × 10^2	W/m·K
1 Btu/ft^2·s		= 1.135 × 10^4	W/m^2
1 cal/cm^2·min		= 6.973 × 10^2	W/m^2
Miscellaneous			
foot^3/second	1 ft^3/s	= 2.832 × 10^{-2}	m^3/s
foot^3/minute	1 ft^3/min	= 4.719 × 10^{-4}	m^3/s
rad	rad	= 1.00 × 10^{-2}	J/kg
roentgen	R	= 2.580 × 10^{-4}	C/kg
curie	Ci	= 3.70 × 10^{10}	disintegration/s

^{a} atm abs: atmospheres absolute;
  atm (g): atmospheres gauge.

^{b} lbf/in^2 (g) (= psig): gauge pressure;
  lbf/in^2 abs (= psia): absolute pressure.

HOW TO ORDER IAEA PUBLICATIONS

Exclusive sales agents for IAEA publications, to whom all orders
and inquiries should be addressed, have been appointed
in the following countries:

UNITED KINGDOM	Her Majesty's Stationery Office, P.O. Box 569, London SE 1 9NH
UNITED STATES OF AMERICA	UNIPUB, Inc., P.O. Box 433, Murray Hill Station, New York, N.Y. 10016

In the following countries IAEA publications may be purchased from the
sales agents or booksellers listed or through your
major local booksellers. Payment can be made in local
currency or with UNESCO coupons.

ARGENTINA	Comisión Nacional de Energía Atómica, Avenida del Libertador 8250, Buenos Aires
AUSTRALIA	Hunter Publications, 58 A Gipps Street, Collingwood, Victoria 3066
BELGIUM	Service du Courrier de l'UNESCO, 112, Rue du Trône, B-1050 Brussels
CANADA	Information Canada, 171 Slater Street, Ottawa, Ont. K 1 A OS 9
C.S.S.R.	S.N.T.L., Spálená 51, CS-110 00 Prague
	Alfa, Publishers, Hurbanovo námestie 6, CS-800 00 Bratislava
FRANCE	Office International de Documentation et Librairie, 48, rue Gay-Lussac, F-75005 Paris
HUNGARY	Kultura, Hungarian Trading Company for Books and Newspapers, P.O. Box 149, H-1011 Budapest 62
INDIA	Oxford Book and Stationery Comp., 17, Park Street, Calcutta 16
ISRAEL	Heiliger and Co., 3, Nathan Strauss Str., Jerusalem
ITALY	Libreria Scientifica, Dott. de Biasio Lucio "aeiou", Via Meravigli 16, I-20123 Milan
JAPAN	Maruzen Company, Ltd., P.O.Box 5050, 100-31 Tokyo International
NETHERLANDS	Marinus Nijhoff N.V., Lange Voorhout 9-11, P.O. Box 269, The Hague
PAKISTAN	Mirza Book Agency, 65, The Mall, P.O.Box 729, Lahore-3
POLAND	Ars Polona, Centrala Handlu Zagranicznego, Krakowskie Przedmiescie 7, PL-00-068 Warsaw
ROMANIA	Cartimex, 3-5 13 Decembrie Street, P.O.Box 134-135, Bucarest
SOUTH AFRICA	Van Schaik's Bookstore, P.O.Box 724, Pretoria
	Universitas Books (Pty) Ltd., P.O.Box 1557, Pretoria
SPAIN	Nautrónica, S.A., Pérez Ayuso 16, Madrid-2
SWEDEN	C.E. Fritzes Kungl. Hovbokhandel, Fredsgatan 2, S-103 07 Stockholm
U.S.S.R.	Mezhdunarodnaya Kniga, Smolenskaya-Sennaya 32-34, Moscow G-200
YUGOSLAVIA	Jugoslovenska Knjiga, Terazije 27, YU-11000 Belgrade

Orders from countries where sales agents have not yet been appointed and
requests for information should be addressed directly to:

Publishing Section,
International Atomic Energy Agency,
Kärntner Ring 11, P.O.Box 590, A-1011 Vienna, Austria